Seventh Workshop on Social Media Mining for Health Applications, Workshop & Shared Task (#SMM4H 2022)

Gyeongju, South Korea
12-17 October 2022

ISBN: 978-1-7138-6195-9

Printed from e-media with permission by:

Curran Associates, Inc.
57 Morehouse Lane
Red Hook, NY 12571

Some format issues inherent in the e-media version may also appear in this print version.

Printed with permission by Curran Associates, Inc. (2023)

Association for Computational Linguistics
209 N. Eighth Street
Stroudsburg, Pennsylvania 18360

Phone: 1-570-476-8006
Fax: 1-570-476-0860

acl@aclweb.org

Additional copies of this publication are available from:

Curran Associates, Inc.
57 Morehouse Lane
Red Hook, NY 12571 USA
Phone: 845-758-0400
Fax: 845-758-2633
Email: curran@proceedings.com
Web: www.proceedings.com

COLING 2022
Volume 29 (2022), No. 18

Proceedings of The Seventh Workshop on Social Media Mining for Health Applications, Workshop & Shared Task (#SMM4H 2022)

The 29th International Conference on Computational Linguistics

October 12 - 17, 2022
Gyeongju, Republic of Korea

Preface

Welcome to the 7th Social Media Mining for Health Applications (#SMM4H) Workshop & Shared Task 2022, co-located at the 29th International Conference on Computational Linguistics. Held as a hybrid event in its seventh iteration, #SMM4H 2022 continues to serve as a unique venue for bringing together data mining researchers interested in building and sharing solutions for utilizing social media data for health informatics. For #SMM4H 2022, we accepted 6 workshop papers and 47 shared task system description papers. Each submission was peer-reviewed by two or three reviewers.

The accepted papers proposed advanced models to detect or extract health-related information in posts written in various languages. Pais et al. report the performance of baseline transformers on a new corpus of Romanian micro-blogging posts annotated with 9 entity classes. Their corpus is made available to the community. Zanwar et al. detect 6 mental health conditions on Reddit posts with an interpretable neural network by combining a feature-based model with a transformer model. Chan et al. describe the collection and annotation process to create a corpus of Dutch Facebook comments. These posts comment on news articles about COVID-19 vaccination where Facebook users shared the knowledge they acquired through their personal experiences. Adhikari et al. detailed their GUI to improve, with incremental learning, their classifier of 8 topics related to COVID-19 in Nepali tweets. Finally, Davydova & Tutubalina and Gasco Sánchez et al. expand the description and the analysis of the results of the #SMM4H shared tasks 2 and 10.

The #SMM4H 2022 shared tasks sought to advance the use of user-generated social media data for pharmacovigilance, epidemiology, patient-centered outcomes, and tracking beliefs and impacts of COVID-19. #SMM4H 2022 included re-runs of three tasks about adverse drug events, changes in medication treatments, and COVID-19 symptoms. In addition, #SMM4H 2022 included seven new tasks on detecting stances toward COVID-19 health mandates, COVID-19 vaccination status, the age of social media users, intimate partner violence, chronic stress, and diseases. The ten tasks required methods for multi-class classification, and named entity recognition and normalization. With 117 teams that registered from 28 countries and 54 teams that participated, the interest in the #SMM4H shared tasks continues to grow. Among the 47 system description papers that were accepted, 10 teams were invited for an oral presentation.

The organizing committee of #SMM4H 2022 would like to thank the program committee, the additional reviewers of system description papers, the organizers of COLING 2022 (especially the workshop co-chairs), the annotators of the shared task data, and, of course, everyone who submitted a paper or participated in the shared tasks. #SMM4H 2022 would not have been possible without them.

Graciela, Davy, Arjun, Ari, Ivan, Karen, Raul, Lucia, Juan, Abeed, Yuting, Yao, Elena, Luis, Darryl, and Martin.

Organizing and Program Committees

Organizing Committee:

Graciela Gonzalez-Hernandez, Cedars-Sinai Medical Center, USA
Davy Weissenbacher, Cedars-Sinai Medical Center, USA
Arjun Magge, University of Pennsylvania, USA
Ari Z. Klein, University of Pennsylvania, USA
Ivan Flores, Cedars-Sinai Medical Center, USA
Karen O'Connor, University of Pennsylvania, USA
Raul Rodriguez-Esteban, Roche Pharmaceuticals, Switzerland
Lucia Schmidt, Roche Pharmaceuticals, Switzerland
Juan M. Banda, Georgia State University, USA
Abeed Sarker, Emory University, USA
Yuting Guo, Emory University, USA
Yao Ge, Emory University, USA
Elena Tutubalina, Insilico Medicine, Hong Kong
Luis Gasco, Barcelona Supercomputing Center, Spain
Darryl Estrada, Barcelona Supercomputing Center, Spain
Martin Krallinger, Barcelona Supercomputing Center, Spain

Program Committee:

Cecilia Arighi, University of Delaware, USA
Natalia Grabar, French National Center for Scientific Research, France
Thierry Hamon, Paris-Nord University, France
Antonio Jimeno Yepes, Royal Melbourne Institute of Technology, Australia
Jin-Dong Kim, Database Center for Life Science, Japan
Corrado Lanera, University of Padova, Italy
Robert Leaman, US National Library of Medicine, USA
Kirk Roberts, University of Texas Health Science Center at Houston, USA
Yutaka Sasaki, Toyota Technological Institute, Japan
Pierre Zweigenbaum, French National Center for Scientific Research, France

Conference Program

#SMM4H'22 Monday, October 17, 2022

9:00–9:15 *Introduction*
Graciela Gonzalez-Hernandez

09:15–10:15 **Oral Presentations Q&A Session 1**

Romanian micro-blogging named entity recognition including health-related entities
Vasile Pais, Verginica Barbu Mititelu, Elena Irimia, Maria Mitrofan, Carol Luca Gasan and Roxana Micu

The Best of Both Worlds: Combining Engineered Features with Transformers for Improved Mental Health Prediction from Reddit Posts
Sourabh Satish Zanwar, Daniel Wiechmann, Yu Qiao and Elma Kerz

Overview of the Seventh Social Media Mining for Health Applications (#SMM4H) Shared Tasks at COLING 2022
Davy Weissenbacher, Juan Banda, Vera Davydova, Darryl Johan Estrada Zavala, Luis Gasco Sánchez, Yao Y. Ge, Yuting Guo, Ari Klein, Martin Krallinger, Mathias Leddin, Arjun Magge, Raul Rodriguez-Esteban, Abeed Sarker, Lucia L. Schmidt, Elena Tutubalina and Graciela Gonzalez-Hernandez

PingAnTech at SMM4H task1: Multiple pre-trained model approaches for Adverse Drug Reactions
Xi Liu, Han Zhou and Chang Su

10:15–10:30 *Break*

10:30–11:30 **Oral Presentations Q&A Session 2**

COVID-19-related Nepali Tweets Classification in a Low Resource Setting
Rabin Adhikari, Safal Thapaliya, Nirajan Basnet, Samip Poudel, Aman Shakya and Bishesh Khanal

Leveraging Social Media as a Source for Clinical Guidelines: A Demarcation of Experiential Knowledge
Jia-Zhen Michelle Chan, Florian Kunneman, Roser Morante, Lea Lösch and Teun Zuiderent-Jerak

Zhegu@SMM4H-2022: The Pre-training Tweet & Claim Matching Makes Your Prediction Better
Pan He, Chen YuZe and Yanru Zhang

CSECU-DSG@SMM4H'22: Transformer based Unified Approach for Classification of Changes in Medication Treatments in Tweets and WebMD Reviews
Afrin Sultana, Nihad Karim Chowdhury and Abu Nowshed Chy

#SMM4H'22 Monday, October 17, 2022 (continued)

11:30–12:15 **Poster Session**

12:15–12:30 *Break*

12:30–13:30 **Oral Presentations Q&A Session 3**

Yet@SMM4H'22: Improved BERT-based classification models with Rdrop and PolyLoss
Yan Zhuang and Yanru Zhang

AIR-JPMC@SMM4H'22: Identifying Self-Reported Spanish COVID-19 Symptom Tweets Through Multiple-Model Ensembling
Adrian Garcia Hernandez, Leung Wai Liu, Akshat Gupta, vineeth ravi, Saheed O. Obitayo, Xiaomo Liu and Sameena Shah

AILAB-Udine@SMM4H'22: Limits of Transformers and BERT Ensembles
Beatrice Portelli, Simone Scaboro, Emmanuele Chersoni, Enrico Santus and Giuseppe Serra

AIR-JPMC@SMM4H'22: Classifying Self-Reported Intimate Partner Violence in Tweets with Multiple BERT-based Models
Alec Louis Clemente Candidato, Akshat Gupta, Xiaomo Liu and Sameena Shah

13:30–13:45 *Break*

13:45–14:30 **Oral Presentations Q&A Session 4**

zydhjh4593@SMM4H'22: A Generic Pre-trained BERT-based Framework for Social Media Health Text Classification
Chenghao Huang, Xiaolu Chen, Yuxi Chen, Yutong Wu, Weimin Yuan, Yan Wang and Yanru Zhang

Fraunhofer FKIE @ SMM4H 2022: System Description for Shared Tasks 2, 4 and 9
Daniel Claeser and Samantha Kent

CASIA@SMM4H'22: A Uniform Health Information Mining System for Multilingual Social Media Texts
Jia Fu, Sirui Li, hui ming yuan, Zhucong Li, zhen gan, Yubo Chen, kang liu, Jun Zhao and Shengping Liu

14:30–14:45 *Break*

14:45–15:25 **Keynote**
Raul Rodriguez-Esteban

15:25–15:40 *Conclusion and Closing Remarks*
Graciela Gonzalez-Hernandez

Table of Contents

RACAI@SMM4H'22: Tweets Disease Mention Detection Using a Neural Lateral Inhibitory Mechanism

Andrei-Marius Avram, Vasile Păis and **Maria Mitrofan**
Research Institute for Artificial Intelligence "Mihai Drăgănescu"
{andrei.avram,vasile,maria}@racai.ro

Abstract

This paper presents our system employed for the Social Media Mining for Health (SMM4H) 2022 competition Task 10 - SocialDisNER. The goal of the task was to improve the detection of diseases in tweets. Because the tweets were in Spanish, we approached this problem using a system that relies on a pre-trained multilingual model and is fine-tuned using the recently introduced lateral inhibition layer. We further experimented on this task by employing a conditional random field on top of the system and using a voting-based ensemble that contains various architectures. The evaluation results outlined that our best performing model obtained 83.7% F1-strict on the validation set and 82.1% F1-strict on the test set.

1 Motivation

Social media data (e.g. Twitter, Facebook) analysis for health informatics is an active area of research that aims to understand public opinion of health-related topics. The Social Media Mining for Health Applications (#SMM4H) workshop is a venue that brings together researchers that want to contribute to this area, including, but not limited to subjects such as: deriving health trends from social media, health-related message classification or disease monitoring from messages on social media.

We chose to participate at the tenth task of the #SMM4H competition - SocialDisNER - whose aim was to improve the disease detection in Spanish tweets by identifying `ENFERMEDAD` entities in the given text (Gasco et al., 2022). We approached this problem using the XLM-RoBERTa multilingual pre-trained model (Conneau et al., 2020) on which we incorporated the recently introduced lateral inhibitory (LI) layer (Păiș, 2022) in the fine-tuning process, together with a conditional random field (CRF) (Lafferty et al., 2001) layer on top of the model and an ensemble system.

The interest in this task came from our involvement within the CURLICAT project for the CEF AT action (Váradi et al., 2022), where named entity recognition systems need to be developed for various domains (health-related included), in Romanian language.

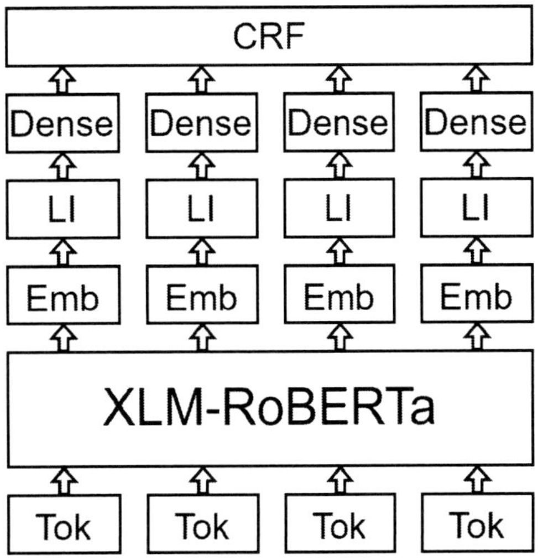

Figure 1: Model architecture that contains XLM-RoBERTa, the LI layer and the CRF layer (in this figure, "Tok" stands for "Token" and "Emb" stands for "Embedding").

2 System Description

We treat the SocialDisNER task as a sequence tagging problem, so we try to predict a label for each token given as input. The architecture of our system consists of using the XLM-RoBERTa pre-trained model that produces an embedding for each token in the input sequence. Then, we employ the LI layer on each token embedding, followed by a dense layer to produce the output probabilities for each of the three possible classes: `O`, `B-ENFERMEDAD` or `I-ENFERMEDAD`[1]. We further use a CRF layer on top of the predicted logits to

[1]We transform the input data to match the Inside–Outside–Beginning format (Ramshaw and Marcus, 1999).

Proceedings of the 29th International Conference on Computational Linguistic, pages 1–3
October 12–17, 2022.

Model	F1-Overlap			F1-Strict		
	P	R	F1	P	R	F1
XLM-RoBERTa	**95.7**	88.3	**91.8**	86.2	75.4	80.4
XLM-RoBERTa+LI*	94.7	88.4	91.4	87.5	**80.3**	**83.7**
XLM-RoBERTa+CRF	95.3	87.8	91.4	**87.8**	79.6	83.5
XLM-RoBERTa+LI+CRF	95.0	88.4	91.6	87.4	80.0	83.5
Ensemble*	94.9	**88.6**	91.6	87.5	80.2	**83.7**

Table 1: The F1-overlap and F1-strict evaluation results obtained by our models on the SocialDisNER task validation set. *These models were sent for the final evaluation.

learn to better predict the most probable sequence of classes for a given input. The overall system architecture is depicted in Figure 1.

It must be noted that we also experimented with models that do not include the LI and/or the CRF layers. In addition, we created a voting-based ensemble system that incorporates the predictions of all four architectures developed in this work: XLM-RoBERTa, XLM-RoBERTa+LI, XLM-RoBERTa+CRF and XLM-RoBERTa+LI+CRF.

We fine-tune each model for 80 epochs using a batch size of 16 and a gradient accumulation of 8, resulting in a total batch size of 128. We use the Adam optimizer (Kingma and Ba, 2014) with a learning rate of 2e-5 and the default values for the weight decay, epsilon and the betas in PyTorch (Paszke et al., 2019). During training, we save only the checkpoint that obtained the highest performance on the validation set.

3 Evaluation

We evaluate and present the results of our models on the validation and test sets. The results of the XLM-RoBERTa with different layers and of the ensemble system are listed in Table 1. Here we compute both the F1-overlap and F1-strict, of which the latter is the main metric used in the competition. The highest F1-overlap was obtained by the XLM-RoBERTa model that contains no special layers with 91.8%. However, when considering the F1-strict as the evaluation metric, this model underperformed all the other architectures by more than 3%, obtaining a F1-strict score of 80.4%. This result shows that, without any additional layers, the model is not capable of predicting exact boundaries for the entities. The highest F1-strict score of 83.7% was obtained by both XLM-RoBERTa+LI and the ensemble system, these two being the ones that we sent for the final evaluation.

Model	P	R	F1
XLM-RoBERTa+LI	**86.8**	**77.9**	**82.1**
Ensemble	86.7	77.9	82.0
Mean	68.0	67.7	67.5
Median	75.8	78.0	76.1

Table 2: The F1-strict evaluation results obtained by our models on the SocialDisNER test set, together with the mean and median results of the task.

The results on the test set are outlined in Table 2. The two models we sent for evaluation, XLM-RoBERTa+LI and the ensemble, obtained similar results, with a 82.1% F1-strict for the former and 82.0% for the later. This puts our solution 14.6% F1-strict above the mean and 6% F1-strict above the median.

4 Conclusion

Analysing social media data remains an important area of research for health informatics, offering a better understanding of public opinion in health-related topics. This work describes our system used to participate at the SMM4H competition SocialDisNER shared task. We treated the problem as a sequence tagging task and employed the XLM-RoBERTa multilingual pre-trained model to process the Spanish tweets. In addition, we analysed the capabilities of the LI layer on this task in combination with a CRF module. We sent to evaluation the XLM-RoBERTa model with the LI layer, as well as an ensemble system. The two systems obtained 82.1% and 82.0% F1-strict scores on the evaluation set, respectively. No additional resources were employed. In the future we plan to apply this experience also on Romanian micro-blogging text (Păiș et al., 2022).

References

Alexis Conneau, Kartikay Khandelwal, Naman Goyal, Vishrav Chaudhary, Guillaume Wenzek, Francisco Guzmán, Édouard Grave, Myle Ott, Luke Zettlemoyer, and Veselin Stoyanov. 2020. Unsupervised cross-lingual representation learning at scale. In *Proceedings of the 58th Annual Meeting of the Association for Computational Linguistics*, pages 8440–8451.

Luis Gasco, Darryl Estrada-Zavala, Eulàlia Farré-Maduell, Salvador Lima-López, Antonio Miranda-Escalada, and Martin Krallinger. 2022. Overview of the SocialDisNER shared task on detection of diseases mentions from healthcare related and patient generated social media content: methods, evaluation and corpora. In *Proceedings of the Seventh Social Media Mining for Health (#SMM4H) Workshop and Shared Task*.

Diederik P Kingma and Jimmy Ba. 2014. Adam: A method for stochastic optimization. *arXiv preprint arXiv:1412.6980*.

John Lafferty, Andrew McCallum, and Fernando CN Pereira. 2001. Conditional random fields: Probabilistic models for segmenting and labeling sequence data.

Adam Paszke, Sam Gross, Francisco Massa, Adam Lerer, James Bradbury, Gregory Chanan, Trevor Killeen, Zeming Lin, Natalia Gimelshein, Luca Antiga, et al. 2019. Pytorch: An imperative style, high-performance deep learning library. *Advances in neural information processing systems*, 32.

Vasile Păiș. 2022. RACAI at SemEval-2022 Task 11: Complex named entity recognition using a lateral inhibition mechanism. In *Proceedings of the 3International Workshop on Semantic Evaluation (SemEval)*.

Vasile Păiș, Maria Mitrofan, Verginica Barbu Mititelu, Elena Irimia, Roxana Micu, and Carol Luca Gasan. 2022. Challenges in creating a representative corpus of romanian micro-blogging text. In *Proceedings of the Workshop on Challenges in the Management of Large Corpora (CMLC-10)*, pages 1–7, Marseille, France. European Language Resources Association.

Lance A Ramshaw and Mitchell P Marcus. 1999. Text chunking using transformation-based learning. In *Natural language processing using very large corpora*, pages 157–176. Springer.

Tamás Váradi, Marko Tadić, Svetla Koeva, Maciej Ogrodniczuk, Dan Tufiş, Radovan Garabík, Simon Krek, and Andraž Repar. 2022. Curated multilingual language resources for cef at (curlicat): Overall view. In *Proceedings of the 23rd Annual Conference of the European Association for Machine Translation*, pages 339–340.

PingAnTech at SMM4H task1: Multiple pre-trained model approaches for Adverse Drug Reactions

Xi Liu[1], Han Zhou[1,2], Chang Su[1]
[1] PingAn Technology
[2] Chengdu University of Technology
doufuxixi3@163.com
zhouhan@stu.cdut.edu.cn
SHUCHUANG254@pingan.com.cn

Abstract

This paper describes the solution for the Social Media Mining for Health (SMM4H) 2022 Shared Task. We participated in Task1a., Task1b. and Task1c. To solve the problem of the presence of Twitter data, we used a pre-trained language model. We used training strategies that involved: adversarial training, head layer weighted fusion, etc., to improve the performance of the model. The experimental results show the effectiveness of our designed system. For task 1a, the system achieved an F1 score of 0.68; for task 1b Overlapping F1 score of 0.65 and a Strict F1 score of 0.49. Task 1c yields Overlapping F1 and Strict F1 scores of 0.36 and 0.30, respectively.

1 Introduction

Mining adverse drug events (ADEs) from social media is one of the most researched topics in the field of social media pharmacovigilance. To promote the research on this topic, the Health Language Processing Lab of the University of Pennsylvania organized Social Media Mining for Health Applications (SMM4H) shared tasks. This year, the SMM4H shared tasks included ten subtasks(Davy Weissenbacher, 2022). Our team focused on three subtasks in Task 1, which are (1) classifying tweets reporting ADEs (Adverse Drug Events); (2) detect ADE spans in the tweets; (3) map these colloquial mentions to their standard concept IDs in the MedDRA vocabulary. The main challenges of this task are as follows: (1) how to handle unbalanced data and (2) the presence of a large amount of noise in the data, e.g., emojis, redundant punctuation, desensitized usernames, some link addresses, etc. In addition, medical expressions in the text are often expressed in non-professional colloquial expressions, which are common problems in social media data and usually mislead the trained model. To address these issues, we use pre-trained language models as the basis; many text pre-processing and adversarial training are described in detail in the following sections, with corresponding experimental results.

2 Task 1a: Classify Tweets Reporting ADEs

This subtask aims to identify whether sentences contain adverse drug reactions, which can be modeled as a classification task. There are a total of 17385 samples in this dataset, where the positive to negative ratio is 1:11 (mentions of ADEs are labeled as 0), a total of 915 samples in the validation set (87 ADE labels and 828 NoADE labels), and 10,984 samples in the test set(Magge et al., 2021).

2.1 Method

Preprocessing Due to the spoken Twitter data, we first cleaned the data. (1) We remove the "@USER" and the placeholder "_" carried after it, and we consider that the URL information does not bring additional information to ADE-related content, so we choose to remove it. (2) We remove the redundant symbolic expressions and keep only one symbol, for example, "!!!", "???" transformed to "!","?". However, since the "...... " symbol is used as a label in task 1b, so we keep this symbol for all tasks. (3) We remove the emoticons from the text. (4) Finally, we remove the extra space symbols from the text. (5) We use an oversampling strategy to increase the number of positive samples. We use the following method: using the cleaned data, we construct duplicate positive samples by randomly combining positive samples with positive sample text or positive samples with negative sample text so that the model focuses more on the text with ADE label information. It is worth noting that we join two samples together in a random order, and the length of the new samples does not exceed the length of the pre-trained model, so we do not need additional processing.

Proceedings of the 29th International Conference on Computational Linguistic, pages 4–6
October 12–17, 2022.

Model set-up	Precision	Recall	F1
Deberta + over-sampling + FGM	0.79	0.59	0.67
Deberta + over-sampling + FGM + weighted-fusion	0.79	0.61	0.69
Average scores	0.65	0.50	0.56

Table 1: Results of Task 1a on the test set.

Model We use the Deberta-v3-large (He et al., 2021) model as a text encoder and then use adversarial training and a learning rate cosine transform strategy to aid the training. The [CLS] vectors of all hidden layers are weighted and fused, and the weights keep increasing as the number of layers increases. Finally, binary classification is output by linear layer mapping.

2.2 Experiments and Results

We set the batch size to 64 and use the Adamw (Loshchilov and Hutter, 2017) optimizer for training. For the Deberta parameter, We set the learning rate of the pre-trained model to 2e-6 and set the learning rate of the other layers to 1e-3. The learning rate decay strategy uses the Cosine Annealing Warm Restarts (Loshchilov and Hutter, 2016) method, and the weight decay factor is set to 0.001. The adversarial training is performed using FGM (Miyato et al., 2016). The number of iterations is 20 rounds using five-fold cross-validation training. The experimental results are shown in Table 1. As can be seen, the results show that our use of [CLS] weight summation does improve the robustness of the model and our model effect exceeds the average score.

3 Task 1b: ADE Span Detection

This subtask task extracts the location of ADE entities from the text and can be considered a sequence annotation task. Data distribution is the same as task 1a.

3.1 Method

Preprocessing We process the data using the method described in Section 2.2. We only use the data with label information for training, so the model of task 1a is used to identify the text labeled ADE as the input of task 1b during inference.

Model The pre-training model selection is the same as task 1a, and we use the W2NER (Li et al., 2022) model architecture instead of the traditional BERT+CRF architecture. We combine character information and location information at the encoding side, use inflated convolution at the decoding side to obtain entity information of different sizes, and finally use matrix location decoding instead of CRF (Lafferty et al., 2001).

3.2 Experiments and Results

We set the batch size to 8 and used the Adamw optimizer for training. The learning rate size is 1e-5, the learning rate decay strategy uses cosine decay, ten epochs are trained, the decay factor is 0.001, the size of the expanded convolution is set to [1,2,3,4], the dropout size is 0.5, and the character information and location information is set to 50. The experimental results are shown in Table 2. The results show that the solution we use exceeds the average score in all three metrics: accuracy, recall, and F1, indicating the effectiveness of our strategy.

Model set-up	Precision	Recall	F1
Ours	0.68	0.62	0.65
Average scores	0.54	0.52	0.53

Table 2: Results of Task 1b on the test set.

4 Task 1c: Normalization

The task is to extract the ADE keywords in tweet data and match these ADE keywords to the correct standard ADE terms. There are two difficulties in this task. First, there are only 1000+ tweet data containing ADE keywords, which constrains us not from using overly complex models. Secondly, there are more than 30,000 standard ADE terms, and most of the standard ADE terms are not match the tweet data.

4.1 Method

Preprocessing We use the pipeline model for task1c. We use the recall model(Huang et al., 2013) to recall the standard ADE terms, and then we use the ranking model to rank the standard ADE terms and select the best matching standard ADE term. In the recalled model, we found that the model tends to confuse similar terms, so we set the labels of these similar terms to 0.5 and added them to the training data as pseudo-data.

Model In the recalled model, we extract ADE keywords from the tweet data by task1b and tokenize the keywords, output word vectors by the pre-trained DeBERTa model, and average pool these word vectors into a keyword vector. We do the same for the standard ADE term to get the corresponding word vector. We calculate the dot product on these two vectors, sort the results, and set a threshold to filter the standard ADE terms. In the ranking model, we believe that the contextual information of the ADE keywords also contributes to word matching. We use [SEP] to concatenate the ADE keyword with the corresponding tweet sentence, feed it into the DeBERTa model, and adopt the output vector corresponding to [CLS] as the vector for that keyword. We do the same for the candidate standard ADE terms to obtain the vectors of candidate words. We dot the product keyword vector with candidate word vectors and get the highest scored standard ADE term predicted by the model.

4.2 Experiments and Results

We train and test on three pre-trained models, the Bert-base model, the ALBERT model, DeBERTa model, and the best result is obtained on the DeBERTa pre-trained model. Moreover, by adjusting the learning rate of CLS layer to 1e-2, dropout rate to 0.3, label smooth rate to 0.1, adding FGM perturbation, R-Drop loss function(Wu et al., 2021), warn up epoch to 0.3, the model achieves the best result on the seventh epoch. The experimental results are shown in Table 3.

Model set-up	Precision	Recall	F1
Ours	0.40	0.34	0.37
Average scores	0.12	0.11	0.12

Table 3: Results of Task 1c on the test set.

5 Conclusion

In this paper, we use task-specific methods for each of the three subtasks of classification, extraction, and normalization. We enhance the effectiveness of our model through various strategies and demonstrate the effectiveness of the model. In future work, we will specifically target pre-training tasks in the negative drug effect vertical, such as how to make the model acquire prior knowledge of drugs. .

References

Luis Gascó Darryl Estrada-Zavala Martin Krallinger Yuting Guo Yao Ge Abeed Sarker Ana Lucia Schmidt Raul Rodriguez-Esteban Mathias Leddin Arjun Magge Juan M. Banda Vera Davydova Elena Tutubalina Graciela Gonzalez-Hernandez Davy Weissenbacher, Ari Z. Klein. 2022. Overview of the seventh social media mining for health applications smm4h shared tasks at coling 2022. In *In Proceedings of the Seventh Social Media Mining for Health (SMM4H) Workshop and Shared Task.*

Pengcheng He, Jianfeng Gao, and Weizhu Chen. 2021. Debertav3: Improving deberta using electra-style pre-training with gradient-disentangled embedding sharing. *arXiv preprint arXiv:2111.09543.*

Po-Sen Huang, Xiaodong He, Jianfeng Gao, Li Deng, Alex Acero, and Larry Heck. 2013. Learning deep structured semantic models for web search using clickthrough data. In *Proceedings of the 22nd ACM international conference on Information & Knowledge Management*, pages 2333–2338.

John Lafferty, Andrew McCallum, and Fernando Pereira. 2001. Conditional random fields: Probabilistic models for segmenting and labeling sequence data. *international conference on machine learning.*

Jingye Li, Hao Fei, Jiang Liu, Shengqiong Wu, Meishan Zhang, Chong Teng, Donghong Ji, and Fei Li. 2022. Unified named entity recognition as word-word relation classification.

Ilya Loshchilov and Frank Hutter. 2016. Sgdr: Stochastic gradient descent with warm restarts. *arXiv preprint arXiv:1608.03983.*

Ilya Loshchilov and Frank Hutter. 2017. Decoupled weight decay regularization. *arXiv preprint arXiv:1711.05101.*

Arjun Magge, Elena Tutubalina, Zulfat Miftahutdinov, Ilseyar Alimova, Anne Dirkson, Suzan Verberne, Davy Weissenbacher, and Graciela Gonzalez-Hernandez. 2021. Deepademiner: a deep learning pharmacovigilance pipeline for extraction and normalization of adverse drug event mentions on twitter. *Journal of the American Medical Informatics Association*, 28(10):2184–2192.

Takeru Miyato, Andrew M. Dai, and Ian Goodfellow. 2016. Adversarial training methods for semi-supervised text classification. *Learning.*

Lijun Wu, Juntao Li, Yue Wang, Qi Meng, Tao Qin, Wei Chen, Min Zhang, Tie-Yan Liu, et al. 2021. R-drop: Regularized dropout for neural networks. *Advances in Neural Information Processing Systems*, 34:10890–10905.

dezzai@SMM4H'22: Tasks 5 & 10 - Hybrid models everywhere

Miguel Ortega-Martín
dezzai, UCM
m.ortega@dezzai.com
m.ortega@ucm.es

Alfonso Ardoiz
dezzai, UCM
alfonso.ardoiz@dezzai.com
aardoiz@ucm.es

Jorge Álvarez
dezzai
jorge.alvarez@dezzai.com

Óscar García-Sierra
dezzai, UCM
oscar.garcia@dezzai.com
oscarg02@ucm.es

Adrián Alonso
dezzai, URJC, Data Science Lab URJC
a.alonso@dezzai.com
adrian.barriuso@urjc.es

Abstract

This paper presents our approaches to SMM4H'22 task 5 - Classification of tweets of self-reported COVID-19 symptoms in Spanish, and task 10 - Detection of disease mentions in tweets – SocialDisNER (in Spanish). We have presented hybrid systems that combine Deep Learning techniques with linguistic rules and medical ontologies, which have allowed us to achieve outstanding results in both tasks.

1 Introduction

The Social Media Mining for Health Applications (SMM4H) (Weissenbacher et al., 2022) workshop aims to promote automatic methods for mining social media data for health informatics. In order to support Spanish Natural Language Processing (NLP), we have focused on both Spanish tasks.

The SMM4H 2022 task 5 focuses on classification of tweets containing self-reports of COVID-19 symptoms in Spanish. It consists of a triple classification task in which participants have to distinguish personal symptoms from symptoms reported by others and references to news articles or other sources, giving rise to three labels: non-personal reports, news and literature mentions, and self-reports.

In the case of SMM4H task 10 - Detection of disease mentions in tweets – SocialDisNER (in Spanish) (Gasco et al., 2022), the task is focused on the recognition of disease mentions in tweets written in Spanish, with the aim of using social media to better understand societal perception of diseases. Therefore, we have tackled this task as a NER offset detection and token classification problem.

Our contributions to these tasks are the following:

- remarkable results for both tasks.
- an analysis of the tasks and the datasets used in them.
- a comparison of several approaches, finding that combining Deep Learning models with linguistic and clinical knowledge achieves the best results.

2 System description

2.1 Task 5

Our pipeline is based on a linguistic preprocessing that allows us to split the tweets in two subgroups, with each of which we use a different RoBERTa (Liu et al., 2019) classifier.

In first place we apply a filter where two subsets of the training set are created. For each tweet we analyze the number of "first person flags", that is, the number of first person verbs and pronouns, and the number of "other person flags", that is, the number of verbs which are not in first person. Then we use a threshold to compare this values: if at least a third of the total flags correspond "first person flags" the tweet is added to subset 1, and else to subset 2. This way only the tweets which the system filter as subset 1 can be labeled as self-report.

In second place, we trained 2 XLM Roberta base classifiers: one ternary classifier trained on all 3 labels from training set, and one binary classifier with only the "news and literature mentions" and the "non-personal report" labeled tweets from the training set. Subsequently, during inference, we apply the previously explained filter on the test dataset. In the end the 3-label classifier is used for subset 1 and the binary classifier for subset 2. This

Proceedings of the 29th International Conference on Computational Linguistic, pages 7–10
October 12–17, 2022.

Model	Post-processing	Precision	Recall	F1 score
RoBERTa	no	0.8434	0.8434	0.8434
RoBERTa	yes	0.8445	0.8445	0.8445
Hybrid	yes	0.8487	0.8487	0.8487

Table 1: Task 5 validation results

Model	Nº predicted	Overlap P	Overlap R	Overlap F1	Strict P	Strict R	Strict F1
RoBERTa	4270	0.951	0.948	0.949	0.852	0.856	0.854
Hybrid	4364	0.939	0.957	0.948	0.856	0.874	0.865

Table 2: Task 10 validation results

way we achieved slightly better results than using just one ternary classifier for all the tweets.

As seen in table 1, we found that cleaning emojis and hashtags slightly improved validation results.

2.2 Task 10

We face this task as a token classification problem with a hybrid system where we combine the outputs from a Transformer model and a medical rule-based NER model. However, this architecture focus heavily in the Transformer side and use the rule model just as a complement. Therefore, we have separately submitted the hybrid system approach results and Transformer only approach results.

As Transformer model we use multilingual XLM-RoBERTa-base (Conneau et al., 2019) as the base model in order to address the multilingual annotation stance of some tweets. This model is fine-tuned in order to adapt it to the token classification task using the IOB tag format strategy (Ramshaw and Marcus, 1995). To perform this strategy, a pre-processing was needed in order to convert all the tokens in the training dataset to this tag set format.

Additionally, a medical rule-based system is built using spaCy utilities and UMLS annotations (Bodenreider, 2004). In particular a pre-built Spanish spaCy model is used and a matcher using all the terms associated to the semantic type T047 "Disease or Syndrome" is built. This technique allows us to find more rare diseases that the Transformer model is not able to recognize since it has not been trained on them. However, this rule-based matcher cannot deal with orthography errors or entities in other languages.

2.3 Model hyperparameters

Models for both tasks were fine-tuned using 4 Nvidia Tesla v100 32GB. Our experiments showed that task 5 models converge after 3 epoch and task 10 model after 5 epochs. Regarding the hyperparameters, the following optimized the evaluation on the development sets for both tasks: batch size = 32, Adam epsilon = 1e-8, learning rate = 5e-5, warm up ratio = 0.1, weight decay = 0.0.

3 Dataset, results and error analysis

3.1 Task 5

Task 5 dataset contains 10,052 training tweets (1,654 labeled as self-reports, 2,413 labeled as non-personal reports and 5,985 as literature/news mentions); 3,578 validation tweets and 6,851 test tweets. Task was evaluated using precision, recall and F1 for the positive class (self-report). Evaluation on validation set showed results seen in table 1.

We consider the quality and inconsistency in the labeling of the dataset to be one of the main problems we have faced, which in many cases leads to confusion between the self-report and the non-personal-report labels. In this regard, we have identified three groups of errors:

(i) Incorrect inclusion of tweets with no symptoms in the training, validation and test sets.

Model	Post-processing	Precision	Recall	F1 score
RoBERTa	yes	0.8464	0.8464	0.8464
Hybrid	yes	0.8521	0.8521	0.8521

Table 3: Task 5 test results

Model	Nº Predict	Strict P	Strict R	Strict F1
RoBERTa	30066	0.821	0.826	0.824
Hybrid	31297	0.828	0.845	0.836

Table 4: Task 10 test results

(ii) Not a clear answer to how to handle the indirect speech annotations (self-reports and non-personal-reports are assigned unclearly in these cases).

(iii) Inconsistencies when labeling tweets describing other people COVID-19 symptoms (in this case, there are more samples labeled as non-personal reports than self-reports).

We developed a filtering system in order to overcome this annotations problem. The main idea behind it is to reinforce the weak barrier between self-reports and non-personal reports that the training data carries, and to reduce the quantity of false positives that the final model could outcome. As stated before, this way only the tweets which the system filter as group 1 can be labeled as self-report.

As seen in table 3, in the end the model got a good grasp of the task, achieving a 0.8521 test F1 score. We consider that better annotated data could lead the model to a better performance.

3.2 Task 10

Task 10 dataset contains 5000 train tweets, 2500 validation tweets and 23430 (2000) test tweets. Task was evaluated with Strict Precision, Recall and F1-score as Leaderboard Scores, and with Overlapping Precision, Recall and F1-score as Additional Scores.

We have encountered some minor issues of wrong entity labeling through the dataset. For instance, there are tweets where a term mainly related with diseases is used with another meaning and therefore should not be labeled as a disease; some tweets in both Spanish and English which contain annotated diseases in both languages; or incoherence in the annotation of non-strict diseases like "fumar" (in English, "smoke") or "enfermedad" (in English, "disease") which are only annotated in some cases.

Table 2 shows our results in the validation dataset. As it can be seen, there is a huge contrast between strict and overlapping scores. This is mainly caused by our approach as a token-based classification problem, which heavily relies on the RoBERTa tokenization process, where the first part of many entities is not classified as entity but as not-entity, therefore the strict score decreases, but the overlapping score keeps its robustness. Table 4 contains our final test results for task 10.

4 Conclusions

In this paper we have reviewed our approaches to SMM4H'22 tasks 5 and 10. For both tasks we have developed hybrid systems combining Deep Learning techniques with rule-based modules. We have analysed our results as well as both datasets, and presented some insights of each one. We also reviewed the limitations of our models and highlighted their weaknesses and strengths. As seen in table 3, in the end we achieved a score of 0.8521 in the test set for task 5. For Task 10, as seen in table 4, we achieved a Strict Precision of 0.828, a Strict Recall of 0.845 and a Strict F1 of 0.836. For both tasks these represents remarkable and top-of-the-chart results.

We consider this two tasks specially relevant due to the need to promote NLP in Spanish, and even more so when applied to the clinical domain using real data.

References

Olivier Bodenreider. 2004. The unified medical language system (umls): integrating biomedical terminology. *Nucleic Acids Res.*, 32(Database-Issue):267–270.

Alexis Conneau, Kartikay Khandelwal, Naman Goyal, Vishrav Chaudhary, Guillaume Wenzek, Francisco Guzmán, Edouard Grave, Myle Ott, Luke Zettlemoyer, and Veselin Stoyanov. 2019. Unsupervised cross-lingual representation learning at scale. *CoRR*, abs/1911.02116.

Luis Gasco, Darryl Estrada-Zavala, Eulàlia Farré-Maduell, Salvador Lima-López, Antonio Miranda-Escalada, and Martin Krallinger. 2022. Overview of the SocialDisNER shared task on detection of diseases mentions from healthcare related and patient generated social media content: methods, evaluation and corpora. In *Proceedings of the Seventh Social Media Mining for Health (#SMM4H) Workshop and Shared Task.*

Yinhan Liu, Myle Ott, Naman Goyal, Jingfei Du, Mandar Joshi, Danqi Chen, Omer Levy, Mike Lewis, Luke Zettlemoyer, and Veselin Stoyanov. 2019. Roberta: A robustly optimized bert pretraining approach. *arXiv preprint arXiv:1907.11692.*

Lance A. Ramshaw and Mitchell P. Marcus. 1995. Text chunking using transformation-based learning. *CoRR*, cmp-lg/9505040.

Davy Weissenbacher, Ari Klein, Luis Gascó, Darryl Estrada-Zavala, Martin Krallinger, Yuting, Yao Ge, Abeed Sarker, Ana Lucia Schmidt, Raul Rodriguez-Esteban, Mathias Leddin, Arjun Magge, Juan M. Banda, Vera Davydova, Elena Tutubalina, and Graciela Gonzalez-Hernandez. 2022. Overview of the Seventh Social Media Mining for Health Applications #SMM4H Shared Tasks at COLING 2022. In *Proceedings of the Seventh Social Media Mining for Health (#SMM4H) Workshop and Shared Task.*

zydhjh4593@SMM4H'22: A Generic Pre-trained BERT-based Framework for Social Media Health Text Classification

Chenghao Huang[1], Xiaolu Chen[1], Yuxi Chen[1], Yutong Wu[2],
Weimin Yuan[1], Yan Wang[1], and Yanru Zhang[1,3]

[1]University of Electronic Science and Technology of China
[2]University of Nottingham Ningbo China
[3]Shenzhen Institute of Advanced Study, UESTC
Email: zydhjh4593@gmail.com yanruzhang@uestc.edu.cn

Abstract

This paper describes our proposed framework for the 10 text classification tasks of Task 1a, 2a, 2b, 3a, 4, 5, 6, 7, 8, and 9, in the Social Media Mining for Health (SMM4H) 2022. According to the pre-trained BERT-based models, various techniques, including regularized dropout, focal loss, exponential moving average, 5-fold cross-validation, ensemble prediction, and pseudo-labeling, are applied for further formulating and improving the generalization performance of our framework. In the evaluation, the proposed framework achieves the **1st place** in Task 3a with a 7% higher F_1-score than the median, and obtains a 4% higher averaged F_1-score than the median in all participating tasks except Task 1a.

1 Introduction

Social media platforms such as Twitter and Reddit are full of various statements made by users, and there is abundant social media data that can be used for mining health-related information. Natural language processing (NLP) plays an important role in social media data manipulation, which helps solve issues including informal expressions, misspelling of terms, noise, and multiple languages in the tweets. The intersection of social media data mining and health applications is concerned in the 7th Social Media Mining for Health Applications (SMM4H) workshop (Weissenbacher et al., 2022). Our team participates in 10 classification tasks of the 7th SMM4H shared tasks, which are Task 1a (Magge et al., 2021), 2a, 2b (Davydova and Tutubalina, 2022), 3a, 4-9.

In this paper, a generic framework is formulated to fine-tune pre-trained bidirectional encoder representations from transformers (BERT)-based models on text classification datasets of SMM4H 2022 shared tasks. For obtaining higher accuracy, we adopt 5-fold cross-validation (CV) to get 5 models and use an ensemble learning technique to obtain the result. Furthermore, considering the evaluation performance on test datasets, we fine-tune our models using regularized dropout (R-drop) (Wu et al., 2021) and exponential moving average (EMA) to increase the models' generalization capability. To deal with data imbalance, focal loss (Lin et al., 2017) is adopted to assign higher weights to samples with minority classes. Additionally, pseudo-labeling (Lee et al., 2013) is adopted as an attempt for improvement. As a result of applying the above techniques, our framework obtains a 4% higher averaged F_1-score than the median in all participating tasks except Task 1a, and a 7% higher F_1-score than the median in Task 3a.

2 Preliminaries

2.1 Problem Description and Datasets

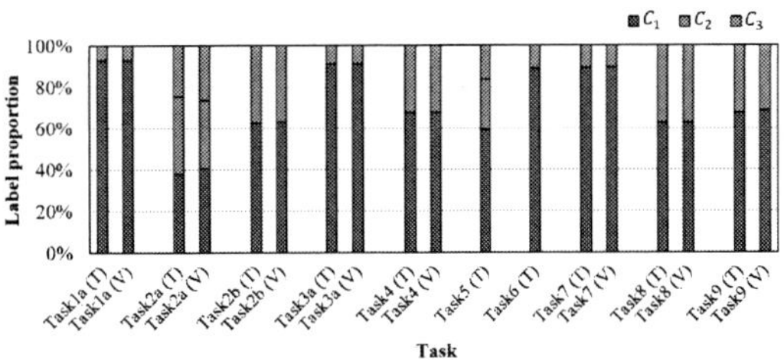

Figure 1: The proportion of different labels in the SMM4H 2022 Task 1a, 2a, 2b, 3a, 4-9, where T and V represent each corresponding task's training dataset and validation dataset, respectively. For Task 1a, C_1 is *noADE* and C_2 is *ADE*. For Task 2a, C_1, C_2, and C_3 are *FAVOR, NONE*, and *AGAINST*, respectively. For Task 5, C_1, C_2, and C_3 are *lit-news_mention*, *non-personal_report*, and *self_report*. For Task 6, C_1 is *Vaccine_chatter* and C_2 is *Self_reports*. C_1 and C_2 of the rest tasks are 0 and 1, respectively.

We participate in 10 text classification tasks, including 8 binary classification tasks and 2 three-way classification tasks summarized as follows.

Binary classification tasks:

- **Task 1a** aims to classify if there is an adverse

Proceedings of the 29th International Conference on Computational Linguistic, pages 11–15
October 12–17, 2022.

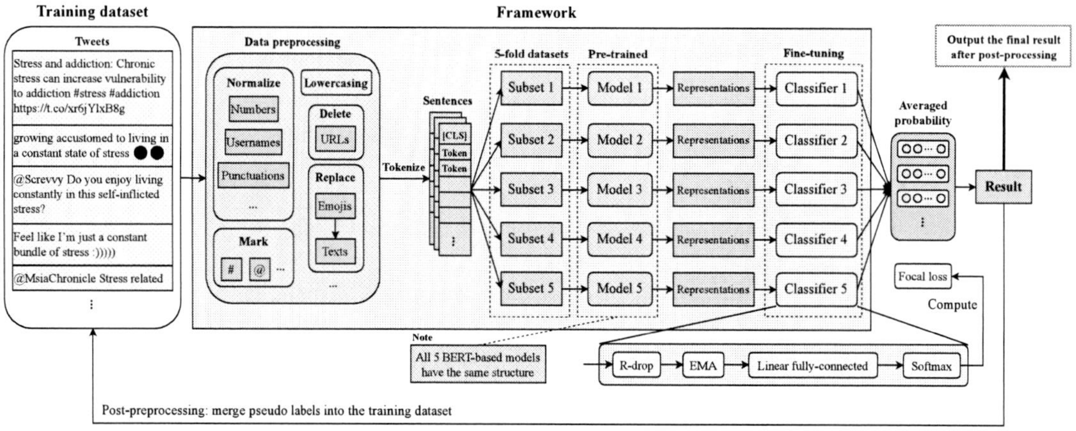

Figure 2: Our framework architecture for the SMM4H 2022 Task 1a, 2a, 2b, 3a, 4-9.

drug event (ADE) in an English tweet.

- **Task 2b** aims to classify if at least one premise/argument is mentioned in a COVID-related English tweet.
- **Task 3a** aims to classify if there is a medication treatment change in an English tweet.
- **Task 4** aims to determine if a self-reporting age is exact in an English tweet.
- **Task 6** aims to classify if there is a COVID-19 vaccination confirmation or just related chatter in an English tweet.
- **Task 7** aims to classify if intimate partner violence exists in an English tweet.
- **Task 8** aims to classify if chronic stress exists in an English tweet (Yang et al., 2022).
- **Task 9** aims to determine if a self-reporting age is exact in an English posting on Reddit.

Three-way classification tasks:

- **Task 2a** aims to determine if a stance is positive, negative, or neutral in an English tweet.
- **Task 5** aims to distinguish if a Spanish tweet about COVID-19 symptoms is a self-report, a non-personal report, or a literature/news mention.

For further analysis, we count the label proportions of each task, illustrated in Figure 1, where C_1, C_2, and C_3 represents different labels in each task. Note that validation datasets of Task 5 and 6 have no labels before the release of the final results.

2.2 Related Work

In recent years, BERT (Devlin et al., 2018), a pre-trained transformer-based model for language representation, has become increasingly successful in text classification. An upgraded version of BERT called RoBERTa (Liu et al., 2019) produces better processing outcomes for NLP tasks. The first pre-trained language model for English tweets, BERTweet (Nguyen et al., 2020), which is built on BERT and was trained using the RoBERTa pre-training procedure, performs better on Twitter NLP tasks. In previous SMM4H workshops, a number of participants used BERT-based models with encouraging results in text classification tasks (Valdes et al., 2021; Ramesh et al., 2021; Sakhovskiy et al., 2021).

3 Framework

A generic framework is formulated for all classification tasks, which is shown in Figure 2. Before model training, simple data preprocessing is conducted. Then, to encode tweet texts, BERT-based classifiers are trained using 5-fold CV. During the inference phase, ensemble learning is applied for more accurate prediction over individual classifiers. In an attempt to further improve the performance, pseudo-labeling based on the inferred test dataset is used.

3.1 Preprocessing

Firstly, to reduce tweets' noise, data preprocessing methods, such as lowercasing, deleting URLs, replacing emojis with their text strings, normalization of numbers, punctuations, and usernames, and marking special characters (#, @, ...), are generally applied on tweets in all participating tasks. Then the training dataset is shuffled and divided evenly into 5 subsets for 5-fold CV during the training phase. After training, 5 models are obtained for the ensemble prediction during the inference phase. Particularly, tweets in validation datasets of Task 5 and 6 have no labels. For each of them, we simply use the training dataset for 5-fold CV.

3.2 Model Training and Inference

Considering the increasing prevalence of BERT-based model usage in text classification, we propose to adopt pre-trained BERT-based models and fine-tune them using different downstream tasks, i.e., shared tasks in SMM4H 2022. By inputting tokenized tweet texts and other information, such as claims in Task 2a and 2b, to BERT-based models, token representations are obtained. Then, we pass these representations through a linear fully-connected layer and a softmax activation function to get the probability of each class. For acquiring more accurate and robust predictions, 5 models are individually trained through 5-fold CV.

During the training phase, R-drop is adopted to alleviate the model overfitting. By constraining the parameter space through symmetric Kullback-Leibler divergence, R-drop is also capable of reducing randomness caused by the traditional dropout technique. Furthermore, in order to further raise the performance on the test dataset, EMA, which averages model parameters in the latest few steps, is applied to improve the generalization capability of our models. In addition, Figure 1 illustrates that the label proportions show different degrees of imbalance. As BERT-based models and fully-connected layers are constructed for our framework, focal loss, which is designed for neural networks, is adopted to prioritize samples with minority labels that are hard to be classified.

During the inference phase, an ensemble technique is used to combine 5 models' results for better performance. The outputs of the 5 models, i.e., all 5 probability vectors, are averaged. Then, the result is obtained according to the highest values in the averaged probability vector.

3.3 Post-processing

To further improve our models' performance, pseudo-labeling is adopted in the post-processing phase. After inference, prediction on test dataset is set as the pseudo labels. Then, training dataset, validation dataset, and test dataset are merged into an entire dataset for another training. Finally, the latest 5 models conduct ensemble inference and output the final prediction for submission.

4 Experiments

4.1 Setup

To compare the performance of different pre-trained models, 3 BERT-based models pre-trained on English corpora are experimented with: (i) BERT, (ii) RoBERTa, and (iii) BERTweet. Note that validation datasets' labels of Task 5 and 6 are released as well as the results. For the offline evaluation, we conduct inference on the validation datasets to sum the micro F_1-score and the macro F_1-score of each task and calculate the mean value. All simulations and approaches are coded in Python and conducted on a personal computer with an NVIDIA GeForce RTX 3090 Ti GPU. Adam optimizer is used (Kingma and Ba, 2014). The hyperparameters are listed in Appendix A.

4.2 Online Evaluation

As shown in Table 1, our framework obtains a 4% higher averaged F_1-score than the median in all participating tasks except Task 1a, and a 7% higher F_1-score than the median in Task 3a. Besides, our framework's precision values of Task 3a and 4 are 6% and 5% higher than the median, respectively. Our framework's recall values of Task 3a, 7, and 8 are 7%, 8%, and 10% higher than the median, respectively. In Task 1a, though our framework's F_1-score and recall value are 2.5% and 7% lower than the mean, respectively, our precision is about 8% higher than the mean.

Task	F_1-score		Precision		Recall	
	Ours	Median	Ours	Median	Ours	Median
2a	**0.586**	0.550	/	/	/	/
2b	**0.701**	0.647	/	/	/	/
3a	**0.655**	0.586	**0.682**	0.617	**0.630**	0.557
4	**0.914**	0.869	**0.921**	0.869	**0.908**	0.889
5	**0.860**	0.840	**0.860**	0.840	**0.860**	0.840
6	**0.800**	0.770	**0.900**	0.900	**0.710**	0.680
7	**0.795**	0.763	**0.795**	0.790	**0.795**	0.716
8	**0.792**	0.750	**0.734**	0.720	**0.859**	0.760
9	**0.891**	0.891	0.893	**0.896**	0.889	**0.919**

Table 1: Our results and the median results on the test datasets of all participating tasks, where the best results are in bold.

4.3 Offline Evaluation

We evaluate the effects of different pre-trained models and techniques on our framework. As shown in Table 2, when not using pseudo labels, our framework achieves the highest F_1-score in Task 1a, 2a, 2b, and 3a. In the rest of the tasks, the use of pseudo labels shows the best performance.

Models	Task1a	Task2a	Task2b	Task3a	Task4	Task5	Task6	Task7	Task8	Task9
BERT	0.558	0.534	0.660	0.587	0.824	0.806	0.737	0.730	0.732	0.814
RoBERTa	0.585	0.551	0.682	0.619	0.858	0.833	0.764	0.755	0.759	0.847
BERTweet (BT)	0.606	0.574	0.707	0.645	0.897	0.869	0.798	0.784	0.791	0.883
BT+OS w/o FL	0.579	0.566	0.699	0.611	0.882	0.854	0.757	0.749	0.776	0.874
BT+Dropout	0.639	0.618	0.751	0.692	0.945	0.908	0.852	0.822	0.827	0.930
BT+RD	0.654	0.630	0.761	0.705	0.951	0.917	0.862	0.833	0.836	0.928
BT+RD+EMA	**0.662**	**0.642**	**0.768**	**0.718**	0.955	0.924	0.872	0.841	0.843	0.934
BT+RD+EMA+PS	0.638	0.625	0.752	0.713	**0.971**	**0.932**	**0.876**	**0.842**	**0.851**	**0.945**

Table 2: Offline F_1-scores of different models with various techniques on the validation dataset of each participating task, where oversampling, focal loss, R-drop, exponential moving average, and pseudo-labeling are abbreviated as OS, FL, RD, EMA, and PS, respectively, and the best results are in bold. Note that 5-fold CV and focal loss are applied in all models.

5 Result Analysis

Through the results shown in Table 2, we can see that BERTweet outperforms BERT and RoBERTa in all participating tasks, indicating the benefit of pre-training on the right corpora. Because the data of Task 9 is from Reddit, our framework's online F_1-score, precision, and recall in Task 9 are in line with the median, shown in Table 1. Besides, the performance on Task 5 is slightly worse than other tasks, which can be attributed to the Spanish data.

In terms of training techniques, the 3rd and 5th rows of Table 2 show that in all participating tasks, the application of dropout improves around 5% on the BERTweet model, indicating that dropout is crucial to help BERT-based models avoid overfitting. Moreover, the effectiveness of R-drop is proved as the performance of R-drop is around 1% higher than the traditional dropout. In addition, EMA helps with the fine-tuning convergence, improving around 0.7% on R-drop-based BERTweet in all participating tasks.

To verify the effectiveness of focal loss, we compare focal loss with oversampling, which randomly duplicates samples with minority classes until achieving data balance. As shown in the 3rd and 4th rows of Table 2, F_1-scores of using focal loss are all higher than oversampling, especially on validation datasets of Task 1a, 3a, 6, and 7, where labels are significantly imbalanced. The reason is that though oversampling enlarges the sizes of training data, overfitting may occur in samples with minority classes. Besides, as the dataset sizes in all participating tasks are not large, undersampling, which will lead to overfitting by reducing the training datasets, is discarded.

Additionally, Table 2 shows that pseudo-labeling is not effective on all participating tasks. We count labeled and unlabeled data, which are shown in Figure 3. As for Task 1a, 2a, 2b, and 3a, where the prediction performance is not as expected, and the size of each test dataset is fairly large. In such a situation, pseudo-labeling will weaken the performance of models. On the contrary, in the other tasks, pseudo-labeling is effective. Therefore, it is concluded that pseudo-labeling only contributes to models which have already obtained outstanding performance.

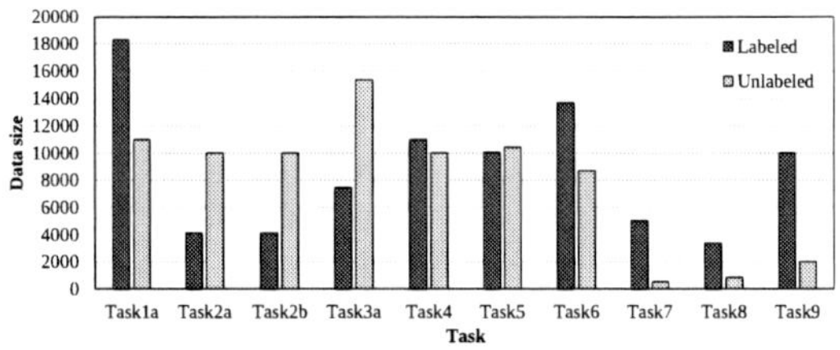

Figure 3: The statistics of labeled data and unlabeled data in the SMM4H 2022 Task 1a, 2a, 2b, 3a, 4-9.

6 Conclusion

In this work, A generic framework based on pre-trained BERT-based models is formulated for 10 text classification tasks, which are Task 1a, 2a, 2b, 3a, 4-9 in the SMM4H 2022. Through 5-fold CV, 5 models are trained and fine-tuned with R-drop, EMA, and focal loss. Then, during the inference, ensemble prediction from 5 models is outputted. In addition, pseudo-labeling is applied for the better fitting capability of our models. As a result, our framework obtains a 4% higher averaged F_1-score than the median in all participating tasks except Task 1a, and achieves the 1st place in Task 3a with a 7% higher F_1-score than the median.

For future work, in order to sufficiently utilize labeled data, contrastive learning (Khosla et al., 2020) will be studied between tweet texts and other information, such as claims in Task 2.

References

Vera Davydova and Elena Tutubalina. 2022. Smm4h 2022 task 2: Dataset for stance and premise detection in tweets about health mandates related to covid-19. In *Proceedings of the Seventh Social Media Mining for Health Applications (#SMM4H) Workshop & Shared Task*, pages –.

Jacob Devlin, Ming-Wei Chang, Kenton Lee, and Kristina Toutanova. 2018. Bert: Pre-training of deep bidirectional transformers for language understanding. *arXiv preprint arXiv:1810.04805*.

Prannay Khosla, Piotr Teterwak, Chen Wang, Aaron Sarna, Yonglong Tian, Phillip Isola, Aaron Maschinot, Ce Liu, and Dilip Krishnan. 2020. Supervised contrastive learning. *Advances in Neural Information Processing Systems*, 33:18661–18673.

Diederik P Kingma and Jimmy Ba. 2014. Adam: A method for stochastic optimization. *arXiv preprint arXiv:1412.6980*.

Dong-Hyun Lee et al. 2013. Pseudo-label: The simple and efficient semi-supervised learning method for deep neural networks. In *Workshop on challenges in representation learning, ICML*, page 896.

Tsung-Yi Lin, Priya Goyal, Ross Girshick, Kaiming He, and Piotr Dollár. 2017. Focal loss for dense object detection. In *Proceedings of the IEEE international conference on computer vision*, pages 2980–2988.

Yinhan Liu, Myle Ott, Naman Goyal, Jingfei Du, Mandar Joshi, Danqi Chen, Omer Levy, Mike Lewis, Luke Zettlemoyer, and Veselin Stoyanov. 2019. Roberta: A robustly optimized bert pretraining approach. *arXiv preprint arXiv:1907.11692*.

Arjun Magge, Elena Tutubalina, Zulfat Miftahutdinov, Ilseyar Alimova, Anne Dirkson, Suzan Verberne, Davy Weissenbacher, and Graciela Gonzalez-Hernandez. 2021. Deepademiner: a deep learning pharmacovigilance pipeline for extraction and normalization of adverse drug event mentions on twitter. *Journal of the American Medical Informatics Association*, 28(10):2184–2192.

Dat Quoc Nguyen, Thanh Vu, and Anh Tuan Nguyen. 2020. Bertweet: A pre-trained language model for english tweets. *arXiv preprint arXiv:2005.10200*.

Sidharth Ramesh, Abhiraj Tiwari, Parthivi Choubey, Saisha Kashyap, Sahil Khose, Kumud Lakara, Nishesh Singh, and Ujjwal Verma. 2021. BERT based transformers lead the way in extraction of health information from social media. In *Proceedings of the Sixth Social Media Mining for Health (#SMM4H) Workshop and Shared Task*, pages 33–38, Mexico City, Mexico. Association for Computational Linguistics.

Andrey Sakhovskiy, Zulfat Miftahutdinov, and Elena Tutubalina. 2021. KFU NLP team at SMM4H 2021 tasks: Cross-lingual and cross-modal BERT-based models for adverse drug effects. In *Proceedings of the Sixth Social Media Mining for Health (#SMM4H) Workshop and Shared Task*, pages 39–43, Mexico City, Mexico. Association for Computational Linguistics.

Alberto Valdes, Jesus Lopez, and Manuel Montes. 2021. UACH-INAOE at SMM4H: a BERT based approach for classification of COVID-19 Twitter posts. In *Proceedings of the Sixth Social Media Mining for Health (#SMM4H) Workshop and Shared Task*, pages 65–68, Mexico City, Mexico. Association for Computational Linguistics.

Davy Weissenbacher, Ari Z. Klein, Luis Gascó, Darryl Estrada-Zavala, Martin Krallinger, Yuting Guo, Yao Ge, Abeed Sarker, Ana Lucia Schmidt, Raul Rodriguez-Esteban, Mathias Leddin, Arjun Magge, Juan M. Banda, Vera Davydova, Elena Tutubalina, and Graciela Gonzalez-Hernandez. 2022. Overview of the seventh social media mining for health applications #smm4h shared tasks at coling 2022. In *Proceedings of the Seventh Social Media Mining for Health Applications (#SMM4H) Workshop & Shared Task*, pages –.

Lijun Wu, Juntao Li, Yue Wang, Qi Meng, Tao Qin, Wei Chen, Min Zhang, Tie-Yan Liu, et al. 2021. R-drop: Regularized dropout for neural networks. *Advances in Neural Information Processing Systems*, 34:10890–10905.

Yuan-Chi Yang, Angel Xie, Sangmi Kim, Jessica Hair, Ali-Mohammed Al-Garadi, and Abeed Sarker. 2022. Automatic detection of twitter users who express chronic stress experiences via supervised machine learning and natural language processing. *Computers, Informatics, Nursing*.

A Hyperparameters

Module	Hyperparameter	Value
Classifier	Batch size	16
	Updata frequency	10 (epochs)
	Learning rate (LR)	1e-4
	LR decay rate	0.99
	Focal loss factor	0.3
	Adam epsilon	1e-6
	EMA start	0
	EMA decay	0.99
	R-drop rate	0.5
	R-drop weight	5
BERT	Embedding size	768
	Hidden size	768
	Sequence length	128
	Learning rate	2e-5

Table 3: Hyperparameters of the proposed framework.

MANTIS at SMM4H'2022: Pre-Trained Language Models Meet a Suite of Psycholinguistic Features for the Detection of Self-Reported Chronic Stress

Sourabh Zanwar
RWTH Aachen University
sourabh.zanwar@rwth-aachen.de

Daniel Wiechmann
University of Amsterdam
d.wiechmann@uva.nl

Yu Qiao
RWTH Aachen University
yu.qiao@rwth-aachen.de

Elma Kerz
RWTH Aachen University
elma.kerz@ifaar.rwth-aachen.de

Abstract

This paper describes our submission to the Social Media Mining for Health (SMM4H) 2022 Shared Task 8, aimed at detecting self-reported chronic stress on Twitter. Our approach leverages a pre-trained transformer model (RoBERTa) in combination with a Bidirectional Long Short-Term Memory (BiLSTM) network trained on a diverse set of psycholinguistic features. We handle the class imbalance issue in the training dataset by augmenting it by another dataset used for stress classification in social media.

1 Introduction

The global increase in social media use over the past decade has afforded researchers new opportunities to mine health-related information that can ultimately be used to improve public health. The Social Media Mining for Health Applications (SMM4H) Shared Task involved ten natural language processing challenges of using social media data for health research (Weissenbacher et al., 2022). In our submission to the task targeting the classification of self-reported chronic stress on Twitter (Task 8), we built hybrid models that combine pre-trained transformer language models with Bidirectional Long Short-Term Memory (BiLSTM) networks trained on a diverse set of psycholinguistic features.

2 Data

The Twitter data provided by the organizers of Task 8 comprised of a total of 4,195 tweets whose distributions over training, development and testing sets are shown in Table 3 of supplementary material[1]. About 37% of the tweets are positive (self-disclosure of chronic stress, Pos) and 63% are negative (non-self-disclosure of chronic stress, Neg). The only preprocessing step that was applied was the removal of HTML and links from the text. To address the class imbalance in the data, we augmented the data using 1000 items with positive labels and 200 ones with negative labels from the Dreaddit dataset (Turcan and McKeown, 2019).

[1] Supplementary material

2.1 Measurement of Psycholinguistic Features

A set of 435 psycholinguistic features used in our approach fall into the following four categories: (1) features of morpho-syntactic complexity (N=19), (2) features of lexical richness, diversity and sophistication (N=77), (3) readability features (N=14), and (4) lexicon features designed to detect sentiment, emotion and/or affect (N=325). Measurements of these features were obtained using an automated text analysis system (for its recent applications, see e.g. Wiechmann et al. (2022) for predicting eye-moving patterns during reading and Kerz et al. (2022) for detection of Big Five personality traits and Myers–Briggs types). Tokenization, sentence splitting, part-of-speech tagging, lemmatization and syntactic PCFG parsing were performed using Stanford CoreNLP (Manning et al., 2014).

3 Description of System Architecture

We conducted experiments with a total of five models: (1) a fine-tuned Bidirectional Encoder Representations from Transformers (BERT) model (Devlin et al., 2018), (2) a fine-tuned RoBERTa model (Robustly Optimized BERT pre-training Approach) (Liu et al., 2019), (3) a bidirectional neural network classifiers trained on measurements of psycholinguistic features described in Section 2.2, and (4) and (5) two hybrid models integrating BERT and RoBERTa predictions with the psycholinguistic features. For each model, we performed experiments with and without data augmentation.

For (1) and (2) we used the pretrained 'bert-base-uncased' and 'roberta-base' models from the Huggingface Transformers library (Wolf et al., 2020) each with an intermediate BiLSTM layer with 256

Proceedings of the 29th International Conference on Computational Linguistic, pages 16–18
October 12–17, 2022.

Figure 1: Proposed hybrid model architecture

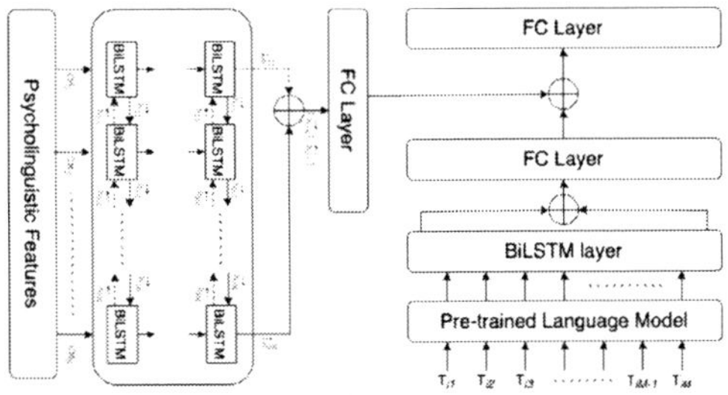

hidden units (Al-Omari et al., 2020). For (3) - the model based solely on psycholinguistic features, we constructed a 2-layer BiLSTM with a hidden state dimension of 32. The input to that model is a sequence $CM_1^N = (CM_1, CM_2 \ldots, CM_N)$, where CM_i, the output of CoCoGen for the ith sentence of a document, is a 435 dimensional vector and N is the sequence length. To predict the labels of a sequence, we concatenate the last hidden states of the last layer in forward ($\overrightarrow{h_n}$) and backward directions ($\overleftarrow{h_n}$). The result vector of concatenation $h_n = [\overrightarrow{h_n}|\overleftarrow{h_n}]$ is then transformed through a 2-layer feedforward neural network, using the Rectifier linear unit (ReLU) is as an activation function. The output of this network is then passed to a Dense Fully Connected (FC) Layer with dropout of 0.2 and is finally passed on to a terminal, fully connected layer. The output is a K dimensional vector, where K is the number of emotion labels. The architecture of the hybrid classification models - models (4) and (5) - consists of two parts: (i) a pre-trained Transformer-based model with a BiLSTM layer and FC layer on top of it and (ii) the psycholinguistic features of the text fed into a BiLSTM layer and a FC layer. We concatenate the outputs of these layers before passing them into a final FC layer with a sigmoid activation function. Since the evaluation is based on the sensitivity of the system – i.e., here the classification of positive labels is more important – , we reduced the threshold of positive labels from 0.5 to 0.3. The model used to generate predictions for the test set was the RoBERTa-PsyLing hybrid model with the following configuration: BiLSTM-PsyLing: 2-layers, hidden size of 512 and dropout 0.2. We trained this model for 12 epochs, saving the model with the best performance from the development set. The optimizer used is AdamW with a learning rate of 2e-5 and a weight decay of 1e-4. We trained 5 models with 5 random 80% splits of the training data and superimposed a meta-learner to get the final predictions.

Table 1: Results on the validation set

Model	Aug	Acc	Prec	Rec	F1
PsyLing	No	69	67	67	67
BERT	No	75	75	77	74
RoBERTa	No	71	77	76	71
BERT-Hybrid	No	78	77	79	78
RoBERTa-Hyb.	No	79	78	79	78
PsyLing	Yes	71	69	67	68
BERT	Yes	79	78	80	78
RoBERTa	Yes	78	79	81	78
BERT-Hybrid	Yes	81	80	81	80
RoBERTa-Hyb.	Yes	82	81	83	**81**

4 Results

An overview of the performance metrics on the validation set is presented in Table 1. We found that the proposed hybrid models consistently outperformed the standard transformer-based baseline models, with an improvement in F1 of up to 3%. Training the models on the augmented data led to an average increase in performance of 3.8% F1 over the non-augmented data. Highest performance on the validation set (F1 = 81) was achieved by the RoBERTa hybrid model. Error inspection indicated that the the majority of errors are False Positives (see Table 5[2]). This behavior was intentionally evoked by lowering the threshold for positive labels, as task scoring focused on the F1 assessment of these labels. Manual inspection of the errors also revealed that some predictions were incorrect due to labeling errors in the development set. Examples of such cases can be found in Table 6[2].

Table 2: Results on the test set (courtesy of challenge organizers)

Model	Acc	Prec	Rec	F1
RoBERTa-Hybrid	79.8	72	76	75

5 Conclusion

We developed hybrid classification systems for the detection of self-reported chronic stress that integrate pre-trained transformer language models with BiLSTM networks trained on a diverse set of psycholinguistic features. Our experiments show that such hybrid models significantly outperform base transformer models for both augmented and non-augmented data.

[2]Tables 5,6 in supplementary material

References

Hani Al-Omari, Malak A. Abdullah, and Samira Shaikh. 2020. Emodet2: Emotion detection in english textual dialogue using bert and bilstm models. In *2020 11th International Conference on Information and Communication Systems (ICICS)*, pages 226–232.

Jacob Devlin, Ming-Wei Chang, Kenton Lee, and Kristina Toutanova. 2018. Bert: Pre-training of deep bidirectional transformers for language understanding. *arXiv preprint arXiv:1810.04805*.

Elma Kerz, Yu Qiao, Sourabh Zanwar, and Daniel Wiechmann. 2022. Pushing on personality detection from verbal behavior: A transformer meets text contours of psycholinguistic features. *arXiv preprint arXiv:2204.04629*.

Yinhan Liu, Myle Ott, Naman Goyal, Jingfei Du, Mandar Joshi, Danqi Chen, Omer Levy, Mike Lewis, Luke Zettlemoyer, and Veselin Stoyanov. 2019. Roberta: A robustly optimized bert pretraining approach. *arXiv preprint arXiv:1907.11692*.

Christopher Manning, Mihai Surdeanu, John Bauer, Jenny Finkel, Steven Bethard, and David McClosky. 2014. The stanford corenlp natural language processing toolkit. In *Proceedings of 52nd annual meeting of the association for computational linguistics: system demonstrations*, pages 55–60.

Elsbeth Turcan and Kathy McKeown. 2019. Dreaddit: A Reddit dataset for stress analysis in social media. In *Proceedings of the Tenth International Workshop on Health Text Mining and Information Analysis (LOUHI 2019)*, pages 97–107. Association for Computational Linguistics.

Davy Weissenbacher, Ari Z. Klein, Luis Gascó, Darryl Estrada-Zavala, Martin Krallinger, Yuting Guo, Yao Ge, Abeed Sarker, Ana Lucia Schmidt, Raul Rodriguez-Esteban, Mathias Leddin, Arjun Magge, Juan M. Banda, Vera Davydova, Elena Tutubalina, and Graciela Gonzalez-Hernandez. 2022. Overview of the seventh social media mining for health applications smm4h shared tasks at coling 2022. In *Proceedings of the Seventh Social Media Mining for Health (SMM4H) Workshop and Shared Task*, pages 33–40.

Daniel Wiechmann, Yu Qiao, Elma Kerz, and Justus Mattern. 2022. Measuring the impact of (psycho-) linguistic and readability features and their spill over effects on the prediction of eye movement patterns. *arXiv preprint arXiv:2203.08085*.

Thomas Wolf, Julien Chaumond, Lysandre Debut, Victor Sanh, Clement Delangue, Anthony Moi, Pierric Cistac, Morgan Funtowicz, Joe Davison, Sam Shleifer, et al. 2020. Transformers: State-of-the-art natural language processing. In *Proceedings of the 2020 Conference on Empirical Methods in Natural Language Processing: System Demonstrations*, pages 38–45.

A Supplementary material

Supplementary material can be found here https://bit.ly/3xq68bx.

NLP-CIC-WFU at SocialDisNER: Disease Mention Extraction in Spanish Tweets Using Transfer Learning and Search by Propagation

Antonio Tamayo
Instituto Politécnico Nacional
Centro de Investigación en Computación
Av. Juan de Dios Batiz,
s/n, 07320, Mexico City, Mexico
atamayoh2019@cic.ipn.mx

Diego Burgos
Wake Forest University
1834 Wake Forest Road,
Winston-Salem,
NC 27109,
Winston Salem, USA
burgosda@wfu.edu

Alexander Gelbukh
Instituto Politécnico Nacional
Centro de Investigación en Computación
Av. Juan de Dios Batiz,
s/n, 07320, Mexico City, Mexico
gelbukh@cic.ipn.mx

Abstract

Named entity recognition (e.g., disease mention extraction) is one of the most relevant tasks for data mining in the medical field. Although it is a well-known challenge, the bulk of the efforts to tackle this task have been made using clinical texts commonly written in English. In this work, we present our contribution to the SocialDisNER competition, which consists of a transfer learning approach to extracting disease mentions in a corpus from Twitter written in Spanish. We fine-tuned a model based on mBERT and applied post-processing using regular expressions to propagate the entities identified by the model and enhance disease mention extraction. Our system achieved a competitive strict F1 of 0.851 on the testing data set.

1 Motivation

Although there are several works for disease mention extraction, the bulk of them has been carried out for clinical texts written in English (Eftimov et al., 2017; Patra and Saha, 2013; Peng et al., 2019; Sachan et al., 2018; Lee et al., 2020; Akhtyamova, 2020; Alsentzer et al., 2019; Yasunaga et al., 2022; Gu et al., 2021). In this work, we present our contribution to the SocialDisNER competition (Gasco et al., 2022) at the SMM4H workshop, task 10 (Weissenbacher et al., 2022). Our system is focused on disease mention extraction from Twitter messages in Spanish. The nature of the texts written in this social network presents new challenges to the disease extraction task because misspellings are frequent (Magumba et al., 2018; Magge et al., 2021). Additionally, many disease mentions can be subsumed in hashtags, urls, or user names (Magumba et al., 2018), which, together with the above, makes it more difficult to identify entities than in common clinical texts as shown in Xiong et al. (2020); García-Pablos et al. (2020); Wang et al. (2019).

2 System description

In this work, we present a transfer learning approach using the model proposed by Tamayo et al. (2022), which is a version of multilingual BERT (Devlin et al., 2019) fine-tuned for disease mention extraction from clinical texts and we apply post-processing rules to extract diseases mentioned in a corpus of tweets in Spanish. Our system tackles the problem in three steps, namely, pre-processing, transfer learning, and post-processing. Below we describe each of them.

2.1 Pre-processing

To implement the fine-tuning process, the BIO scheme (Begin, Inside, Outside) (Ramshaw and Marcus, 1995) was used. Since the dataset provided by SocialDisNER is formatted in a different way, pre-processing was needed to take it to the BIO scheme. We used the disease mentions in the provided structured dataset as a reference to annotate disease mentions in each tweet with their corresponding labels in the BIO scheme. Tokenization was carried out using SpaCy (Honnibal and Montani, 2017) instead of a NER dedicated library such as SciSpacy (Neumann et al., 2019) because the former works for Spanish.

2.2 Transfer learning

We tackled disease mention extraction as a sequence labeling problem using the whole tweet as input, and the labels mentioned above as output. We randomly split partitions of the training dataset into training (75%) and validation (25%) sets. This partition was done iteratively five times with random seeds. Additionally, we carried out a hyperparameter tuning searching for the best model's configuration using a grid search for the epochs (3, 5, 7) and the learning rate (5e-03, 5e-05, 5e-07). 7 epochs and a learning rate of 5e-05 yielded the best results. With regard to the rest of hyperparameters, default values were kept. For this process,

Proceedings of the 29th International Conference on Computational Linguistic, pages 19–22
October 12–17, 2022.

we used a transformer library and the model available at Hugging Face[1]. Google Colab Pro with a GPU Tesla P100 with 27.3 gigabytes of available RAM was used to run all the experiments. The data we used for our training process together with the source code to replicate this work are available at a GitHub repository[2].

2.3 Post-processing plus search by propagation

Post-processing was carried out through a custom Python script to clean up and format the output as follows: 1) Because mBERT works with a subword tokenization system, we decoded the output that contained subwords. 2) We concatenated all the named entities detected by the model one after the other. This means that if the model detected a named entity whose final character position (or final character position plus one) concurred with the first position of the next named entity detected, our system considered that these two entities were part of one single entity. This was necessary because the model extracts parts of some entities separately. 3) We also applied simple but effective post-processing based on some orthographic and grammatical rules which are detailed in Table 1. 4) Under the assumption that SocialDisNER participants were required to extract all the mentions of a disease mention occurring in a tweet, we used the entities extracted by the model to identify and extract any repetitions of said entities in the same document. In order to retrieve misspelled mentions or mentions subsumed by hashtags, urls, or user names, we carried out a search by propagation applying the following steps: a) lowercase both the entity identified by the model and the tweet, b) concatenate multi-word entities, c) delete accents, and d) search entity occurrences throughout the tweet. Lastly, since we work with the BIO scheme, the last post-processing step consisted of decoding the predictions to put them in the data format required by SocialDisNER.

If the disease mention detected ...	**... then apply this rule**
1. Starts with punctuation mark	1. Delete the match and adjust the entity's beginning index
2. Contains a mark of new line	2. Replace the match with a space
3. Contains a space before and/or after a hyphen or a parenthesis	3. Delete the space(s) and adjust the entity's ending index
4. Ends with non-content words or punctuation marks	4. Delete the match and adjust the entity's ending index
5. Concurs with non-content words or punctuation/hashtag marks	5. Leave out of the entities detected

Table 1: Post-processing rules

Model	P	R	F1
mBERT + post-processing	0.861	0.876	0.868

Table 2: Results (5-iteration mean) on the development dataset

3 Results and error analysis

The results achieved by our model on the validation dataset can be seen in Table 2.

Likewise, for future reference and improvement, we present the following brief error analysis. First, our system extracts false positives. They are meaningful entities, but they are not in the gold standard (e.g., EFyC, *formación diabetológica*). Second, the model truncates some entities (e.g., *problemas de apre* insted of *problemas de aprendizaje*, *respiratorias crónicas* instead of *Enfermedades respiratorias crónicas*). The first example of the latter type of error is caused by the nature of the mBERT model which works with subwords tokenization. Finally, we consider that there are some errors resulting from an incorrect tagging of the dataset (e.g., our model extracts the entity *cáncer* but in the gold standard appears *niñas y niños que hay hoy con cáncer*).

4 Conclusions

In this work, we presented a system based on mBERT following a fine-tuning approach plus simple post-processing and search by propagation to extract disease mentions from Tweets in Spanish. We achieved competitive results with a strict F1 of 0.851, Precision of 0.842, and Recall of 0.860 on the test dataset of the SocialDisNER competition.

[1]The model is available at: https://bit.ly/3zGlxWy

[2]https://github.com/ajtamayoh/NLP-CIC-WFU-Contribution-to-SocialDisNER-shared-task-2022.git

Acknowledgements

This work was done with partial support from the Mexican Government through the grant A1-S-47854 of CONACYT, Mexico, grants 20220852 and 20220859 of the Secretaría de Investigación y Posgrado of the Instituto Politécnico Nacional, Mexico. The authors thank the CONACYT for the computing resources brought to them through the Plataforma de Aprendizaje Profundo para Tecnologías del Lenguaje of the Laboratorio de Supercómputo of the INAOE, Mexico and acknowledge the support of Microsoft through the Microsoft Latin America PhD Award.

References

Liliya Akhtyamova. 2020. Named entity recognition in spanish biomedical literature: Short review and bert model. In *2020 26th Conference of Open Innovations Association (FRUCT)*, pages 1–7. IEEE.

Emily Alsentzer, John R Murphy, Willie Boag, Wei-Hung Weng, Di Jin, Tristan Naumann, and Matthew McDermott. 2019. Publicly available clinical bert embeddings. *arXiv preprint arXiv:1904.03323*.

Jacob Devlin, Ming-Wei Chang, Kenton Lee, and Kristina Toutanova. 2019. BERT: Pre-training of deep bidirectional transformers for language understanding. In *Proceedings of the 2019 Conference of the North American Chapter of the Association for Computational Linguistics: Human Language Technologies, Volume 1 (Long and Short Papers)*, pages 4171–4186, Minneapolis, Minnesota. Association for Computational Linguistics.

Tome Eftimov, Barbara Koroušić Seljak, and Peter Korošec. 2017. A rule-based named-entity recognition method for knowledge extraction of evidence-based dietary recommendations. *PloS one*, 12(6):e0179488.

Aitor García-Pablos, Naiara Perez, and Montse Cuadros. 2020. Vicomtech at cantemist 2020. In *Proceedings of the Iberian Languages Evaluation Forum (IberLEF 2020), CEUR Workshop Proceedings*, volume 17, page 25.

Luis Gasco, Darryl Estrada-Zavala, Eulàlia Farré-Maduell, Salvador Lima-López, Antonio Miranda-Escalada, and Martin Krallinger. 2022. Overview of the SocialDisNER shared task on detection of diseases mentions from healthcare related and patient generated social media content: methods, evaluation and corpora. In *Proceedings of the Seventh Social Media Mining for Health (#SMM4H) Workshop and Shared Task*.

Yu Gu, Robert Tinn, Hao Cheng, Michael Lucas, Naoto Usuyama, Xiaodong Liu, Tristan Naumann, Jianfeng Gao, and Hoifung Poon. 2021. Domain-specific language model pretraining for biomedical natural language processing. *ACM Transactions on Computing for Healthcare (HEALTH)*, 3(1):1–23.

Matthew Honnibal and Ines Montani. 2017. spacy 2: Natural language understanding with bloom embeddings, convolutional neural networks and incremental parsing. *To appear*, 7(1):411–420.

Jinhyuk Lee, Wonjin Yoon, Sungdong Kim, Donghyeon Kim, Sunkyu Kim, Chan Ho So, and Jaewoo Kang. 2020. Biobert: a pre-trained biomedical language representation model for biomedical text mining. *Bioinformatics*, 36(4):1234–1240.

Arjun Magge, Davy Weissenbacher, Karen O'Connor, Matthew Scotch, and Graciela Gonzalez-Hernandez. 2021. Seed: Symptom extraction from english social media posts using deep learning and transfer learning. *medRxiv*.

Mark Abraham Magumba, Peter Nabende, and Ernest Mwebaze. 2018. Ontology boosted deep learning for disease name extraction from twitter messages. *Journal of Big Data*, 5(1):1–19.

Mark Neumann, Daniel King, Iz Beltagy, and Waleed Ammar. 2019. ScispaCy: Fast and robust models for biomedical natural language processing. In *Proceedings of the 18th BioNLP Workshop and Shared Task*, pages 319–327, Florence, Italy. Association for Computational Linguistics.

Rakesh Patra and Sujan Kumar Saha. 2013. A kernel-based approach for biomedical named entity recognition. *The Scientific World Journal*, 2013.

Yifan Peng, Shankai Yan, and Zhiyong Lu. 2019. Transfer learning in biomedical natural language processing: An evaluation of BERT and ELMo on ten benchmarking datasets. In *Proceedings of the 18th BioNLP Workshop and Shared Task*, pages 58–65, Florence, Italy. Association for Computational Linguistics.

Lance Ramshaw and Mitch Marcus. 1995. Text chunking using transformation-based learning. In *Third Workshop on Very Large Corpora*.

Devendra Singh Sachan, Pengtao Xie, Mrinmaya Sachan, and Eric P Xing. 2018. Effective use of bidirectional language modeling for transfer learning in biomedical named entity recognition. In *Machine learning for healthcare conference*, pages 383–402. PMLR.

Antonio Tamayo, Diego A. Burgos, and Alexander Gelbukh. 2022. mbert and simple post-processing: A baseline for disease mention detection in spanish. In *Working Notes of Conference and Labs of the Evaluation (CLEF) Forum. CEUR Workshop Proceedings*.

Zhiguo Wang, Patrick Ng, Xiaofei Ma, Ramesh Nallapati, and Bing Xiang. 2019. Multi-passage bert: A globally normalized bert model for open-domain question answering. *arXiv preprint arXiv:1908.08167*.

Davy Weissenbacher, Ari Z. Klein, Luis Gascó, Darryl Estrada-Zavala, Martin Krallinger, Yuting Guo, Yao Ge, Abeed Sarker, Ana Lucia Schmidt, Raul Rodriguez-Esteban, Mathias Leddin, Arjun Magge, Juan M. Banda, Vera Davydova, Elena Tutubalina, and Graciela Gonzalez-Hernandez. 2022. Overview of the seventh social media mining for health applications (#smm4h) shared tasks at coling 2022. In *In Proceedings of the Seventh Social Media Mining for Health (#SMM4H) Workshop and Shared Task.*

Ying Xiong, Yuanhang Huang, Qingcai Chen, Xiaolong Wang, Yuan Nic, and Buzhou Tang. 2020. A joint model for medical named entity recognition and normalization. *Proceedings http://ceur-ws. org ISSN*, 1613(0073):17.

Michihiro Yasunaga, Jure Leskovec, and Percy Liang. 2022. LinkBERT: Pretraining language models with document links. In *Proceedings of the 60th Annual Meeting of the Association for Computational Linguistics (Volume 1: Long Papers)*, pages 8003–8016, Dublin, Ireland. Association for Computational Linguistics.

yiriyou@SMM4H'22: Stance and Premise Classification in Domain Specific Tweets with Dual-View Attention Neural Networks

Huabin Yang[1], Zhongjian Zhang[1], Yanru Zhang[1,2]*
[1]University of Electronic Science and Technology of China, Chengdu
[2]Shenzhen Institute for Advanced Study, UESTC
202121080420@std.uestc.edu.cn
zhongjianzhang6972@gmail.com
yanruzhang@uestc.edu.cn

Abstract

The paper introduces the methodology proposed for the shared Task 2 of the Social Media Mining for Health Application (SMM4H) in 2022. Task 2 consists of two subtasks: Stance Detection and Premise Classification, named Subtask 2a and Subtask 2b, respectively. Our proposed system is based on dual-view attention neural networks and achieves an F1 score of 0.618 for Subtask 2a (0.068 more than the median) and an F1 score of 0.630 for Subtask 2b (0.017 less than the median). Further experiments show that the domain-specific pretrained model, cross-validation, and pseudo-label techniques contribute to the improvement of system performance.

1 Introduction

Since the outbreak of the COVID-19 pandemic, there has been a large amount of research on COVID-19 in the natural language processing (NLP) arena. Meanwhile, social media platforms such as Twitter are widely used to share and spread users' views on various issues, so the Social Media Mining for Health Application (SMM4H) Shared Task in 2022 (Weissenbacher et al., 2022) focuses on leveraging COVID-related tweets for health research. In this paper, we describe our methodology for the shared Task 2: Classification of stance and premise in tweets about health mandates related to COVID-19 (in English). We are provided with labeled training set containing texts from Twitter about three health mandates (claims) related to COVID-19 pandemic: *school closures*, *stay at home orders* and *face masks*[1]. Subtask 2a aims at detecting the stance of the text's author concerning the given claim (e.g., *school closures*), while Subtask 2b aims at identifying whether there are premises in an argument.

*Corresponding author

[1]see Davydova and Tutubalina (2022) for more details

2 Methodology

2.1 Data

We use the dataset provided by the organizers of Task 2. For Subtask 2a, the tweets in the training and validation sets are annotated for stance towards the given claim according to three categories: FAVOR, AGAINST, and NONE. For Subtask 2b, the tweets in the training and validation sets are annotated as 1 for containing a premise (argument) and 0 otherwise. In addition, we note that the data in both the training and validation sets are in English. And the emoji in the tweets have been replaced with the corresponding textual representations (e.g., :face_with_medical_mask). However, we find that the language of tweets in the test set is variant. Most tweets are in English, but there are tweets in other languages, such as Hindi or Urdu. Even for tweets in the same language, some of them contain emoticons that will affect the overall meaning, indicating that data preprocessing on the test set is essential before predicting the results. Table 1 gives the statistics of the Task 2 dataset.

Training	Validation	Test
3556	600	2000(10000)

Table 1: Statistics of the Task 2 dataset. Note that only 2000 tweets in the test set were included in the metrics computing, other 8000 tweets were added to avoid manual annotation.

2.2 Preprocessing

In order to reduce the noise of tweets in the test set, we preprocess the test set as follows:

- For those tweets are not in English, we use Google Translation to translate them into English, considering that multilingual text can affect the final results. However, we note that Google Translation does not always translate the sentence as its intended meaning, so this

Proceedings of the 29th International Conference on Computational Linguistic, pages 23–26
October 12–17, 2022.

kind of operation may result in unexpected effects.

- To unify the impact of emoticons in the test set, we perform emoji codification to convert the emoticons into textual representations using the python emoji package [2].

Figure 1 shows examples of data preprocessing. (a) shows a good example of preprocessing while (b) proves that Google Translation doesn't always work well.

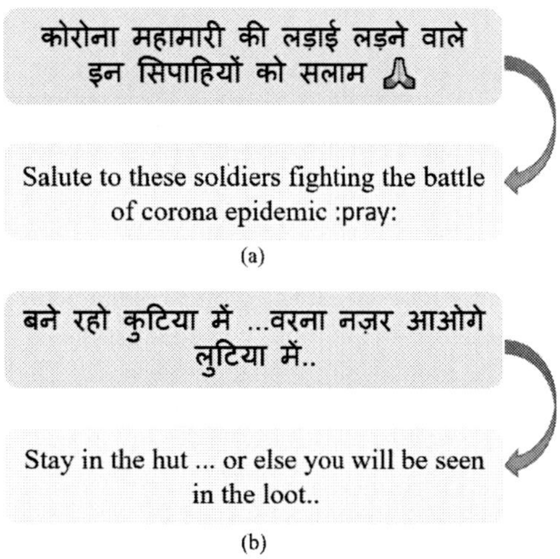

Figure 1: Examples of preprocessing. (a) Good preprocessing result. (b) Bad preprocessing result.

2.3 Modeling

Dual-view Attention Neural Networks (Xu et al., 2020) learn the representations of subjective and objective features of texts. Motivated by (Xu et al., 2020; Glandt et al., 2021), we focus on exploring the impact of subjective and objective information from tweets on stance detection and premise classification. Figure 2 shows the architecture of our proposed system. We use BERT-based models to extract subjective and objective features from texts, then we concatenate the two [CLS] embeddings (i.e., subjective and objective features, denoted by $\boldsymbol{f}_{subj}$ and $\boldsymbol{f}_{obj}$ respectively) and use the result as input to feed into a fully connected layer, followed by a sigmoid activation to produce a fusion vector:

$$\boldsymbol{g} = sigmoid(\boldsymbol{W}_u[\boldsymbol{f}_{subj} \oplus \boldsymbol{f}_{obj}] + \boldsymbol{b}_u),$$

where $\oplus$ denotes vector concatenation and $\boldsymbol{W}_u, \boldsymbol{b}_u$ are trainable parameters (Xu et al., 2020). We aim to attain an optimal combination between $\boldsymbol{f}_{subj}$ and $\boldsymbol{f}_{obj}$, so vector $\boldsymbol{g}$ serves as a weight vector as follows:

$$\boldsymbol{f}_{dual} = \boldsymbol{g} \odot \boldsymbol{f}_{subj} + (\boldsymbol{1} - \boldsymbol{g}) \odot \boldsymbol{f}_{obj},$$

where $\odot$ denotes the element-wise product (Xu et al., 2020; Glandt et al., 2021). Vector $\boldsymbol{f}_{dual}$ stands for dual-view features of texts and is used to train our classifier, which consists of 2 linear layers with 1024 hidden units respectively and a ReLU activation function. For each subtask of Task 2, the model was trained for 50 epochs with a batch size of 16, a dropout probability of 0.15, and a learning rate of 0.00003.

2.4 Cross-validation and Pseudo-label

Cross-validation is a data resampling method used to evaluate models and prevent overfitting. The basic form of cross-validation is k-fold cross-validation (Refaeilzadeh et al., 2009). Pseudo-label (Lee et al., 2013) uses semi-supervised learning to increase the training data to improve the robustness of the model, which can be regarded as a method to increase the fitting ability of the model by expanding the decision boundary. This technique trains a model by labeled data and uses the trained model to predict labels for unlabeled data, then adds these newly obtained pseudo-labeled data to the training set and retrains the model.

3 Results and Discussion

	F1 score	
	Subtask 2a	**Subtask 2b**
Model$_{roberta}$	0.750	0.636
Model$_{v2}$	0.801	0.747
Model$_{v2}$ + cv	0.802	0.750
Model$_{v2}$ + cv + pl	**0.805**	**0.754**

Table 2: Performance of different approaches on the SMM4H 2022 Task 2 **Validation Set**. *roberta* denotes RoBERTa-large pre-trained model, *v2* denotes COVID-Twitter-BERT-v2 pre-trained model, *cv* denotes cross-validation, and *pl* denotes pseudo-label.

Table 2 shows the performance of the suggested approaches on the validation set for both subtasks. We can see that COVID-Twitter-BERT-v2 based model (Müller et al., 2020) performs better than RoBERTa-large based model (Liu et al., 2019). Additionally, we applied techniques including pseudo-label and 5-folds cross-validation with a majority

[2] https://pypi.org/project/emoji/

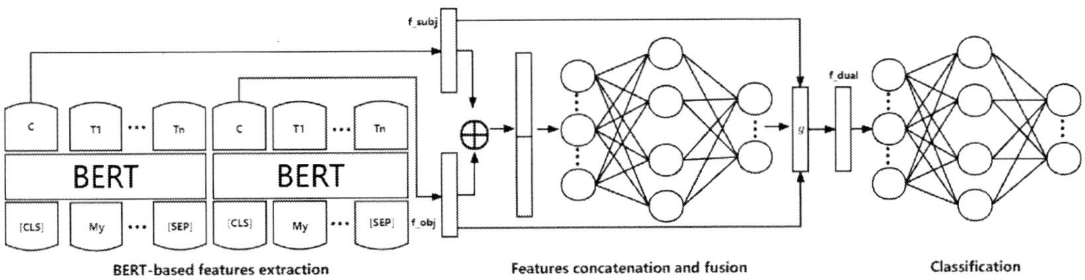

Figure 2: Architecture of our proposed system

voting strategy to our experiments, both methods proved to be useful in improving the performance of the system. We conducted an error analysis and found that, in the Stance Detection task, positive or negative words such as "support" or "don't" might lead to misclassification to some extent, especially when distinguishing NONE from the other two stances (e.g., "Schools are doing their best to ensure that education for children is not disrupted. I support my children's school and all self financed schools wholeheartedly" is predicted as FAVOR label while the truth label is NONE, we note that it is also difficult for humans to distinguish between FAVOR and NONE). As for the Premise Classification task, it also comes with challenges of how to identify a given statement that can be used as an argument in a discussion, it is also necessary to distinguish between sentiment polarity and argumentation.

In the evaluation phase, our submitted systems including Model_{v2}, $\text{Model}_{v2} + cv$, and $\text{Model}_{v2} + cv + pl$ (*v2* denotes COVID-Twitter-BERT-v2 pre-trained model, *cv* denotes 5-folds cross-validation, and *pl* denotes pseudo-label. As for the pseudo-label method, we selected 2000 English tweets in the test set to predict labels and added them to the training set to train our final submitted model), and the last one achieved the highest score for all of our submissions. Table 3 shows the results of the $\text{Model}_{v2} + cv + pl$ compared to the mean and median of all competing submissions.

4 Conclusion

Our proposed methods achieve an F1 score of 0.618 for Subtask 2a (the mean score is 0.491 and the median score is 0.550) and an F1 score of 0.630 for Subtask 2b (the mean score is 0.574 and the median score is 0.647) on the test set. We observe that the domain-specific pre-trained model can significantly outperform the general pre-trained model. Besides, we also notice that the system's performances on the test set are lower than on the validation set, which can be attributed to overfitting. In terms of future work, stochastic weight averaging, adversarial training, and exponential moving averaging can be investigated to improve the generalizability and robustness of our system.

	F1 score	
	Subtask 2a	**Subtask 2b**
Ours	**0.618**	0.630
Median	0.550	**0.647**
Mean	0.491	0.574

Table 3: Comparison of our best performing system (i.e., $\text{Model}_{v2} + cv + pl$) with the mean and median of all competing systems in the evaluation phase.

Acknowledgements

We would like to thank the reviewers for their insightful suggestions.

References

Vera Davydova and Elena Tutubalina. 2022. SMM4H 2022 Task 2: Dataset for stance and premise detection in tweets about health mandates related to COVID-19. In *Proceedings of the Seventh Social Media Mining for Health Applications (#SMM4H) Workshop & Shared Task*, pages –.

Kyle Glandt, Sarthak Khanal, Yingjie Li, Doina Caragea, and Cornelia Caragea. 2021. Stance detection in COVID-19 tweets. In *Proceedings of the 59th Annual Meeting of the Association for Computational Linguistics and the 11th International Joint Conference on Natural Language Processing (Volume 1: Long Papers)*, pages 1596–1611.

Dong-Hyun Lee et al. 2013. Pseudo-label: The simple and efficient semi-supervised learning method for deep neural networks. In *Workshop on challenges in representation learning, ICML*, volume 3, page 896.

Yinhan Liu, Myle Ott, Naman Goyal, Jingfei Du, Mandar Joshi, Danqi Chen, Omer Levy, Mike Lewis, Luke Zettlemoyer, and Veselin Stoyanov. 2019. Roberta: A robustly optimized BERT pretraining approach. *CoRR*, abs/1907.11692.

Martin Müller, Marcel Salathé, and Per E Kummervold. 2020. COVID-Twitter-BERT: A Natural Language Processing Model to Analyse COVID-19 Content on Twitter. *arXiv preprint arXiv:2005.07503*.

Payam Refaeilzadeh, Lei Tang, and Huan Liu. 2009. Cross-validation. *Encyclopedia of database systems*, 5:532–538.

Davy Weissenbacher, Ari Z. Klein, Luis Gascó, Darryl Estrada-Zavala, Martin Krallinger, Yuting Guo, Yao Ge, Abeed Sarker, Ana Lucia Schmidt, Raul Rodriguez-Esteban, Mathias Leddin, Arjun Magge, Juan M. Banda, Vera Davydova, Elena Tutubalina, and Graciela Gonzalez-Hernandez. 2022. Overview of the Seventh Social Media Mining for Health Applications #SMM4H Shared Tasks at COLING 2022. In *Proceedings of the Seventh Social Media Mining for Health Applications (#SMM4H) Workshop & Shared Task*, pages –.

Chang Xu, Cécile Paris, Surya Nepal, Ross Sparks, Chong Long, and Yafang Wang. 2020. DAN: Dual-view representation learning for adapting stance classifiers to new domains. *arXiv preprint arXiv:2003.06514*.

SINAI@SMM4H'22: Transformers for biomedical social media text mining in Spanish

Mariia Chizhikova[1], Pilar López-Úbeda[2], Manuel C. Díaz-Galiano[1], L. Alfonso Ureña-López[1], M. Teresa Martín-Valdivia[1]
[1]Department of Computer Science. University of Jaén. Jaén, Spain
[2]R+D+I department. HT medica. Jaén, Spain
mc000051@red.ujaen.es, p.lopez@htmedica.com,
{mcdiaz, laurena, maite}@ujaen.es

Abstract

This paper covers participation of the SINAI team in Tasks 5 and 10 of the Social Media Mining for Health (#SSM4H) workshop at COLING-2022. These tasks focus on leveraging Twitter posts written in Spanish for healthcare research. The objective of Task 5 was to classify tweets reporting COVID-19 symptoms, while Task 10 required identifying disease mentions in Twitter posts. The presented systems explore large RoBERTa language models pre-trained on Twitter data in the case of tweet classification task and general-domain data for the disease recognition task. We also present a text pre-processing methodology implemented in both systems and describe an initial weakly-supervised fine-tuning phase alongside with a submission post-processing procedure designed for Task 10. The systems obtained 0.84 F1-score on the Task 5 and 0.77 F1-score on Task 10.

1 Introduction

Social media networks are widely used as a base for public health related research. Among other websites, Twitter constitutes a useful data source for researches due to the real time nature of the content and the ease to accessing publicly available information (Sinnenberg et al., 2017).

However, extraction of information from tweets remains challenging due to some particularities of the language variety used in the social network. Spelling mistakes, shortenings, abbreviations, a-grammatical word truncation alongside with frequent use of emoji, hashtags, emoticons and even usage of digits and mathematical symbols to represent phonetic sounds are some characteristics that distinguish linguistic register typically adopted by Twitter users (Jalbuena, 2012). This fact restricts the efficiency of rule-based information retrieval approaches, such as exact matching, used to extract data in many researches (Chew and Eysenbach, 2010; McNeil et al., 2012; Lyles, 2013).

Named entity recognition (NER) is a task in Natural Language Processing (NLP) which has as objective structuring free-text data by identifying relevant terms in it. Text classification is another way of structuring data which improves information retrieval by assigning standardized labels to texts(Du et al., 2019).

In this paper we present the systems developed by the SINAI team for Shared Tasks 5 and 10 at the Social Media Mining for Health (#SMM4H) workshop at COLING 2022. On the one hand, task 5 brings the community effort to design systems that would automatically classify Spanish tweets reporting COVID-19 in 3 categories: self reports, non-personal reports and literature/news mentions. The presented tweet classification systems relies on RoBERTiuto (Pérez et al., 2021), a language model pre-trained on social media text. On the other hand, the purpose of Task 10 (SocialDisNER) is to design systems that would recognize all kinds of disease mentions in tweets written in Spanish. With this objective, we developed a system based on a model pre-trained on general-domain text (Gutiérrez-Fandiño et al., 2021). In order to leverage large scale additional Silver Standard data with automatically generated labels provided by task's organizers we designed a two-stage fine-tuning framework.

Concerning data processing, we describe a submission post-processing procedure aimed to improve system's performance and we enhanced our systems with additional text pre-processing.

2 Data

The organizers of both tasks made available corpora of tweets in Spanish that represent first-hand experience of diseases and other health-related content.

Proceedings of the 29th International Conference on Computational Linguistic, pages 27–30
October 12–17, 2022.

2.1 Task 5

The annotated dataset for this classification task is a collection of 10,052 tweets characterized with class imbalance, as its major part is labeled as literature/news mentions (5,985). We randomly split this set in order to use 30% of labelled data as development set.

An unannotated collection of 3,578 was provided as validation data and the test set consisted of 6,851 tweets.

2.2 Task 10

The Gold Standard SocialDisNER corpus is a collection of 9,500 manually annotated tweets (Gasco et al., 2022a). The corpus was randomly split in train, validation and test sets by the organizers. To prevent participants from manually annotating tweets, final predictions were made on a set of 23,430 tweets (test+background data), while only 2,000 of these were used to evaluate the presented systems.

As an additional resource, a set of large-scale Silver Standard data which contained mentions automatically extracted from 85,000 tweets was released. Its subset labelled with disease mentions consists of 84,988 tweets with 116,260 annotations 16,034 from which are unique (Gasco et al., 2022b).

3 Pre-processing

The peculiarities of the linguistic register adopted by Twitter users make text pre-processing an essential step in order to get a language model to perform robustly. For both tasks, we followed the same rule-based pre-processing procedure which consists of the following steps:

1. Newline characters are replaced with spaces.
2. Character repetitions are limited to max of 3.
3. Hashtag symbol (#) is removed and its text is split into words when possible.
4. Emoji are replaced by their text representations between two 'emoji' tokens.

The steps 2-4 were carried out by using the `pysentimiento` Python package (Pérez et al., 2021).

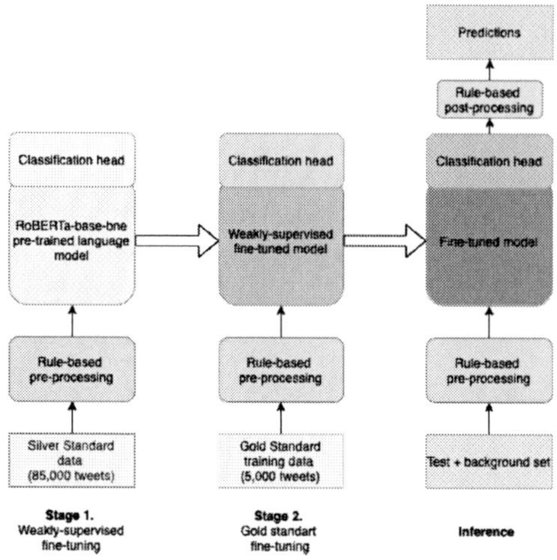

Figure 1: Task 10 model fine-tuning process overview

4 Task 5. Classification of tweets containing self-reported COVID-19 symptoms

To address the tweet classification task we opted for fine-tuning two variants (*cased* and *uncased-deacc*, which lowercases and removes accents) of RoBERTuito - a RoBERTa architecture language model (Liu et al., 2019) pre-trained on social media text (Pérez et al., 2021).

5 Task 10- Detection of disease mentions in tweets – SocialDisNER

The core element of the system presented for SocialDisNER task is the RoBERTa-base-bne: RoBERTa pre-trained on Spanish general-domain corpus (Gutiérrez-Fandiño et al., 2021).

Weak supervision In order to leverage the additional large-scale data provided by SocialDisNER organizers, we designed a two-step model fine-tuning process in which we first feed the model automatically generated (*weak*) Silver Standard data and fine-tune the model on it using as validation data the corresponding split of the Gold Standard dataset. Secondly, we take the resulting model to the second fine-tuning stage using the Gold Standard training set. Figure 1 provides a summary of the described process.

The aim of weakly supervised modelling is to bootstrap system performance without the need of manually annotating more data (Lison et al., 2020).

Submission post-processing To improve overall quality of NER performed by our system, we incorporated a rule-based post-processing which is used in both single-stage and two-stage fine-tuning. At this step, the entities detected by the transformer model are subjected to the following procedures:

- Strip all punctuation marks and whitespaces from detected mentions.
- Retrieve a list of relevant terms from train, validation and Silver Standard and perform an exact match.
- Select the longest mention if two entities occur within the same offset.

6 Experimental setup

All models were fine-tuned on a single NVIDIA Tesla V100 GPU by adding a linear classifier layer preceded by a 0.1 dropout layer on top of the original architecture using the Hugging Face Transformers Python library (Wolf et al., 2019). In addition, to take the maximum advantage of the pre-trained model, we performed hyperparameter optimization that relied on the Optuna framework (Akiba et al., 2019).

In the two-stage learning process, we performed optimization at each step within the same hyperparameter space which was also used for a single-stage fine-tuning.

7 Results

Our team submitted a total of two runs per task. In Task 5 these corresponded to predictions made with *cased* model (Run 1) and *uncased-deacc* model (Run 2). Only the best run was kept after the official evaluation.

As for Task 10, Run 1 corresponds to the system based on the fine-tuned RoBERTa model and Run 2 to the one based on the same pre-trained model, but subjected to the-two stage fine-tuning procedure described in Section 5.

The metrics selected to evaluate performance of the participant systems are precision, recall and F1-score for the self-report class or for each mention extracted where the spans overlap exactly in Task 5 and Task 10 respectively. Table 1 displays the results for each of presented systems.

8 Error analysis

8.1 Task 5

The best performing system reached 0.77 F-1 score during in-house evaluation. While carrying out the evaluation and analysis of the results, we were able to determine that many miss-classified tweets contained an large number of spelling errors or incorporated direct speech citations.

8.2 Task 10

During the development process, the best performing model reached 0.81 strict F1-score on the validation set. Most common false negatives were related to detection of mentions that formed part of complex hashtags, for example, *Depresión* in *#VivirConDepresión* (*eng. #LiveWithDepression*). Also were observed some cases of incorrect extraction of complex disease mentions, where the system tended to shorten complex entities: *reacciones dermatológicas* (*eng. dermatologic reactions*) instead of *reacciones dermatológicas tras la vacuna* (*eng. dermatological reactions after the vaccine*).

9 Conclusions and future work

This paper presents the participation of the SINAI team in #SMM4H workshop at COLING 2022. Firstly, we describe rule-based tweet preprocessing which was used in both tasks and aimed to normalize the peculiarities of the linguistic variety used in Twitter,.

For tweet classification task we compared transfer learning potential of two versions of RoBERTa model pre-trained on social media text: *cased* and *uncased-deacc*. The latter performed better achieving 0.84 F1-score.

To address the automatic disease mention recognition in tweets we employed a large pre-trained model and compared a state-of-the art fine-tuning approach with a weak-supervision enhanced two-stage fine-tuning. For the tweet classification task, we compared transfer learning performance of two versions of language models pre-trained on social media text.

The official evaluation for Task 10 reveals that the model subjected to weak-supervision fine-tuning performs slightly better in terms of precision (0.739 vs 0.756) and F1-score (0.769 vs 0.775), which, however, cannot be considered a significant performance boost.[1]

[1]You can access to the best performing model trained for this task at `https://huggingface.co/chizhikchi/spanish-SM-disease-finder`

		Precision	Recall	F1-score
Task 5	Run 2	0.84	0.84	0.84
	Median	0.84	0.84	0.84
Task10	Run 1	0.739	**0.802**	0.769
	Run 2	**0.756**	0.795	**0.77**
	Mean (STD)	0.680 (0.245)	0.677 (0.254)	0.675 (0.246)

Table 1: Official evaluation results for systems presented by the SINAI team

Acknowledgements

This work has been partially supported by Big Hug project (P20_00956, PAIDI 2020) and WeLee project (1380939, FEDER Andalucía 2014-2020) funded by the Andalusian Regional Government, LIVING-LANG project (RTI2018-094653-B-C21) funded by MCIN/AEI/10.13039/501100011033

References

Takuya Akiba, Shotaro Sano, Toshihiko Yanase, Takeru Ohta, and Masanori Koyama. 2019. Optuna: A next-generation hyperparameter optimization framework. In *Proceedings of the 25rd ACM SIGKDD International Conference on Knowledge Discovery and Data Mining*.

Cynthia Chew and Gunther Eysenbach. 2010. Pandemics in the age of twitter: content analysis of tweets during the 2009 h1n1 outbreak. *PloS one*, 5(11):e14118.

Jingcheng Du, Qingyu Chen, Yifan Peng, Yang Xiang, Cui Tao, and Zhiyong Lu. 2019. Ml-net: multi-label classification of biomedical texts with deep neural networks. *Journal of the American Medical Informatics Association*, 26(11):1279–1285.

Luis Gasco, Darryl Estrada, Eulàlia Farré, and Martin Krallinger. 2022a. SocialDisNER corpus: gold standard annotations for detection of disease mentions in Spanish tweets. Type: dataset.

Luis Gasco, Darryl Estrada-Zavala, Eulàlia Farré-Maduell, Salvador Lima-López, Antonio Miranda-Escalada, and Martin Krallinger. 2022b. Overview of the SocialDisNER shared task on detection of diseases mentions from healthcare related and patient generated social media content: methods, evaluation and corpora. In *Proceedings of the Seventh Social Media Mining for Health (#SMM4H) Workshop and Shared Task*.

Asier Gutiérrez-Fandiño, Jordi Armengol-Estapé, Marc Pàmies, Joan Llop-Palao, Joaquín Silveira-Ocampo, Casimiro Pio Carrino, Carme Armentano-Oller, Carlos Rodriguez-Penagos, Aitor Gonzalez-Agirre, and Marta Villegas. 2021. MarIA: Spanish Language Models. page 22.

Maria Cecilia M. Jalbuena. 2012. Linguistic Features of English in Twitter. *MSEUF Research Studies*, 14(1):1–1.

Pierre Lison, Aliaksandr Hubin, Jeremy Barnes, and Samia Touileb. 2020. Named entity recognition without labelled data: A weak supervision approach. *arXiv preprint arXiv:2004.14723*.

Yinhan Liu, Myle Ott, Naman Goyal, Jingfei Du, Mandar Joshi, Danqi Chen, Omer Levy, Mike Lewis, Luke Zettlemoyer, and Veselin Stoyanov. 2019. Roberta: A robustly optimized bert pretraining approach. *arXiv preprint arXiv:1907.11692*.

CR Lyles. 2013. Ló pez a, pasick r, sarkar u."5 mins of uncomfyness is better than dealing with cancer 4 a lifetime": an exploratory qualitative analysis of cervical and breast cancer screening dialogue on twitter. *Journal of Cancer Education*, 28(1):127–33.

Karen McNeil, Paula M Brna, and Kevin E Gordon. 2012. Epilepsy in the twitter era: a need to re-tweet the way we think about seizures. *Epilepsy & behavior*, 23(2):127–130.

Juan Manuel Pérez, Damián A Furman, Laura Alonso Alemany, and Franco Luque. 2021. Robertuito: a pre-trained language model for social media text in spanish. *arXiv preprint arXiv:2111.09453*.

Juan Manuel Pérez, Juan Carlos Giudici, and Franco Luque. 2021. pysentimiento: A Python Toolkit for Sentiment Analysis and SocialNLP tasks. Technical Report arXiv:2106.09462, arXiv. ArXiv:2106.09462 [cs] type: article.

Lauren Sinnenberg, Alison M. Buttenheim, Kevin Padrez, Christina Mancheno, Lyle Ungar, and Raina M. Merchant. 2017. Twitter as a Tool for Health Research: A Systematic Review. *American Journal of Public Health*, 107(1):e1–e8.

Thomas Wolf, Lysandre Debut, Victor Sanh, Julien Chaumond, Clement Delangue, Anthony Moi, Pierric Cistac, Tim Rault, Rémi Louf, Morgan Funtowicz, et al. 2019. Huggingface's transformers: State-of-the-art natural language processing. *arXiv preprint arXiv:1910.03771*.

BOUN-TABI@SMM4H'22: Text-to-Text Adverse Drug Event Extraction with Data Balancing and Prompting

Gökçe Uludoğan* and **Zeynep Yirmibeşoğlu***

Department of Computer Engineering
Bogazici University
Istanbul 34342, Turkey
{gokce.uludogan, zeynep.yirmibesoglu}@boun.edu.tr

Abstract

This paper describes models developed for the Social Media Mining for Health 2022 Shared Task. We participated in two subtasks: classification of English tweets reporting adverse drug events (ADE) (Task 1a) and extraction of ADE spans in such tweets (Task 1b). We developed two separate systems based on the T5 model, viewing these tasks as sequence-to-sequence problems. To address the class imbalance, we made use of data balancing via over- and undersampling on both tasks. For the ADE extraction task, we explored prompting to further benefit from the T5 model and its formulation. Additionally, we built an ensemble model, utilizing both balanced and prompted models. The proposed models outperformed the current state-of-the-art, with an F1 score of 0.655 on ADE classification and a Partial F1 score of 0.527 on ADE extraction.

1 Introduction

This paper describes the models developed for the classification and extraction of Adverse Drug Event (ADE) mentions in English tweets (Tasks 1a and 1b) in the Social Media Mining for Health (SMM4H) 2022 Shared Task (Weissenbacher et al., 2022). We viewed the ADE classification and extraction tasks as sequence-to-sequence problems and fine-tuned the Text-to-Text Transfer Transformer (T5) model (Raffel et al., 2019), which is a pretrained model where the input and the output are both represented as text rather than class labels or spans.

Observing a significant class imbalance in the training data, we made use of over- and undersampling to balance out positive and negative class instances. After this operation, with an equal number of positive and negative instances (1:1 ratio), we obtained an 8% improvement in F1 score for ADE classification, and a 2.2% improvement in Strict F1 score for ADE extraction, showing the power of a balanced class distribution.

Prompt-based learning, or prompting is a method where the input and output text is modified with a string template such that training resembles language model training, where the model fills out missing information (finds the label) (Liu et al., 2021). We have made use of prompting for the ADE extraction task with three different templates, obtaining higher partial, but lower strict F1 scores compared to raw and balanced models. In the end, we have also put together the strengths of our balanced and prompted models in ensemble for the ADE extraction task.

2 Methodology

2.1 Data and Preprocessing

The task organizers provided the second version of the dataset presented in (Magge et al., 2021; Weissenbacher et al., 2022). For both tasks, we split the training data into a training set and a development set with a 80:20 ratio and used the validation set provided by the organizers to evaluate the models. We cleaned tweets with `preprocessor` library[1] and converted emojis to their text aliases using `emoji` library[2].

2.2 Model

In this work, the T5 model suggested by Raffel et al. (Raffel et al., 2019) is employed for ADE classification and extraction. This is an encoder-decoder model pre-trained on multiple tasks, whose architecture resembles that of the original Transformer (Vaswani et al., 2017), and the pre-training objective is to reconstruct randomly corrupted spans. The model can be fine-tuned on multiple tasks by adding a task-denoting prefix in front of the input text, such as "assert ade", or "ner ade". We fine-tuned the T5 model separately for the ADE clas-

*Equal contribution: order determined by a coin flip.

[1] https://github.com/s/preprocessor

[2] https://github.com/carpedm20/emoji

Proceedings of the 29th International Conference on Computational Linguistic, pages 31–34
October 12–17, 2022.

sification and extraction tasks with raw, balanced (over- and undersampled) and prompted datasets.

2.3 Data Balancing

Only 7.06% of the tweets in our training set contained ADE mentions (positive instances). To eliminate this imbalance between positive (ADE) and negative (noADE) instances, we oversampled the 982 ADE tweets up to 6463, and undersampled the 12926 noADE tweets by half using the `imbalanced-learn` library (Lemaître et al., 2017), such that there's a 1:1 ratio. As a second approach, we once again oversampled the 982 ADE tweets up to 6463, and used all of the noADE tweets, obtaining a 2:1 (noADE:ADE) ratio.

2.4 Prompting

Prompting is a method where templates are applied to input and output text for prediction tasks, so that the probability of text can directly be modeled (Liu et al., 2021). This method shows useful in low-resource scenarios with few labeled data (Schick and Schütze, 2021; Gao et al., 2021). Since there are only 982 instances of ADE mentions in the training set, we explored prompting for the ADE extraction task where the goal is to identify ADE mentions in tweets. To this end, we used three templates illustrated in Table 1 to transform data.

2.5 Ensemble Modeling

In neural network training, a change in seed, initialization of parameters or a slight change in the training set alters the outcome of the model. Therefore, we ensembled the predictions of our ADE extraction models, such that the strengths and weaknesses of the models can be compensated by each other. We first applied majority voting and chosen the span predicted at least half of the models for each tweet if there exists such a span. We combined the predictions of different models for the rest of tweets by taking the intersection of the predicted spans.

2.6 Implementation Details

We adapted the implementation[3] of Raval et al. (2021), and fine-tuned the T5 model separately for the classification and extraction tasks. Our models were trained on a single 12 GB GPU for 12 epochs. We took the maximum sequence length as 130, and batch size as 16. We made use of the AdamW optimizer (Loshchilov and Hutter, 2019) with an initial learning rate of *1e-4*. The source code is available at `https://github.com/gokceuludogan/boun-tabi-smm4h22`.

[3]https://github.com/shivamraval98/MultiTask-T5_AE

3 Results

3.1 Task 1a: ADE Classification

Model	Precision	Recall	F1
Raw data	0.75	0.69	0.72
Balanced data (1:1 ratio)	0.75	0.86	0.80
Balanced data (2:1 ratio)	0.73	0.86	0.79

Table 2: Comparison of models on ADE classification task (i.e. Task 1a) on the validation set

Table 2 presents the performance of different models on the validation set with respect to precision, recall and F1 metrics for the ADE class. Unsurprisingly, the models with balanced data outperform the model trained on raw data where tweets mentioning ADEs are rare. The model trained on the balanced data with 1:1 positive/negative ratio performed the best with an F1 score of 0.80 on the validation set. This model was then used to produce predictions for our official submission. Our model obtained higher scores compared to the mean of all submissions across metrics and achieved the state-of-the art result exceeding the performance reported in Magge et al. (2021) (see Table 3).

Model	Precision	Recall	F1
Our model	0.688	0.625	0.655
Mean	0.646	0.497	0.562
Magge et al. (2021)	0.61	0.64	0.63

Table 3: Comparison of our model with the mean of all submissions and the state-of-the-art (Magge et al., 2021) on ADE classification task (i.e. Task 1a) on the test set.

3.2 Task 1b: ADE Extraction

Results for the ADE extraction task on the validation set are reported in Table 4. The models are evaluated with two metrics, Strict F1 score where only perfect matches are considered as true matches and Partial F1 score in which partial matches (i.e. overlapping spans) are also taken into account. The models trained with balanced data achieved the best Strict F1 scores, yet they lag behind all prompting models in terms of Partial F1 score. The highest score in Partial F1 metric is obtained with the

Templates		Input	Output
T1	ADE noADE	Is there a negative drug effect in : [X]	[Y] is a negative drug effect. There isn't a negative drug effect.
T2	ADE noADE	Did the patient suffer from a side effect? [X]	Yes, the patient suffered from [Y]. No, the patient didn't suffer from a side effect.
T3	ADE noADE	[X] Did the patient suffer from a side effect?	Yes, the patient suffered from [Y]. No, the patient didn't suffer from a side effect.

Table 1: Prompting templates for the ADE extraction task given input [X] and output ADE span [Y]. T1, T2 and T3 denote Templates 1, 2 and 3.

Model	Partial F1	Strict F1
Raw data	0.605	0.481
Balanced data (1:1)	0.612	<u>0.503</u>
Balanced data (2:1)	0.639	0.482
Prompt/T1	0.636	0.424
Prompt/T2	<u>0.662</u>	0.408
Prompt/T3	0.638	0.393
Ensemble	0.657	0.500

Table 4: Comparison of models on ADE extraction task (i.e. Task 1b) on the validation set

model using the second prompt template. However, this model's Strict F1 score is significantly lower than the best performing model (0.408 vs 0.503). The ensemble model addressing the performance gap between the best models with respect to Strict F1 and Partial F1 metrics combined the predictions of Balanced data (1:1) and Prompt/T2, and obtained the second best score across metrics. Observing the success of the ensemble model on both overlapping and strict metrics, we used its predictions as our official submission to Task 1b. As seen in Table 5, the model beats the state-of-the-art (Magge et al., 2021) with respect to Partial Recall and F1 metrics as well as obtaining significantly higher strict scores than the average of all official submissions on the test set. Yet, its Partial F1 score was at par with the other submissions, suggesting that our model achieves acceptable results for overlapping spans, but the strength of our model lies rather in its detection of perfect spans.

3.3 Analysis

We analyzed the performance of our submissions on the validation set provided by the organizers. We observed that our ADE classification model, trained on the balanced data with 1:1 ratio, per-

Model	Partial			Strict		
	P	R	F1	P	R	F1
Our model	0.507	**0.549**	**0.527**	**0.384**	**0.412**	**0.398**
Mean	**0.539**	0.517	**0.527**	0.344	0.339	0.341
Magge et al. (2021)	0.53	0.38	0.44	-	-	-

Table 5: Comparison of our model with the mean of all submissions and the state-of-the-art (Magge et al., 2021) on ADE extraction task (i.e. Task 1b) on the test set. P and R denote precision and recall, respectively.

formed fairly well in detecting ADE mentions with 56 correct predictions out of 65 ADE mentions. However, the model also made 19 incorrect ADE predictions (i.e. false positives). Similar results were also observed in our ensemble ADE extraction model. The model predictions included a lot of false positives (42) in contrast to relatively few false negatives (14).

4 Conclusion

In this study, we applied the T5 model and its problem formulation (i.e. viewing tasks as sequence-to-sequence problems) to ADE classification and extraction tasks. We used over- and undersampling to address class imbalance, a major challenge in ADE extraction from social media. In addition, we explored prompting methods to further benefit from the T5 language model and its sequence-to-sequence formulation. Our official submission for the ADE classification task made use of data balancing via over- and undersampling, and achieved an F1 score of 0.655, outperforming the state-of-the-art. For the ADE extraction task, we ensembled our balanced and prompted models to increase generalization for both partial and exact matches. Our submission of the ensemble model achieved a Partial F1 score of 0.527, beating the current state-of-the art and performing at par with the mean of all submissions. In the future, we intend to experiment with other methods of dealing with imbalanced data, such as data augmentation. The ADE classification and extraction tasks is also planned to be used in a multi-task learning scenario with the T5 model.

Acknowledgements

The models trained in this paper were fully performed at Boğaziçi University Telecommunications and Informatics Technologies Research Center servers (TETAM resources).

References

Tianyu Gao, Adam Fisch, and Danqi Chen. 2021. Making pre-trained language models better few-shot learners. In *Proceedings of the 59th Annual Meeting of the Association for Computational Linguistics and the 11th International Joint Conference on Natural Language Processing (Volume 1: Long Papers)*, pages 3816–3830, Online. Association for Computational Linguistics.

Guillaume Lemaître, Fernando Nogueira, and Christos K. Aridas. 2017. Imbalanced-learn: A python toolbox to tackle the curse of imbalanced datasets in machine learning. *Journal of Machine Learning Research*, 18(17):1–5.

Pengfei Liu, Weizhe Yuan, Jinlan Fu, Zhengbao Jiang, Hiroaki Hayashi, and Graham Neubig. 2021. Pre-train, prompt, and predict: A systematic survey of prompting methods in natural language processing. *arXiv preprint arXiv:2107.13586*.

Ilya Loshchilov and Frank Hutter. 2019. Decoupled weight decay regularization. In *7th International Conference on Learning Representations, ICLR 2019, New Orleans, LA, USA, May 6-9, 2019*. OpenReview.net.

Arjun Magge, Elena Tutubalina, Zulfat Miftahutdinov, Ilseyar Alimova, Anne Dirkson, Suzan Verberne, Davy Weissenbacher, and Graciela Gonzalez-Hernandez. 2021. Deepademiner: a deep learning pharmacovigilance pipeline for extraction and normalization of adverse drug event mentions on twitter. *Journal of the American Medical Informatics Association*, 28(10):2184–2192.

Colin Raffel, Noam Shazeer, Adam Roberts, Katherine Lee, Sharan Narang, Michael Matena, Yanqi Zhou, Wei Li, and Peter J Liu. 2019. Exploring the limits of transfer learning with a unified text-to-text transformer. *arXiv preprint arXiv:1910.10683*.

Shivam Raval, Hooman Sedghamiz, Enrico Santus, Tuka Alhanai, Mohammad Ghassemi, and Emmanuele Chersoni. 2021. Exploring a unified Sequence-To-Sequence Transformer for medical product safety monitoring in social media. In *Findings of the Association for Computational Linguistics: EMNLP 2021*, pages 3534–3546, Punta Cana, Dominican Republic. Association for Computational Linguistics.

Timo Schick and Hinrich Schütze. 2021. Few-shot text generation with natural language instructions. In *Proceedings of the 2021 Conference on Empirical Methods in Natural Language Processing*, pages 390–402, Online and Punta Cana, Dominican Republic. Association for Computational Linguistics.

Ashish Vaswani, Noam Shazeer, Niki Parmar, Jakob Uszkoreit, Llion Jones, Aidan N Gomez, Łukasz Kaiser, and Illia Polosukhin. 2017. Attention is all you need. *Advances in neural information processing systems*, 30.

Davy Weissenbacher, Ari Z. Klein, Luis Gascó, Darryl Estrada-Zavala, Martin Krallinger, Yuting Guo, Yao Ge, Abeed Sarker, Ana Lucia Schmidt, Raul Rodriguez-Esteban, Mathias Leddin, Arjun Magge, Juan M. Banda, Vera Davydova, Elena Tutubalina, and Graciela Gonzalez-Hernandez. 2022. Overview of the seventh social media mining for health applications #SMM4H shared tasks at COLING 2022. In *Proceedings of the Seventh Social Media Mining for Health (#SMM4H) Workshop and Shared Task*.

uestcc@SMM4H'22: RoBERTa based Adverse Drug Events Classification on Tweets

Chunchen Wei[1], Ran Bi[1], and Yanru Zhang[1,2]

[1]University of Electronic Science and Technology of China, China
[2]Shenzhen Institute for Advanced Study, UESTC, China

Abstract

This is a description of our participation in the *ADE Mining in English Tweets* shared task, organized by the Social Media Mining for Health *SMM4H* 2022 workshop. We participate in the subtask a of shared Task 1, and the paper introduces the system we developed for solving the task. The task requires classifying the given tweets by whether they mention the Adverse Drug Effects. We utilize RoBERTa model and apply several methods during training and finetuning period. We also try to improve the performance of our system by preprocessing the dataset but improve the precision only. The results of our system on test set are 0.601 in F1-score, 0.705 in precision, and 0.524 in recall.

1 Introduction

Adverse Drug Events (ADEs) refer to negative side effects related to the drug. In the area of Social Media Pharmacovigilance, mining ADEs from social media is one of the most studied topics. In the Social Media Mining for Health 2022 (SMM4H) shared Task 1a (Weissenbacher et al., 2022), we focus on classifying tweets reporting ADEs and labeling them with ADE or noADE. In training process, we input 18000 labeled tweets provided by organizers to RoBERTa model and finetune the model by several practical methods such as Stochastic Weight Averaging (SWA), then output the prediction results. The system is shown in Figure 1.

We also attempt to preprocess the tweets by deleting some stop words and emojis, the performance of all approaches is shown in the paper.

2 Data

The dataset (Magge et al., 2021) provided by the organizer includes a training set of 17,385 tweets, a validation set of 915 tweets, and a test set of 10,984 tweets. It is worth noting that the sample proportion is highly unbalanced, tweets mentioning ADEs are only 7% of the training data.

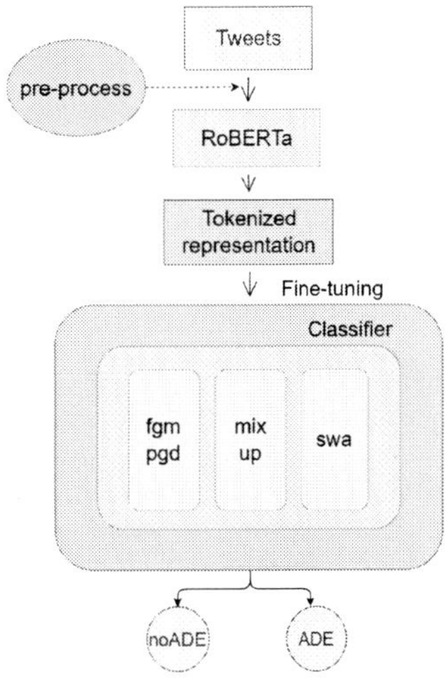

Figure 1: The architecture of system.

For data preprocessing, we save the original training data at first containing the redundant symbols and emojis in the tweets. Then we perform following preprocessing on the original dataset to construct another dataset as a comparison:

a) lowercase all words and remove whitespace.

b) remove instances of '@USER' followed by '_'.

c) remove extranous characters and emojis.

d) remove stopwords[1].

3 Method

In this section, we introduce the methodology we propose to accomplish the task of the competition.

We directly use the original dataset as input of the system at first. Then we pass the input into the BERT-based model RoBERTa (Lee and Toutanova, 2018) and get the token representations. By inputting the representations through the fully connected layers and finetuning the model, we obtain the final model and use it to predict the test set.

[1]`https://www.nltk.org/nltk_data/`

Proceedings of the 29th International Conference on Computational Linguistic, pages 35–37
October 12–17, 2022.

During the period of finetuning, we add several methods to improve the performance of our model:

- Fast Gradient Method (FGM) was first proposed by Miyato et al. (2016), we use this method to add a disturbance to the original input training samples and conduct adversarial training, so as to improve the robustness and generalization ability of the model.
- Project Gradient Descent (PGD) proposed by Madry et al. (2017) is an iterative attack and each iteration will project the disturbance into the specified range, we apply this method to further improve the generalization ability of the model.
- Mix-up is a simple and effective data augmentation method (Zhang et al., 2017), we apply this method on the provided dataset to prevent the model from overfitting.
- Stochastic Weight Averaging (SWA) is a method to improve the generalization ability of deep learning-based model through gradient descent and does not require additional computation (Izmailov et al., 2018), we use it in finetuning phase.

4 Experiments

In the shared Task 1a, the RoBERTa model is trained for 10 epochs with a learning rate 5×10^{-5} using Adam optimizer (Kingma and Ba, 2014). We set the batch size in training duration to 16 and 64 in validation and test duration, respectively. In addition, embedding size, output size, and sequence length in the experiments are set up as 768, 768, and 256.

We also conduct the ablation experiments to test the performance of several methods mentioned in last section. We first set the parameter alpha to 0.2 during the process of mix-up. Then we add FGM and PGD methods into the training process and use SWA with a learning rate 5×10^{-5}. As the results of ablation experiment shown in Table 1, the performance of the model on validation set improves gradually as methods add.

In order to verify the effectiveness of data preprocessing, we use the datasets before and after data preprocessing to train the model respectively. The results on validation set of two datasets are in Table 2, it shows that the application of data preprocessing increases precision but reduces recall.

Method	P	R	F1
RoBERTa	0.844	0.721	0.756
RoBERTa$^+$	**0.891**	0.640	0.745
RoBERTa$^\#$	0.851	0.718	0.779
RoBERTa*	0.859	**0.765**	**0.809**

$^+$RoBERTa with mixup.
$^\#$RoBERTa with mixup and swa.
*RoBERTa with mixup, swa, fgm and pgd.

Table 1: Ablation experiments on the validation set.

Dataset	P	R	F1
Dataset$_{raw}$	0.859	**0.765**	**0.809**
Dataset$_{cleaned}$	**0.873**	0.750	0.761

Table 2: Task1a results on validation set.

As for the prediction results of test set, points of three evaluation metrics significantly drop compared with results on validation set. In our two submitted results of prediction, one uses data preprocessing and another not, both are predicted by the RoBERTa model added with three methods. In addition, full training set (validation set + training set) is used to train the model. Results of our submissions are demonstrated in Table 3, mean results of all submissions by all participants are in the last row of table.

5 Conclusion

In this work, to complete the task of binary classification on tweets, we propose a system based on RoBERTa model applying with several methods to improve the generalization ability of our model, and get the results of 0.731 in precision, 0.524 in recall, and 0.601 in F1 score. We also validate the effectiveness of finetuning methods and data preprocessing in the experiments. The performance of our proposed system on test dataset compared with the mean results indicates the robustness and generalization ability of our system, and we will continue to improve it in the future.

	P	R	F1
Dataset$_{raw}$	0.705	**0.524**	**0.601**
Dataset$_{cleaned}$	**0.731**	0.339	0.463
mean	0.646	0.497	0.562

Table 3: Task1a results on test set.

References

Pavel Izmailov, Dmitrii Podoprikhin, Timur Garipov, Dmitry Vetrov, and Andrew Gordon Wilson. 2018. Averaging weights leads to wider optima and better generalization. *arXiv preprint arXiv:1803.05407*.

Diederik P Kingma and Jimmy Ba. 2014. Adam: A method for stochastic optimization. *arXiv preprint arXiv:1412.6980*.

J Devlin M Chang K Lee and K Toutanova. 2018. Pre-training of deep bidirectional transformers for language understanding. *arXiv preprint arXiv:1810.04805*.

Aleksander Madry, Aleksandar Makelov, Ludwig Schmidt, Dimitris Tsipras, and Adrian Vladu. 2017. Towards deep learning models resistant to adversarial attacks. *arXiv preprint arXiv:1706.06083*.

Arjun Magge, Elena Tutubalina, Zulfat Miftahutdinov, Ilseyar Alimova, Anne Dirkson, Suzan Verberne, Davy Weissenbacher, and Graciela Gonzalez-Hernandez. 2021. Deepademiner: a deep learning pharmacovigilance pipeline for extraction and normalization of adverse drug event mentions on twitter. *Journal of the American Medical Informatics Association*, 28(10):2184–2192.

Takeru Miyato, Andrew M Dai, and Ian Goodfellow. 2016. Adversarial training methods for semi-supervised text classification. *arXiv preprint arXiv:1605.07725*.

Davy Weissenbacher, Ari Z. Klein, Luis Gascó, Darryl Estrada-Zavala, Martin Krallinger, Yuting Guo, Yao Ge, Abeed Sarker, Ana Lucia Schmidt, Rual Rodriguez-Esteban, Mathias Leddin, Arjun Magge, Juan M. Banda, Vera Davydova, Elena Tutubalina, and Graciela Gonzalez-Hernandez. 2022. Overview of the seventh social media mining for health applications (# smm4h) shared tasks at coling 2022. In *In Proceedings of the Seventh Social Media Mining for Health (# SMM4H) Workshop and Shared Task*.

Hongyi Zhang, Moustapha Cisse, Yann N Dauphin, and David Lopez-Paz. 2017. mixup: Beyond empirical risk minimization. *arXiv preprint arXiv:1710.09412*.

Zhegu@SMM4H-2022: The Pre-training Tweet & Claim Matching Makes Your Prediction Better

Pan He[1], YuZe Chen[3], Yanru Zhang[1,2]*
[1]University of Electronic Science and Technology of China, Chengdu
[2]Shenzhen Institute for Advanced Study, UESTC
[3]Chengdu University of Technology
newtonysls@gmail.com
chen.yuze@student.zy.cdut.edu.cn
yanruzhang@uestc.edu.cn

Abstract

SMM4H-2022 (Weissenbacher et al., 2022) Task 2 is to detect whether containing premise in the tweets of users about COVID-19 on the social medias or their stances for the claims. In this paper, we propose **T**weet **C**laim **M**atching (**TCM**), which is a new pre-training task constructed by the tweets and claims similarly to Next Sentence Prediction (NSP). We first continue to pre-train the standard pre-trained language models on the labelled dataset and then fine-tune them for obtaining better performance. Compared with the solid baseline (Glandt et al., 2021), we achieve the absolute improvement of 7.9% in Task 2a and obtain the SOTA results.

1 Introduction

Since its appearance in 2019, COVID-19 has generated a wide range of discussions on the social medias. Users make statements that may contain their stances and arguments, and these Twitter posts are value for estimating the level of cooperation with the mandates. Research on this information has attracted widespread attention, and various automated approaches based on machine learning have been proposed. In this paper, we present the solution of our team Zhegu for identifying the stance and premise of Twitter posts in SMM4H-2022 Task 2. To summarize, our main contribution are:

- We constructed **TCM**. A new tweet & claim matching task for pre-training in SMM4H-2022 Task 2.
- Our system obtains the SOTA performance on Task2a of the stance prediction.

2 Data

All data sets (Davydova and Tutubalina, 2022) contain texts from Twitter about three health mandates related to the COVID-19 pandemic: Face Masks, Stay At Home Orders, and School closures. There are 3,556 of the training data, 600 for validation, and 2,000 for testing. Task2a needs to determine whether the attitude of the authors of the tweets are supported, against or neutral to the claims, which is a triple classification. Task2b is a bi-classification, which aims to figure out whether the tweets contain at least one premise or argument.

3 Methodology

In this section, we firstly present our model and strategies we use. Secondly, we give the basic idea and the construction of our **TCM** pre-training task.

3.1 Fine tune

We selected two pre-trained language models (PLMs) as our backbones: RoBERTa-large (Liu et al., 2019) and COVID-Twitter-BERT-v2 (Müller et al., 2020). They both have 24 hidden layers, with the hidden size of 1024. We also design various strategies to obtain the representation of sentences, including CLS, averaging the last layer of PLMs, attention computation for the last four layers and concatenating the CLS of the last four layers. The small amount of the training dataset and the large scale of backbone can easily lead to over-fitting. In this case, we also used FGSM (Goodfellow et al., 2014), R-Drop (Wu et al., 2021), Exponential Moving Average (EMA) to improve the generalization of our model.

3.2 TCM

It has been proved that continuously pre-training on the domain corpus can improve the performance of fine-tuning (Gururangan et al., 2020). We perform further pre-training on backbone based on the training dataset. BERT (Devlin et al., 2018) contains two pre-training tasks including Masked Language Model (MLM) and NSP, while RoBERTa (Liu et al., 2019) only has the MLM. Consequently, We pre-train the MLM for both PLMs. Given a data set

*Corresponding author

Proceedings of the 29th International Conference on Computational Linguistic, pages 38–41
October 12–17, 2022.

RoBERTa-large (ours)							
FGSM	EMA	R-Drop	MLM	TCM	Pseudo-label	Validation	Test
√	-	-	-	-	-	0.788	-
√	√	-	-	-	-	0.798	-
√	√	√	-	-	-	0.800	-
√	√	√	√	-	-	0.812	-
√	√	√	√	√	-	0.841 (+0.051)	-
COVID-Twitter-BERT-v2 (ours)							
√	√	√	-	-	-	0.853	-
√	√	√	√	√	-	0.868	-
√	√	√	√	√	√	**0.869** (+0.079)	0.634
Baseline (COVID-Twitter-BERT)						0.790	-

Table 1: The average F1 scores of various strategies in the validation and test dataset. We report the performance of baseline from (Glandt et al., 2021) directly. The bolded and underlined: the SOTA results of all methods. The underlined: the second best.

of the Task2a $((\boldsymbol{T}, \boldsymbol{C}), \boldsymbol{Y})$, where $\boldsymbol{T_i}$ denotes the tweet text, $\boldsymbol{C_i}$ represents the corresponding claim, and $\boldsymbol{Y_i}$ is the label. And $\boldsymbol{C_i} \in \mathcal{C} =$ {Face Masks, Stay At Home Orders, School Closures}. Naturally, $\boldsymbol{T_i}$ only corresponds to its unique $\boldsymbol{C_i}$ rather than $\mathcal{C} - \boldsymbol{C_i}$. This is a typical bi-classification as well as NSP. We construct the dataset of TCM by randomly sampling $\boldsymbol{C^{\text{sample}}}$ for $\boldsymbol{T_i}$ from $\mathcal{C}$ in a certain probability of $p = 0.5$. $((\boldsymbol{T_i}, \boldsymbol{C^{\text{sample}}}), \mathbf{1})$ is a positive sample if $\boldsymbol{C^{\text{sample}}} = \boldsymbol{C_i}$, otherwise negative sample $((\boldsymbol{T_i}, \boldsymbol{C^{\text{sample}}}), \mathbf{0})$. This TCM task incorporates the idea of contrastive learning, which enables the PLMs to determine whether the tweet matches the claim, thus improving the performance of the model.

4 Experiments

4.1 Experiments Settings

For fine-tuning, different learning rates are used for the backbone and other layers. The backbone is set to 2e-5 and the other layers are 1e-4. The epochs of the RoBERTa-large and COVID-Twitter-BERT-v2 are set to 11 and 4 respectively. Others detailed hyperparameters can be found in the Table 2. The evaluation metric we use is the average F1 score of the three claims, which is different from the website.

4.2 Main Results & Analysis

Table 1 presents the final results on the validation set and test set of Task2a. Task2b can be found in the Table 4. By comparing the effects of four different sentence representations, we finally adopted averaging the hidden states of the last layer of PLMs, as shown in the Table 3. The pre-training corpus of COVID-Twitter-BERT-v2 are the tweets related to COVID-19, while RoBERTa is pre-trained on the general corpus. Overall, the results of RoBERTa are worse 2.8% than COVID-Twitter-BERT-v2. However, RoBERTa still outperformed 5.1% compared to the baseline(Glandt et al., 2021), which takes COVID-Twitter-BERT (Müller et al., 2020) as the backbone. Meanwhile, FGSM, R-Drop and EMA bring different degrees of improvement. It brings relative small promotion when we merely pre-train the MLM task. The improvement is about 2.9% than the MLM only after adding our TCM task. Similarly, for COVID-Twitter-BERT-v2, the boost of 1.6% from the TCM and MLM task is also impressive. It is worth mentioning that we do not perform K-fold training, nor do we perform any model ensemble. The experimental results strongly prove that the TCM we have constructed is simple yet effective.

5 Conclusion

We present the solution of our team Zhegu in SMM4H-2022 Task 2. We construct a new TCM pre-training task by incorporating the idea of contrastive learning for the relationship between tweets and claims. The effectiveness of our TCM has been demonstrated by two PLMs. Even in RoBERTa with only the MLM, the addition of TCM brings a decent improvement. In summary, absolute im provement of 5.1% and 7.9% are achieved in both backbones compared to baseline. And we propose to investigate the relationship between Twitter text and its claim when fine-tuning in our future work.

References

Vera Davydova and Elena Tutubalina. 2022. Smm4h 2022 task 2: Dataset for stance and premise detection in tweets about health mandates related to covid-19. In *Proceedings of the Seventh Social Media Mining for Health Applications (#SMM4H) Workshop Shared Task*, pages –.

Jacob Devlin, Ming-Wei Chang, Kenton Lee, and Kristina Toutanova. 2018. Bert: Pre-training of deep bidirectional transformers for language understanding. *arXiv preprint arXiv:1810.04805*.

Kyle Glandt, Sarthak Khanal, Yingjie Li, Doina Caragea, and Cornelia Caragea. 2021. Stance detection in covid-19 tweets. In *Proceedings of the 59th Annual Meeting of the Association for Computational Linguistics and the 11th International Joint Conference on Natural Language Processing (Volume 1: Long Papers)*, pages 1596–1611.

Ian J Goodfellow, Jonathon Shlens, and Christian Szegedy. 2014. Explaining and harnessing adversarial examples. *arXiv preprint arXiv:1412.6572*.

Suchin Gururangan, Ana Marasović, Swabha Swayamdipta, Kyle Lo, Iz Beltagy, Doug Downey, and Noah A Smith. 2020. Don't stop pretraining: adapt language models to domains and tasks. *arXiv preprint arXiv:2004.10964*.

Yinhan Liu, Myle Ott, Naman Goyal, Jingfei Du, Mandar Joshi, Danqi Chen, Omer Levy, Mike Lewis, Luke Zettlemoyer, and Veselin Stoyanov. 2019. Roberta: A robustly optimized bert pretraining approach. *arXiv preprint arXiv:1907.11692*.

Martin Müller, Marcel Salathé, and Per E Kummervold. 2020. Covid-twitter-bert: A natural language processing model to analyse covid-19 content on twitter. *arXiv preprint arXiv:2005.07503*.

Davy Weissenbacher, Ari Z. Klein, Luis Gascó, Darryl Estrada-Zavala, Martin Krallinger, Yuting Guo, Yao Ge, Abeed Sarker, Ana Lucia Schmidt, Raul Rodriguez-Esteban, Mathias Leddin, Arjun Magge, Juan M. Banda, Vera Davydova, Elena Tutubalina, and Graciela Gonzalez-Hernandez. 2022. Overview of the seventh social media mining for health applications smm4h shared tasks at coling 2022. In *Proceedings of the Seventh Social Media Mining for Health Applications (#SMM4H) Workshop Shared Task*, pages –.

Lijun Wu, Juntao Li, Yue Wang, Qi Meng, Tao Qin, Wei Chen, Min Zhang, Tie-Yan Liu, et al. 2021. R-drop: Regularized dropout for neural networks. *Advances in Neural Information Processing Systems*, 34:10890–10905.

A Implement Details

All experiments were run on a RTX 3090 GPU with 24G of video memory. The rest of hyperparameters is shown in the Table 2. The main performance metric is calculated according to the following formula:

FGSM	epsilon	0.5
EMA	start step	0
	decay	0.99
R-Drop	alpha	0.5
Max sequence length	128	
Batch size	16	
Pseudo label rate	0.88	
Optimizer	AdamW	
Adam epsilon	1e-6	
Weight decay	0.1	
Scheduler	linear warmup	
Warmup rate	0.1	
Max grad norm	1	

Table 2: The hyperparameters of various strategies for fine-tuning. The same parameters are used for both Task2a and Task2b. Besides, we directly apply the same pre-training hyperparameters of BERT (Devlin et al., 2018).

$$F1 = \frac{1}{3}\sum_{C}\frac{1}{2}(F_1^{macro} + F_1^{micro}) \quad (1)$$

where $C \in \mathcal{C}$. The evaluation metric is different from the official website, so the results we obtained are not the same as the official website. For the sentence representation, we evaluate four common methods in validation dataset of Task2a as it shown in Table 3. There is no the best representation which will always yield the best results for different downstream tasks. And we choose to average the last hidden states as the sentence representations for the model by comparing the results obtained from experiments under the same parameters.

Sentence representation	Validation
CLS	0.818
Mean	**0.841**
Attention	0.797
CLS-4	0.801

Table 3: CLS: the [CLS] representation. Mean: average the hidden states of the last layer. Attention: attention calculation of the hidden states of the last 4 layers. CLS-4: concatenate the [CLS] of the last 4 layers.

B Another Results

Model	Validation
RoBERTa-large	0.820
COVID-Twitter-BERT-v2	0.831

Table 4: The final average F1 scores in the validation dataset of Task2b. It is worth noting that the results we present here obtained by using the same optimization strategy as Task2a.

Task	Validation (ours)	Test (ours)	Test (Median)	Test (Mean)
Task2a	0.869	0.634	0.550	0.491
Task2b	0.831	0.698	0.647	0.574

Table 5: The results of the test dataset. Median: the median score of all teams. Mean: the average score of all teams. Our results are excessively higher than the median and mean score in the test dataset of both Task2a and Task2b.

MaNLP@SMM4H'22: BERT for Classification of Twitter Posts

Keshav Kapur
Manipal Institute of Technology
keshav29kapur@gmail.com

Rajitha Harikrishnan
Manipal Institute of Technology
rajithasuja@gmail.com

Sanjay Singh
Manipal Institute of Technology
sanjay.singh@manipal.edu

Abstract

The reported work is our straightforward approach for the shared task "Classification of tweets self-reporting age" organized by the "Social Media Mining for Health Applications (SMM4H)" workshop. This literature describes the approach that was used to build a binary classification system, that classifies the tweets related to birthday posts into two classes namely, exact age(positive class) and non-exact age(negative class). We made two submissions with variations in the preprocessing of text which yielded F1 scores of 0.80 and 0.81 when evaluated by the organizers.

1 Introduction

Determining the exact age of an individual is crucial to increasing the use of social media data for research purposes. In this contemporary world, adolescents use social media to the extent that it can have some very severe effects on their overall well-being if not monitored. Hence, a few applications like Twitter have set up some age restrictions for the well-being of an individual. They automatically detect the age of an individual trying to protect them from viewing unnecessary and harmful content.

In this work, we determine the exact age of an individual based on their tweets on Twitter. This helps in validating if a particular user has faked their age or not. One of the major challenges we faced while working with the data is that some of them could tweet about their friend's or relative's birthday which was getting misclassified. We have used BERT model which helps in the binary classification of our data. While developing our system for this task, we have discovered BERT that outperforms traditional training models.

2 Methodology

2.1 Pre-Processing

Initially, the organisers provided us with 8,800 training data and 2,200 validation data. This dataset consisted of three fields: tweet id, text of the Tweet Object and annotated binary class label(exact age present/absent). The training and validation data were later combined and it was pre-processed for further development. For preprocessing, we removed URLs, emoticons, hashtags, and mentions using a python package *tweet-preprocessor*. After that we removed: contractions from the tweets, special characters and extra spaces. Then we used python package called *Natural Language Toolkit* for removing the stop words. After these steps, we have further divided the pre-processing into two techniques: pronouns and removing pronouns.

2.2 Model

In our model, we have used BERT (*bert-base-uncased*) from the Hugging Face library as a classifier and Softmax as the activation function. In the BERT model, there is an important special token [CLS] which is used as an input for our choice of classifier. We have used the Adam optimizer to fine-tune our BERT model. We trained the model with 4 to 10 epochs which converged after 10 epochs. Learning rate of the optimizer is given by 5e - 5.The batch size used is 32.

3 Evaluation

In the validation phase, our model produced satisfactory results with about 90%. In the test data, 10,000 tweets were provided by the organizers. We have first pre-processed with pronouns and then removed pronouns in the next round of preprocessing. After evaluation, our models generated an F1-score of 0.80 and 0.81.

Table 1 shows our evaluation scores for Precision,

Proceedings of the 29th International Conference on Computational Linguistic, pages 42–43
October 12–17, 2022.

Model	Precision	Recall	F1-Score
Model 1	0.839	0.780	0.808
Model 2	0.771	0.870	0.818

Table 1: Evaluation scores

Recall, and F1-Score as provided by the organizers. Model 1 shows scores of pre-processing with pronouns and Model 2 shows scores of pre-processing with pronouns removed.

4 Conclusion

We discussed our approach to fine-tuning our BERT model on Task 4 of the 2022 Social Media Mining for Health applications shared task. As we observe from the results, the given training data was inadequate to train on a BERT model. There was an imbalance in the number of positives and negatives given in our dataset(refer to Figure 1). An interesting observation drawn from this work is that BERT models rely on huge and balanced datasets for learning patterns. Future work might consider collecting more data points for training, fine-tuning our BERT model, and applying other state-of-the-art methods like RoBERTa.

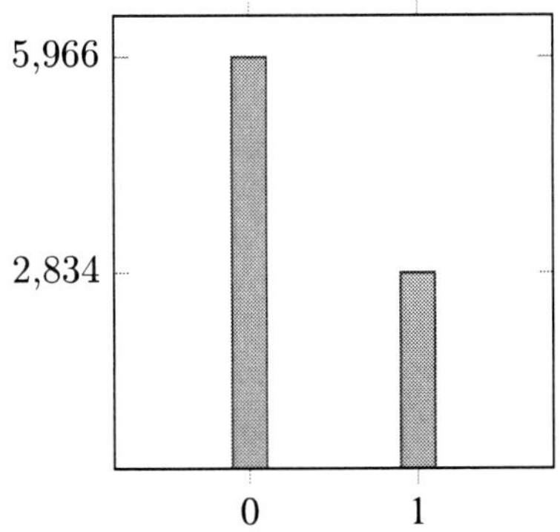

Figure 1: Summary of Dataset Labels

References

Jacob Devlin, Ming-Wei Chang, Kenton Lee, and Kristina Toutanova. 2018. BERT: pre-training of deep bidirectional transformers for language understanding. *CoRR*, abs/1810.04805.

Katikapalli Subramanyam Kalyan, Ajit Rajasekharan, and Sivanesan Sangeetha. 2021. Ammus : A survey of transformer-based pretrained models in natural language processing.

Diederik P. Kingma and Jimmy Ba. 2014. Adam: A method for stochastic optimization.

Yinhan Liu, Myle Ott, Naman Goyal, Jingfei Du, Mandar Joshi, Danqi Chen, Omer Levy, Mike Lewis, Luke Zettlemoyer, and Veselin Stoyanov. 2019. Roberta: A robustly optimized bert pretraining approach.

Arjun Magge, Ari Klein, Antonio Miranda-Escalada, Mohammed Ali Al-garadi, Ilseyar Alimova, Zulfat Miftahutdinov, Eulalia Farre-Maduell, Salvador Lima Lopez, Ivan Flores, Karen O'Connor, Davy Weissenbacher, Elena Tutubalina, Abeed Sarker, Juan M Banda, Martin Krallinger, and Graciela Gonzalez-Hernandez, editors. 2021. *Proceedings of the Sixth Social Media Mining for Health (#SMM4H) Workshop and Shared Task*. Association for Computational Linguistics, Mexico City, Mexico.

John_Snow_Labs@SMM4H'22: Social Media Mining for Health (#SMM4H) with Spark NLP

Veysel Kocaman[1], Cabir Celik[1], Damla Gurbaz[1], Gursev Pirge[1], Bunyamin Polat[1], Halil Saglamlar[1], Meryem Vildan Sarikaya[1], Gokhan Turer[1], David Talby[1]

[1]John Snow Labs Inc., 16192 Coastal Highway, Lewes, DE , USA 19958,
veysel@johnsnowlabs.com

Abstract

Social media has become a major source of information for healthcare professionals but due to the growing volume of data in unstructured format, analyzing these resources accurately has become a challenge. In this study, we trained health related NER and text classification models on different datasets published within the Social Media Mining for Health Applications (#SMM4H 2022) workshop[1]. We utilized transformer based Bert for Token Classification and Bert for Sequence Classification algorithms as well as vanilla NER and text classification algorithms from Spark NLP library during this study without changing the underlying DL architecture. The trained models are available within a production-grade code base as part of the Spark NLP library; can scale up for training and inference in any Spark cluster; has GPU support and libraries for popular programming languages such as Python, R, Scala and Java.

1 Introduction

A primary building block in text mining systems is Named Entity Recognition (NER) - which is regarded as a critical precursor for question answering, topic modeling, information retrieval, etc. (Li et al., 2022). In the medical domain, NER recognizes the first meaningful chunks out of a clinical note, which are then fed down the processing pipeline as an input to subsequent downstream tasks such as clinical assertion status detection (Uzuner et al., 2011), clinical entity resolution (Tzitzivacos, 2007) and de-identification of sensitive data (Uzuner et al., 2007).

Text classification is also one of the important tasks of machine learning and has been extensively used in several areas of NLP. It can be defined as assigning a sentence or document an appropriate category.

In this paper, we share our results on the #SMM4H Shared Task, which involves NLP challenges of using social media data for health research, including informal, colloquial expressions and misspellings of clinical concepts, noise, data sparsity, ambiguity, and multilingual posts. The contest involves classification and NER tasks. As John Snow Labs team, we contributed to many tasks and trained several NER and Classification models on English and Spanish social media data.

2 Background

2.1 Spark NLP Library

All the experiments are run within Spark NLP library, a popular open-source NLP library that has been downloaded more than 30 million times so far[2].

Spark NLP library has two versions: Open source and enterprise. The open-source version has all the features and components that could be expected from any NLP library, using the latest Deep Learning (DL) frameworks and research trends. Enterprise library is licensed (free for academic purposes) and designed towards solving

[1] https://healthlanguageprocessing.org/smm4h-2022/

[2] https://pepy.tech/project/spark-nlp

Proceedings of the 29th International Conference on Computational Linguistic, pages 44–47
October 12–17, 2022.

real world problems in the healthcare domain and extends the open-source version. The licensed version has the following modules to help researchers and data practitioners in various means: Named entity recognition (NER), assertion status (negativity scope) detection, relation extraction, entity resolution (SNOMED, RxNorm, ICD10 etc.), clinical spell checking, contextual parser, deidentification and obfuscation (Kocaman and Talby, 2021).

2.2 Named Entity Recognition in Spark NLP

There are two DL architectures for NER tasks in Spark NLP: *BiLSTM-CNN-Char* and *BertForTokenClassification*. *BiLSTM-CNN-Char* architecture (Kocaman and Talby, 2020) is a modified version of the architecture proposed by Chiu and Nichols (2016). It may be defined as a neural network architecture that automatically detects word and character-level features using a hybrid bidirectional LSTM and CNN architecture, eliminating the need for most feature engineering steps.

NER systems are usually a part of an end-to-end NLP pipeline through which the text is fed and then several text preprocessing steps are applied. Since the DL algorithm we implement is sentence-wise, and the features (embeddings and casing) are token-wise, sentence splitting and tokenization are the most important steps leading to better accuracy. Using a DL based sentence detector module (Scweter and Ahmed, 2019) and a highly customizable rule-based tokenizer in Spark NLP, we ensured that the generated features are more informative.

The second architecture, *BertForTokenClassification (BFTC)*, can load Bert (Peters et al., 2018; Radford et al., 2018) based models with a token classification head on top (a linear layer on top of the hidden-states output) e.g., for NER tasks. Spark NLP lets users utilize this transformer architecture into Spark NLP with just a few lines of code.

2.3 Classification in Spark NLP

To be able to process large volume of data, the text classification model needs to be scalable, and accurate, as it is used to filter out documents, reviews, and twccts that do not contain any indication of adverse events. To achieve this, Spark NLP offers two DL architectures: *ClassifierDL* and *Bert for Sequence Classifier* (BFSC). ClassifierDL uses a Fully Connected Neural Network (FCNN) model that does not require hand-crafted features and relies on a single embedding vector for classification. Given the conversational nature of social media text, we can utilize the entire document to get efficient embeddings (with little text clipping in case of BioBERT embeddings) that we directly feed to the classifier model. Since there is only a single feature vector as input to the model, we test multiple embedding techniques to analyze performance (Haq et al., 2022).

Similar to the token classification problem for NER tasks, *BertForSequenceClassification (BFSC)* is used for text classification tasks based on Bert architecture.

3 Implementation Details

Models for classification tasks were trained by using Bert for Sequence Classifier (BFSC) and Bert for Token Classifier (BTFC) was used for training NER tasks. Embeddings and hyper parameters used for training the documents are shown in Table 1. Once the pre-trained models were loaded within a pipeline, those models were used for prediction. *MedicalBertForTokenClassifier* and *MedicalBertForSequenceClassification* annotators were used to test the performances of the models on the validation datasets.

4 Experimental Results

Using the official training sets from the contest, we trained models and obtained metrics on the test sets used in the challenges. The results are shown in Table 1, where all the tested configurations obtained decent accuracies over the validation dataset. Tasks 1,2,3a, 5 and 10 are not worked on as a team; but some team members made separate individual efforts on them.

Unless otherwise specified, Biobert-base-cased-v1.2 embeddings were used for model training of English texts. Often times, cleaning the social media material (removing emojis, tags, repeating punctuations etc.) did not improve the results.

Task 1a is about classifying tweets reporting Adverse Drug Events (ADEs). BFSC gave the best results. The training dataset was unbalanced; therefore, it has been enriched with different ADE datasets (Haq et al., 2022) to increase the number of ADE entities.

Task 1b involves detecting ADE intervals in tweets. The model was trained using BFTC. As mentioned above, the training dataset was unbalanced and just like Task 1a, the dataset was enriched by using datasets (Haq et al., 2022) with the aim of increasing the number of ADE entities.

Table 1. Detailed information about the competition tasks, embeddings, parameters and model performance.

Task	Additional Data	Model Performance					
		P	R	F1	P	R	F1
1a	Yes	Validation			Test		
		0.97	0.96	0.96	*		
1b	Yes	Validation			Test		
		0.89	0.88	0.88	*		
2a	No	Validation			Test		
		0.72	0.72	0.72	*		
2b	No	Validation			Test		
		0.74	0.72	0.73	*		
3a	No	Validation			Test		
		0.79	0.72	0.75	*		
3b	No	Validation			Test		
		0.86	0.85	0.85	0.84	0.82	0.83
4	No	Validation			Test		
		0.79	0.85	0.82	0.82	0.80	0.81
5	No	Validation			Test		
		0.77	0.77	0.77	*		
6	Yes	Validation			Test		
		0.79	0.78	0.78			
7	Yes	Validation			Test		
		0.80	0.66	0.73	0.88	0.66	0.71
8	No	Validation			Test		
		0.72	0.86	0.78	0.68	0.88	0.76
9	No	Validation			Test		
		0.85	0.90	0.87	0.86	0.85	0.86
10	No	Validation			Test		
		0.69	0.87	0.77	*		

* Tasks 1,2,3a, 5 and 10 are not worked on as a team; but some of the team members did some individual efforts on the side.

In Task 2a, classification of stance in tweets about health mandates is expected. BFSC with biobert-base-cased-v1.2 embeddings produced the best results, whereas bert_base_cased embeddings produced slightly higher metrics.

Task 2b involves binary classification of premise in tweets about health mandates. BFSC produced the better F1 score values.

In Task 3a, detection of tweets where the users self-declare changing their medication treatments is expected. The model trained using BFSC produced higher F1 score values. Training data was unbalanced and using different embeddings and cleaning the tweets did not improve the results.

Task 3b involves binary classification of drug reviews from WebMD.com. BFSC produced around 0.84 F1 score values.

Task 4 involves identification of the self-report exact age in tweet posts. BFSC produced the best results (F1-score of 0.82).

For Task 5, all the tested configurations obtained decent accuracies over the validation dataset. The difference of scores between the two classes is probably due to the imbalance in the dataset.

Task 6 involves identification of self-reported COVID-19 vaccination status in English tweets. The two classes in the provided training dataset are Vaccine_chatter and Self_reports (tweets of users clearly stating that they have been vaccinated). The classes are significantly imbalanced with a ratio of 1 to 8. BFSC produced an F1 score value (for the positive class) of 0.72.

Task 7 involves the identification of self-reported Intimate partner violence (IPV) in English tweets. The dataset is significantly unbalanced, and the negative tweets include non-IPV domestic violence and non-self-reported IPV, which can hardly be distinguished. Where the model could not select between domestic violence and IPV, better results were obtained by introducing family-related words (father, mother, daughter, sister etc.) to the model and enabling it to learn about domestic violence. BFSC produced F1 values around 0.72 (for the positive class) for the validation dataset.

Task 8 is about detecting tweets that are self-disclosures of chronic stress. For the classification task, a pre-trained Roberta For Sequence Classification model from Hugging Face with Roberta embeddings is fine-tuned on the given dataset. Without cleaning the tweets, the model resulted in better scores. The validation F1 score was 0.78 and the test F1 score is 0.76.

Task 9 involves identification of the exact age in social media forum (Reddit) posts. BFSC produced the best results (F1-score of 0.87).

Task 10 requires detection of disease mentions in Spanish tweets, and BFTC was used to produce the models. Huggingface's 5-language embeddings (bert-base-5lang-cased) were used to get the highest F1 score values.

5 Conclusion

In this study, we show through experiments on #SMM4H contest datasets in two languages that models trained by BFSC and BFTC can easily be adapted to the Spark NLP environment and achieve decent scores on social/health-related datasets with zero code changes. We trained BFSC models for classification and BFTC models for NER purposes that can be used as a pretrained model out of the box within Spark NLP for Healthcare library.

Considering the F1 scores, all the models produced decent performances during training. The problem of getting low metrics due to unbalanced datasets was solved by enriching the dataset with data from similar datasets and consequently, there was a substantial increase in the F1 scores.

References

Jason P. C. Chiu and Eric Nichols. 2016. *Named entity recognition with bidirectional LSTM-CNNs. Transactions of the Association for Computational Linguistics*. 4 (2016) 357–370. https://doi.org/10.48550/arXiv.1511.08308

Hasham Ul Haq, Veysel Kocamanand David Talby. 2022. Mining adverse drug reactions from unstructured mediums at scale. W3PHIAI workshop at AAAI-22. https://doi.org/10.48550/arXiv.2201.01405.

Veysel Kocaman and David Talby. 2020. *Improving Clinical Document Understanding on COVID-19 Research with Spark NLP.* arXiv:2012.04005v1

Veysel Kocaman and David Talby. 2021. *Spark NLP: Natural Language Understanding at Scale. Software Impacts*, arXiv:2101.10848v1 [cs.CL].

Jing Li, Aixin Sun, Jianglei Han and Chenliang Li. 2022. *A Survey on Deep Learning for Named Entity Recognition. IEEE Transactions on Knowledge and Data Engineering,* 34(1). https://doi.org/10.48550/arXiv.1812.09449.

Matthew Peters, Mark Neumann, Mohit Iyyer, Matt Gardner, Christopher Clark, Kenton Lee, and Luke Zettlemoyer. 2018. *Deep contextualized word representations.* In *Proceedings of the 2018 Conference of the North American Chapter of the Association for Computational Linguistics: Human Language Technologies, Volume 1 (Long Papers)*. Association for Computational Linguistics, page 2227–2237.

Alec Radford, Karthik Narasimhan, Tim Salimans, and Ilya Sutskever. 2018. *Improving language understanding with unsupervised learning*. Technical report, OpenAI.

Stefan Schweter and Sajawel Ahmed. 2019. Deep-EOS: General-purpose neural networks for sentence boundary detection. In *Proceedings of the 15th Conference on Natural Language Processing (KONVENS 2019)*.

Dimitri Tzitzivacos. 2007. *International Classification of Diseases 10th edition (ICD-10): main article. CME: Your SA Journal of CPD.* 25(1): 8–10.

Ozlem Uzuner, Yuan Luo and Peter Szolovits. 2007. *Evaluating the State-of-the-Art in Automatic De-identification. Journal of the American Medical Informatics Association*. 14(5): 550–563

Ozlem Uzuner, Brett South, Shuying Shen and Scott L. DuVall. 2011. *2010 i2b2/VA challenge on concepts, assertions, and relations in clinical text. Journal of the American Medical Informatics Association* 18(5): 552–556

READ-BioMed@SocialDisNER: Adaptation of an Annotation System to Spanish Tweets

Antonio Jimeno Yepes[1,2] and Karin Verspoor[1,2]

[1]School of Computing Technologies, RMIT University, Melbourne, Australia
[2]School of Computing and Information systems
The University of Melbourne, Melbourne, Australia
[1]{antonio.jose.jimeno.yepes,karin.verspoor}@rmit.edu.au

Abstract

We describe the work of the READ-BioMed team for the preparation of a submission to the SocialDisNER Disease Named Entity Recognition (NER) Task (Task 10) in 2022. We had developed a system for named entity recognition for identifying biomedical concepts in English MEDLINE citations and Spanish clinical text for the LivingNER 2022 challenge (Miranda-Escalada et al., 2022). Minimal adaptation of our system was required to perform named entity recognition in the Spanish tweets in the SocialDisNER task, given the availability of Spanish pre-trained language models and the SocialDisNER training data. Minor additions included treatment of emojis and entities in hashtags and Twitter account names.

1 Motivation

In this paper, we describe the READ-BioMed (Reading, Extraction, and Annotation of Documents in BioMedicine) approach to the 2022 SocialDisNER Task 10 (Weissenbacher et al., 2022). The documents in this task are Spanish-language tweets and the task involves the annotation of disease entities in tweet text.

The READ-BioMed team has extensive experience in the processing of Twitter to identify medical entities (Jimeno-Yepes et al., 2015; Jimeno Yepes and MacKinlay, 2016), the analysis of large sets of tweets for trend analysis (MacKinlay et al., 2015; Jimeno Yepes et al., 2015; Huang et al., 2016), pharmacovigilance using social media (MacKinlay et al., 2017; Li et al., 2020), and for syndromic surveillance (Ofoghi et al., 2016).

Our approach was to adapt a system previously developed for annotation of MEDLINE citations in the English language and clinical text in Spanish. Our system relies on a pre-trained transformer based language model, which was fine-tuned for biomedical concept recognition for the LitCOIN challenge earlier this year[1], where we ranked in the top 5 submissions[2]. More recently, we adapted this system for processing Spanish clinical texts (Jimeno Yepes and Verspoor, 2022) in the LivingNER challenge (Miranda-Escalada et al., 2022).

2 Methods

In this section, we describe the methods that we have used in the challenge. We describe the data and the pre-processing step, the training and annotation of the data and the post-processing steps that we have followed to prepare our submissions using the validation and testing sets.

2.1 Data

We used the data set provided by the task organisers (Gasco et al., 2022). This data set contains 5,000 tweets in the training set, 2,500 tweets in the validation set and 23,430 tweets in the testing set.

2.2 From tweets to BIO

Tweets are messages of up to 280 characters. So, we did not split the tweets into sentences as done in previous challenges (Jimeno Yepes and Verspoor, 2022) to support analysis with BERT-like models (Devlin et al., 2019), instead processing a complete tweet as a single unit.

We define the following list of characters used to identify token boundaries, some of these tokens were defined as Unicode characters, e.g. '...'. All characters except for space (' ') and the new line ('\n') were included as tokens.

```
split_tokens = ['¡','\xa0','_',
'¿','=','+','*',';','&',' ',
'-','.','[',']','...','(',')',
',','/','%',':','#','@','"',
'"','-',"\n",'?',"'",'!','"']
```

[1]https://ncats.nih.gov/funding/challenges/litcoin

[2]https://ncats.nih.gov/funding/challenges/litcoin/winners

Proceedings of the 29th International Conference on Computational Linguistic, pages 48–51
October 12–17, 2022.

In addition to the token list defined above, we also used the python library `emoji`[3], set up for the Spanish language, to identify emoji tokens. Tweets in all the data sets (training, validation and testing) were tokenised using the same process.

The training and validation sets include information about the disease entities annotated per tweet, which are provided in a tab separated values file. For each entity, the tweet id and the offset of the entities are available and are used to identity the tokens and label each one of them using the IOB labeling (Ramshaw and Marcus, 1995). The B label is used to denote the first token of a disease entity, the I label is used to denote any additional token within the entity and the O label is to denote any token that is not part of a disease entity. The IOB labels were used to train our system.

2.3 Training

In our previous work in biomedical Spanish named entity recognition as part of the LivingNER challenge (Miranda-Escalada et al., 2022) and for English language concept recognition in the LitCoin challenge,[4] we evaluated pre-trained language models such as BioBERT (Lee et al., 2020), SciBERT (Beltagy et al., 2019) and PubMedBERT (Gu et al., 2021).

We utilised the NERDA[5] framework for named entity recognition. It consists of a BERT-based system followed by a fully connected layer with as many outputs as labels need to be predicted. Our motivation was to be able to make changes and adapt the software according to (Jimeno Yepes, 2022), to prevent overfitting.

Since the challenge data was in Spanish, we reused the pre-trained language model for Spanish biomedical data (Carrino et al., 2021) that we used in the LivingNER challenge. More specifically, we have used the model *PlanTL-GOB-ES/roberta-base-biomedical-es* available from Huggingface[6]. This model is based on a RoBERTa (Liu et al., 2019) and trained trained on a biomedical-clinical corpus in Spanish collected from several sources.

We used the training data set provided by the organisers (5,000 tweets) for learning the model and the validation set (2,500 documents) was used to control the training process. The IOB labels that our system was trained for included the O label for out entity tokens, representing a total of 3 token labels. We did not consider any of the extended data sets provided by the organisers.

2.4 Annotation

The trained model obtained using the process described above has been used to annotate the tokens in the validation and testing sets with IOB labels. The IOB labels were matched to the tweet text and the positions of these tokens were used to generate the tab separated values submission files. A pipeline of this process is shown in figure 1.

We submitted the annotations on the validation set to the submission system provided by the organisers, which showed that precision was relatively high, while recall could be improved. The analysis of errors in the validation set showed that disease entities in hashtags, identified by the hashtag symbol # and account names @, were not identified. The main reason for this problem is that such terms were not tokenised to allow disease identification.

To solve this problem, the predicted disease entities in the tweet were identified and these entities were matched against these special terms by lower casing the terms and doing an exact match. This method boosted the recall from 0.856 in the strict evaluation to 0.887 and from 0.914 in the relaxed evaluation to 0.947, while suffering only a small decrease in precision. The validation set contains 2,500 tweets and it took over 7 minutes to annotate them using the trained model.

The testing set was processed using the methodology presented above. The testing set contained 23,430 tweets and it took just over one hour to annotate them using the trained model.

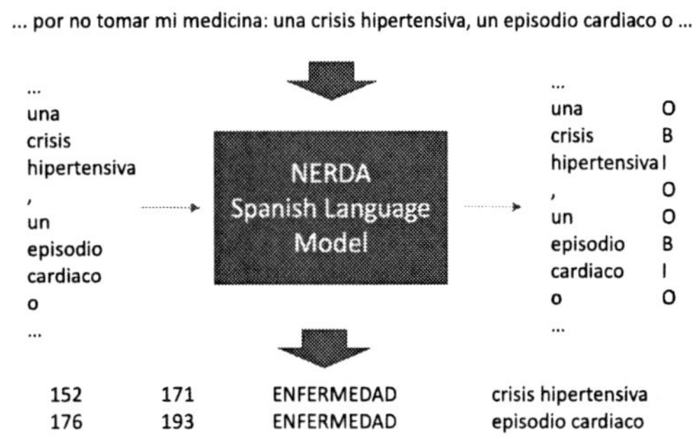

Figure 1: Annotation pipeline. A tweet is tokenised, processed by the trained model in NERDA and the IOB output is converted into the submission format.

[3] https://github.com/carpedm20/emoji

[4] https://bitgrit.net/competition/13#private-leaderboard

[5] https://github.com/ebanalyse/NERDA

[6] https://huggingface.co/PlanTL-GOB-ES/roberta-base-biomedical-clinical-es

3 Results

We generated only one submission for the testing set using the procedure presented above. Results of our submission (READ-BioMed) using strict entity matching are presented in table 1, which also includes the results for the mean and median values of all participants. Our submission achieved a substantially higher performance compared to the mean and median of the challenge participants.

	Precision	Recall	F1
READ-BioMed	0.868	0.875	0.871
Mean	0.680	0.677	0.675
Median	0.758	0.780	0.761

Table 1: Official results on the testing set using strict matching. These results contain out submission (READ-BioMed) and the mean and median of all participants.

Table 2 shows the results of the fine tuned model of the validation set. Both results are reported, the precision, recall and F1 of the B and I labels and the results by the organisers system on a submission built on the validation set. Comparing the results of our submission in table 1 and the results on the validation set using strict matching, we can see that despite the lower result of our submission, the decrease in performance is limited.

Evaluation	Precision	Recall	F1
B label	0.936	0.951	0.944
I label	0.904	0.877	0.890
Strict	0.881	0.887	0.884
Relaxed	0.944	0.947	0.946

Table 2: Evaluation of the fine tuned models on the validation data set for B and I labels and results for the strict and relaxed evaluation.

The B label is predicted with high precision and recall, which might explain the performance on the relaxed evaluation, e.g. being able to find the beginning of an entity (B label) but not being so successful with the following tokens (I label).

Acknowledgements

This work was supported by Commonwealth Innovation Connections Grant ICG001884 from the Australian Department of Industry, Science, Energy and Resources to RMIT. This research used the HPC-GPGPU Facility at the University of Melbourne, supported by LIEF Grant LE170100200.

References

Iz Beltagy, Kyle Lo, and Arman Cohan. 2019. SciBERT: A pretrained language model for scientific text. *arXiv preprint arXiv:1903.10676*.

Casimiro Pio Carrino, Jordi Armengol-Estapé, Asier Gutiérrez-Fandiño, Joan Llop-Palao, Marc Pàmies, Aitor Gonzalez-Agirre, and Marta Villegas. 2021. Biomedical and clinical language models for Spanish: On the benefits of domain-specific pretraining in a mid-resource scenario. *arXiv preprint arXiv:2109.03570*.

Jacob Devlin, Ming-Wei Chang, Kenton Lee, and Kristina Toutanova. 2019. BERT: Pre-training of deep bidirectional transformers for language understanding. In *Proceedings of the 2019 Conference of the North American Chapter of the Association for Computational Linguistics: Human Language Technologies, Volume 1 (Long and Short Papers)*, pages 4171–4186, Minneapolis, Minnesota. Association for Computational Linguistics.

Luis Gasco, Darryl Estrada-Zavala, Eulàlia Farré-Maduell, Salvador Lima-López, Antonio Miranda-Escalada, and Martin Krallinger. 2022. Overview of the SocialDisNER shared task on detection of diseases mentions from healthcare related and patient generated social media content: methods, evaluation and corpora. In *Proceedings of the Seventh Social Media Mining for Health (#SMM4H) Workshop and Shared Task*.

Yu Gu, Robert Tinn, Hao Cheng, Michael Lucas, Naoto Usuyama, Xiaodong Liu, Tristan Naumann, Jianfeng Gao, and Hoifung Poon. 2021. Domain-specific language model pretraining for biomedical natural language processing. *ACM Transactions on Computing for Healthcare (HEALTH)*, 3(1):1–23.

Pin Huang, Andrew MacKinlay, and Antonio Jimeno Yepes. 2016. Syndromic surveillance using generic medical entities on twitter. In *Proceedings of the Australasian Language Technology Association Workshop 2016*, pages 35–44.

Antonio Jimeno Yepes. 2022. Hyperplane bounds for neural feature mappings. *arXiv preprint arXiv:2201.05799*.

Antonio Jimeno Yepes and Andrew MacKinlay. 2016. Ner for medical entities in twitter using sequence to sequence neural networks. In *Proceedings of the Australasian Language Technology Association Workshop 2016*, pages 138–142.

Antonio Jimeno Yepes, Andrew MacKinlay, and Bo Han. 2015. Investigating public health surveillance using twitter. In *Proceedings of BioNLP 15*, pages 164–170.

Antonio Jimeno-Yepes, Andrew MacKinlay, Bo Han, and Qiang Chen. 2015. Identifying diseases, drugs, and symptoms in twitter. In *MEDINFO 2015: eHealth-enabled Health*, pages 643–647. IOS Press.

Antonio Jimeno Yepes and Karin Verspoor. 2022. The read-biomed team in livingner task 1 (2022): Adaptation of an english annotation system to spanish. In *LiverNER challenge, IberLEF 2022*.

Jinhyuk Lee, Wonjin Yoon, Sungdong Kim, Donghyeon Kim, Sunkyu Kim, Chan Ho So, and Jaewoo Kang. 2020. BioBERT: a pre-trained biomedical language representation model for biomedical text mining. *Bioinformatics*, 36(4):1234–1240.

Ying Li, Antonio Jimeno Yepes, and Cao Xiao. 2020. Combining social media and fda adverse event reporting system to detect adverse drug reactions. *Drug safety*, 43(9):893–903.

Yinhan Liu, Myle Ott, Naman Goyal, Jingfei Du, Mandar Joshi, Danqi Chen, Omer Levy, Mike Lewis, Luke Zettlemoyer, and Veselin Stoyanov. 2019. Roberta: A robustly optimized bert pretraining approach. *arXiv preprint arXiv:1907.11692*.

Andrew MacKinlay, Hafsah Aamer, and Antonio Jimeno Yepes. 2017. Detection of adverse drug reactions using medical named entities on twitter. In *AMIA Annual Symposium Proceedings*, volume 2017, page 1215. American Medical Informatics Association.

Andrew MacKinlay, Antonio Jimeno Yepes, and Bo Han. 2015. Identification and analysis of medical entity co-occurrences in twitter. In *Proceedings of the ACM Ninth International Workshop on Data and Text Mining in Biomedical Informatics*, pages 22–22.

Antonio Miranda-Escalada, Eulàlia Farré-Maduell, Salvador Lima-López, Darryl Estrada, Luis Gascó, and Martin Krallinger. 2022. Mention detection, normalization & classification of species, pathogens, humans and food in clinical documents: Overview of livingner shared task and resources. *Procesamiento del Lenguaje Natural*.

Bahadorreza Ofoghi, Meghan Mann, and Karin Verspoor. 2016. Towards early discovery of salient health threats: A social media emotion classification technique. In *Proceedings of the Pacific Symposium on Biocomputing 2016*, pages 504–515. World Scientific.

Lance Ramshaw and Mitch Marcus. 1995. Text chunking using transformation-based learning. In *Third Workshop on Very Large Corpora*.

Davy Weissenbacher, Ari Z. Klein, Luis Gascó, Darryl Estrada-Zavala, Martin Krallinger, Yuting Guo, Yao Ge, Abeed Sarker, Ana Lucia Schmidt, Raul Rodriguez-Esteban, Mathias Leddin, Arjun Magge, Juan M. Banda, Vera Davydova, Elena Tutubalina, and Graciela Gonzalez-Hernandez. 2022. Overview of the seventh social media mining for health applications smm4h shared tasks at coling 2022. In *Proceedings of the Seventh Social Media Mining for Health (SMM4H) Workshop and Shared Task*.

PLN CMM at SocialDisNER: Improving Detection of Disease Mentions in Tweets by Using Document-Level Features

Matías Rojas[1], Jose Barros[1], Kinan Martin[3], Mauricio Araneda[2], and Jocelyn Dunstan[1,4]
[1]Center for Mathematical Modeling (CMM), University of Chile.
[2]National Center for Artificial Intelligence (CENIA), Chile.
[3]Department of Computer Sciences, Massachusetts Institute of Technology, USA.
[4]Initiative for Data & Artificial Intelligence, University of Chile.
{matias.rojas.g, jose.barros.s, mauricio.araneda}@ug.chile.cl
krmkrm@mit.edu, jdunstan@uchile.cl

Abstract

This paper describes our approaches used to solve the SocialDisNER task, which belongs to the Social Media Mining for Health Applications (SMM4H) shared task. This task aims to identify disease mentions in tweets written in Spanish. The proposed model is an architecture based on the FLERT approach. It consists of fine-tuning a language model that creates an input representation of a sentence based on its neighboring sentences, thus obtaining the document-level context. The best result was obtained using an ensemble of six language models using the FLERT approach. The system achieved an F_1 score of 0.862, significantly surpassing the average performance among competitor models of 0.680 on the test partition.

1 Introduction

Understanding the social perception of diseases is crucial for quantifying the effectiveness of public policies on medicine. In particular, analyzing social network data allows us to study this issue in depth. In this context, the SocialDisNER shared task (Gasco et al., 2022) aims to better understand diseases' social perception by using Named Entity Recognition (NER) to identify diseases mentioned in Twitter data: from highly prevalent conditions such as cancer and diabetes to rare immunological and genetic diseases. In this paper, we discuss our approaches used to solve the SocialDisNER task. Specifically, we use several language models trained on both general and domain-specific corpora, exploiting the use of document-level features. Finally, we study the use of ensembles and their impact on performance compared to individual models.

2 Task and Data Description

2.1 Task

Task 10 of SMM4H aims to identify disease mentions in tweets written in Spanish. The identification of entity mentions must be strict, meaning that an entity is considered correct when the system simultaneously identifies both entity types and boundaries.

2.2 Data

Regarding the statistics of the dataset, $10,000$ tweets were released: 50% for training, 25% for validation, and 25% for testing. There are $15,173$ disease mentions in the training partition, while $4,252$ for validation. There were also nested entities within the annotations, which we simplified using the common strategy of keeping only the outermost entity in a nesting, as pointed out in Báez et al. (2022).

3 Methodology

This section describes the models used in our experiments and the chosen baseline.

3.1 Baseline

The baseline model consists of fine-tuning the *xlm-roberta-large-ner-hrl* model for the token classification task. We decided on using this baseline due to its previous NER-specific training on high-resource multilingual corpora, including the Spanish CoNLL 2002 corpus (Tjong Kim Sang, 2002). Unlike our primary FLERT-based approach, the training of this architecture is performed at the sentence level, i.e., only the tokens of the current sentence are used as input to the model. This difference allows comparing the performance of sentence-level context against document-level context models.

3.2 FLERT Model

Our main proposal for the task is based on the FLERT approach (Schweter and Akbik, 2020). This architecture is similar to the baseline since it fine-tunes a language model but differs in that it considers document-level context rather than sentence-level context. To this purpose, we add

Proceedings of the 29th International Conference on Computational Linguistic, pages 52–54
October 12–17, 2022.

a window of 64 tokens from the previous sentence and 64 tokens from the following sentence. We consider the document-level context relevant since there are tweets containing multiple sentences, which implies that two contiguous sentences describe the same idea with a high probability.

The Spanish language models selected to test this approach are the following: the biomedical and clinical versions of RoBERTa (*bsc-bio-es* and *bsc-bio-ehr-es*) (Carrino et al., 2022), the Spanish version of BERT (BETO) (Cañete et al., 2020) and the Spanish version of RoBERTa (*roberta-base-bne*) (Gutiérrez-Fandiño et al., 2022). We also tested concatenating the representations obtained with BETO and Clinical RoBERTa. Finally, although in most of the experiments, we did not use preprocessing on the data, we tested Clinical RoBERTa but converted emojis to text, removed URLs, and unified usernames.

3.3 Ensemble

The second approach is to ensemble all FLERT-based models, regardless of their domains and data preprocessing methods. For this purpose, we used a voting system, which counts the number of times an entity mention was identified and keeps only those that exceed a defined voting threshold.

4 Experiments

The training process for the FLERT-based models was analogous to each language model. We searched for an optimal learning rate between 1e-5, 5e-5, 5e-6, and 1e-6. For the sake of brevity, we only present the best model, trained with a learning rate of 5e-6. We used the Adam optimizer with linear decay and no warm-up steps. The models were trained for 20 epochs using a batch size of 16 sentences with a maximum length of 512 tokens.

In contrast, the baseline model was trained with a learning rate of 2e-5 for 3 epochs with a weight decay of 0.01, which is the setting that gave us the best result. The rest of the hyperparameters are the same. The training of each model took approximately 3 hours using a Tesla V100 GPU.

As shown in Table 1, the baseline model obtained the lowest results, which can be explained since this model was trained on a general domain corpus, lacking both clinical language and sufficient Spanish language specialization. In contrast, the FLERT-based models obtained better results, demonstrating the importance of incorporating the

Model	P	R	F_1
Baseline	0.860	0.763	0.809
FLERT [BETO]	0.865	0.835	0.850
FLERT [RoBERTa]	0.863	0.833	0.848
FLERT [Bio RoBERTa]	0.887	0.854	0.867
FLERT [Clinical RoBERTa]	0.881	0.857	0.870
FLERT [Stacked]	0.880	0.837	0.858
FLERT [Ensemble]	**0.903**	**0.866**	**0.884**

Table 1: Overall results in the validation partition.

Model	P	R	F_1
Other systems (mean)	0.680	0.677	0.675
Other systems (median)	0.758	0.780	0.761
FLERT [Clinical RoBERTa]	0.867	0.831	0.848
FLERT [Ensemble]	**0.882**	**0.843**	**0.862**

Table 2: Overall results in the testing partition.

document-level context. The best results were obtained using the Clinical RoBERTa model, reaching an F_1 score of 0.870, and the ensemble model reaching 0.884 points. Analyzing the ensemble approach, which obtained the best performance, the best results in the validation set were obtained using a threshold of two votes. The Clinical RoBERTa and ensemble-based models were the final systems submitted to the shared task. The source code to reproduce our experiments is freely available to the research community [1].

5 Results and Conclusion

The overall results of our submissions are shown in Table 2. The ensemble approach obtained the best results, achieving a strict F_1 score of 0.862, followed by the Clinical RoBERTa model with 0.848. Both models outperformed the average F_1 score of the rest of the systems submitted (0.675) by 0.187 and 0.173 points, respectively. This demonstrates that using FLERT document-level features delivers great results independent of its simplicity, especially when using domain-specific language models. In future work, we believe that incorporating a more extensive range of domain-specific models into the ensemble architecture, such as models trained on social media, can improve performance further. In addition, we would like to test character-level contextualized embeddings, such as Clinical Flair (Rojas et al., 2022), since it is mentioned that they are helpful for clinical corpora in Spanish. Finally, to measure the impact of the unstructured nature of tweets, we would like to test our models on other datasets created for disease recognition, such as the Chilean Waiting List (Báez et al., 2020).

[1] https://github.com/plncmm/socialdisner

References

Pablo Báez, Felipe Bravo-Marquez, Jocelyn Dunstan, Matías Rojas, and Fabián Villena. 2022. Automatic extraction of nested entities in clinical referrals in spanish. *ACM Trans. Comput. Healthcare*, 3(3).

Pablo Báez, Fabián Villena, Matías Rojas, Manuel Durán, and Jocelyn Dunstan. 2020. The Chilean waiting list corpus: a new resource for clinical named entity recognition in Spanish. In *Proceedings of the 3rd Clinical Natural Language Processing Workshop*, pages 291–300, Online. Association for Computational Linguistics.

Casimiro Pio Carrino, Joan Llop, Marc Pàmies, Asier Gutiérrez-Fandiño, Jordi Armengol-Estapé, Joaquín Silveira-Ocampo, Alfonso Valencia, Aitor Gonzalez-Agirre, and Marta Villegas. 2022. Pretrained biomedical language models for clinical NLP in Spanish. In *Proceedings of the 21st Workshop on Biomedical Language Processing*, pages 193–199, Dublin, Ireland. Association for Computational Linguistics.

José Cañete, Gabriel Chaperon, Rodrigo Fuentes, Jou-Hui Ho, Hojin Kang, and Jorge Pérez. 2020. Spanish pre-trained bert model and evaluation data. In *PML4DC at ICLR 2020*.

Luis Gasco, Darryl Estrada-Zavala, Eulàlia Farrá-Maduell, Salvador Lima-López, Antonio Miranda-Escalada, and Martin Krallinger. 2022. Overview of the SocialDisNER shared task on detection of diseases mentions from healthcare related and patient generated social media content: methods, evaluation and corpora. In *Proceedings of the Seventh Social Media Mining for Health (#SMM4H) Workshop and Shared Task*.

Asier Gutiérrez-Fandiño, Jordi Armengol-Estapé, Marc Pàmies, Joan Llop-Palao, Joaquin Silveira-Ocampo, Casimiro Pio Carrino, Carme Armentano-Oller, Carlos Rodriguez-Penagos, Aitor Gonzalez-Agirre, and Marta Villegas. 2022. Maria: Spanish language models. *Procesamiento del Lenguaje Natural*, 68(0):39–60.

Matías Rojas, Jocelyn Dunstan, and Fabián Villena. 2022. Clinical flair: A pre-trained language model for Spanish clinical natural language processing. In *Proceedings of the 4th Clinical Natural Language Processing Workshop*, pages 87–92, Seattle, WA. Association for Computational Linguistics.

Stefan Schweter and Alan Akbik. 2020. FLERT: Document-level features for named entity recognition.

Erik F. Tjong Kim Sang. 2002. Introduction to the CoNLL-2002 shared task: Language-independent named entity recognition. In *COLING-02: The 6th Conference on Natural Language Learning 2002 (CoNLL-2002)*.

CLaCLab at SocialDisNER: Using Medical Gazetteers for Named-Entity Recognition of Disease Mentions in Spanish Tweets

Harsh Verma, Parsa Bagherzadeh, Sabine Bergler
CLaC Labs, Concordia University
{h_ver, p_bagher, bergler} @cse.concordia.ca

Abstract

This paper summarizes the CLaC submission for SMM4H 2022 Task 10 which concerns the recognition of diseases mentioned in Spanish tweets. Before classifying each token, we encode each token with a transformer encoder using features from Multilingual RoBERTa Large, UMLS gazetteer, and DISTEMIST gazetteer, among others. We obtain a strict F1 score of 0.869, with competition mean of 0.675, standard deviation of 0.245, and median of 0.761. Our code is available here.

1 Motivation

Finding mentions of diseases in tweets in all languages has become an important tool for epidemiologists, especially in times of a pandemic. SocialDisNER (Gasco et al., 2022) at SMM4H 2022 concerns the recognition of disease mentions in Spanish tweets. A disease mention can include both lay and professional language and may be located in hashtags or usernames as well as the tweet text. Disease entities are underlined in Example 1:

(1) *Pasos sábado 23 de mayo! ... Sumemos 1 millón de pasos por la epilepsia, #1MillonDePasos, #Epilepsia, #ADosMetrosDeDistancia #InvestigaEpilepsia @RetoDravet #juntosesmejor @FundacionDravet*

2 System

Our contribution consists of a simple pipeline incorporating four main components: a standard tokenizer for splitting hashtags and username tokens with an ad hoc extension for disease recognition, word embeddings for Spanish and English from RoBERTa Large, four gazetteer lists extracted from relevant domain resources and represented as one hot vectors, and a linear classifier that produces BIO (*beginning, inside, outside*) tags for each token for recognition of multi-token disease mentions. The simplicity of this pipeline and its knowledge injection from readily available domain resources rather than training purely from training data make our system's strength.

Tokenization Sequence analysis classifying each token into BIO tags requires good tokenization. We pre-process the data using ANNIE (Cunningham et al., 2002) and the Twitter English Tokenizer (Bontcheva et al., 2013).

Hashtags and usernames usually include several words strung together without separating characters, e.g. #InvestigaEpilepsia. Since disease mentions can occur in these word composites, their individual word components have to be separated. Camel casing makes splitting this example into *Investiga* and *Epilepsia* easy, but for hashtags like #nosolohaycovid only knowledge of words in Spanish leads to the desired split into *no*, *solo*, *hay* and *covid*. Because disease names are not always part of the general vocabulary of a language, disease gazetteers are a lightweight means to inject such domain knowledge into a general machine learning system.

A gazetteer lists entities of a particular type. For example, a disease gazetteer would contain a list of disease names like COVID, Hepatitis, Epilepsia etc. Ad hoc gazetteer lists that are written for a specific task often suffer from limited coverage. We circumvent our own limitations by using four gazetteer lists compiled from extant resources:

***GoldGaz*:** disease names compiled from the gold annotations of SocialDisNER training and validation data

***DistemistGaz*:** (Miranda-Escalada et al., 2022) Spanish disease gazetteer compiled from Snomed-CT (Donnelly et al., 2006)

***SilverGaz*:** disease gazetteer compiled from silver standard data made available for SocialDisNER (Gasco et al., 2022)

Proceedings of the 29th International Conference on Computational Linguistic, pages 55–57
October 12–17, 2022.

UmlsGaz: disease gazetteer of Spanish and English disease terms we compiled from UMLS (Bodenreider, 2004). Semantic type T047 represents *Disease or Syndrome*, identifying all UMLS concepts representing diseases.

Twitter tokenizer extension The GATE Twitter English Tokenizer (Bontcheva et al., 2013) identifies hashtags and usernames, but doesn't split them. To split usernames and hashtags, we identify the longest substring in a hashtag/username that matches an entity in our gazetteer, then treat that substring as a separate token. The extension matches first against GoldGaz, then DistemistGaz, followed by UmlsGaz, and finally SilverGaz.

Classification We frame the named entity recognition (NER) task as a sequence labeling task with the BIO (*beginning, inside, outside*) tagging scheme. In other words, we classify each token as either `B` (token is the beginning of a disease name), `I` (token lies inside a disease name) or `O` (token lies outside a disease name).

Model Preprocessing produces S, a list of t tokens. S is fed into a pretrained XLM RoBERTa Large (Conneau et al., 2020) model, which returns the matrix $\mathbf{H} \in \mathbb{R}^{t \times d}$, with $d = 1024$. The ith row of $\mathbf{H}$ is a vector of size d and corresponds to the contextualized embedding of the ith token.

We use the GATE ANNIE Gazetteer plugin to find tokens that match terms in UmlsGaz exactly (case-insensitively), creating $\mathbf{G}_{umls} \in \mathbb{R}^{t \times 2}$, a matrix where each row is a one hot vector indicating whether the ith token matches with *UmlsGaz* or not. Similarly, we create $\mathbf{G}_{silver}$ and $\mathbf{G}_{distemist}$ corresponding to *SilverGaz* and *DistemistGaz* respectively. Then, $\mathbf{H}$ is row-wise concatenated with $\mathbf{G}_{umls}$, $\mathbf{G}_{distemist}$, and $\mathbf{G}_{silver}$ to produce $\mathbf{Z} \in \mathbb{R}^{t \times 1030}$. $\mathbf{Z}$ is then fed into a 6-layer Transformer Encoder (Vaswani et al., 2017) with positional encoding and 10 attention heads per layer. The Transformer Encoder outputs $\mathbf{Y} \in \mathbb{R}^{t \times 1030}$ where the ith row represents the encoded representation of the ith token. $\mathbf{Y}$ is fed into a linear classifier which produces $\mathbf{I} \in \mathbb{R}^{t \times 3}$, classifying each token into one of the 3 classes corresponding to the BIO tagging scheme. This represents our submission system M_{sub}.

Training XLM RoBERTa large is fine-tuned on the training data using the Adam optimizer (Kingma and Ba, 2015) with a learning rate of `1e-5` and early stopping. Training for 10 epochs takes 2.5 hours on one Nvidia RTX 3090 gpu.

3 Ablation and test results

The evaluation is by strict F1 score, requiring exactly matching gold spans without partial credit.

Our submission model M_{sub} was described above in Section 2.2. Let $M_{no\text{-}gaz}$ be the same model as M_{sub} but without the one-hot features generated from gazetteer matching. Let $M_{no\text{-}tok}$ be the same model as M_{sub} but without the tokenizer extension using disease gazetteers (hashtags and usernames are tokenized with the GATE Hashtag Tokenizer only). Finally, let $M_{RoBERTa}$ be the baseline model without one-hot features, custom tokenization, and transformer encoder – the output of the language model $\mathbf{H}$ is fed directly into the classifier.

Model	F1 score	Precision	Recall
$M_{RoBERTa}$	0.880	0.856	0.905
$M_{no\text{-}gaz}$	0.888	0.875	0.902
$M_{no\text{-}tok}$	0.888	0.884	0.892
M_{sub}	0.892	0.882	0.900

Table 1: Ablation on development set

We see that M_{sub} improves on $M_{RoBERTa}$ by only 1.2%. Omitting gazetteers or the tokenizer extension each looses only 0.4% over M_{sub}. $M_{no\text{-}gaz}$ looses precision while $M_{no\text{-}tok}$ looses recall against the submission system.

Table 2 shows competition performance of M_{sub} on the test set. We see that performance on the test set decreased by only 2.3%, highlighting the robustness of our technique.

Model	F1 score	Precision	Recall
M_{sub}	0.869	0.851	0.888
Mean	0.675	0.680	0.677
Std. dev.	0.246	0.245	0.254
Median	0.761	0.758	0.780

Table 2: Competition results on test set

4 Conclusion

For the task of detecting disease mentions, a general large, pre-trained language model (XLM RoBERTa large) was enhanced with task-oriented preprocessing (splitting hashtags into component words) and lookup in available quality word lists.

This combination was effective and in competition placed our system nearly 20% above the mean.

References

Olivier Bodenreider. 2004. The unified medical language system (UMLS): integrating biomedical terminology. *Nucleic acids research*, 32(suppl_1):D267–D270.

Kalina Bontcheva, Leon Derczynski, Adam Funk, Mark A Greenwood, Diana Maynard, and Niraj Aswani. 2013. Twitie: An open-source information extraction pipeline for microblog text. In *Proceedings of the international conference recent advances in natural language processing RANLP 2013*, pages 83–90.

Alexis Conneau, Kartikay Khandelwal, Naman Goyal, Vishrav Chaudhary, Guillaume Wenzek, Francisco Guzmán, Edouard Grave, Myle Ott, Luke Zettlemoyer, and Veselin Stoyanov. 2020. Unsupervised cross-lingual representation learning at scale. In *Proceedings of the 58th Annual Meeting of the Association for Computational Linguistics*, pages 8440–8451, Online. Association for Computational Linguistics.

Hamish Cunningham, Diana Maynard, Kalina Bontcheva, and Valentin Tablan. 2002. GATE: A Framework and Graphical Development Environment for Robust NLP tools and applications. In *Proceedings of the 40th Anniversary Meeting of the Association for Computational Linguistics (ACL'02)*.

Kevin Donnelly et al. 2006. Snomed-ct: The advanced terminology and coding system for ehealth. *Studies in health technology and informatics*, 121:279.

Luis Gasco, Darryl Estrada-Zavala, Eulàlia Farré-Maduell, Salvador Lima-López, Antonio Miranda-Escalada, and Martin Krallinger. 2022. Overview of the SocialDisNER shared task on detection of diseases mentions from healthcare related and patient generated social media content: methods, evaluation and corpora. In *Proceedings of the Seventh Social Media Mining for Health (#SMM4H) Workshop and Shared Task*.

Diederik P Kingma and Jimmy Ba. 2015. Adam: A method for stochastic optimization. In *Proceedings of ICLR 2015*.

Antonio Miranda-Escalada, Luis Gascó, Salvador Lima-López, Eulàlia Farré-Maduell, Darryl Estrada, Anastasios Nentidis, Anastasia Krithara, Georgios Katsimpras, Georgios Paliouras, and Martin Krallinger. 2022. Overview of DisTEMIST at BioASQ: Automatic detection and normalization of diseases from clinical texts: results, methods, evaluation and multilingual resources. In *Working Notes of Conference and Labs of the Evaluation (CLEF) Forum. CEUR Workshop Proceedings*.

Ashish Vaswani, Noam Shazeer, Niki Parmar, Jakob Uszkoreit, Llion Jones, Aidan N Gomez, Łukasz Kaiser, and Illia Polosukhin. 2017. Attention is all you need. In *Proceedings of Advances in neural information processing systems NIPS 2017*.

CIC NLP at SMM4H 2022: a BERT-based approach for classification of social media forum posts

Atnafu Lambebo Tonja[*1], **Olumide Ebenezer Ojo**[*2],**Muhammad Arif**[*3], **Abdul Gafar Manuel Meque**[*4],**Olga Kolesnikova**[*5], **Grigori Sidorov**[*6], **Alexander Gelbukh**[*7]

[*]Instituto Politécnico Nacional (IPN), Centro de Investigación en Computación (CIC), Mexico City, Mexico
[1]atnafu.lambebo@wsu.edu.et,{[5]kolesolga,[3] olumideoea}@gmail.com
{[4]mariff2021,[3]gafar_meque,[4]sidorov,[7]gelbukh}@cic.ipn.mx

Abstract

This paper describes our submissions for the Social Media Mining for Health (SMM4H) 2022 shared tasks. We participated in 2 tasks: a) Task 4: Classification of Tweets self-reporting exact age and b) Task 9: Classification of Reddit posts self-reporting exact age. We evaluated the two(BERT and RoBERTa) transformer based models for both tasks. For Task 4 RoBERTa-Large achieved an F1 score of 0.846 on the test set and BERT-Large achieved an F1 score of 0.865 on the test set for Task 9.

1 Introduction

Social media platforms have become more integrated in this digital era, and have impacted various people's perceptions of networking and socializing. The Social Media Mining for Health Applications (SMM4H) Shared Task involves natural language processing (NLP) challenges of using social media data for health research, including informal, colloquial expressions and misspellings of clinical concepts, noise, data sparsity, ambiguity, and multilingual posts (Gasco et al., 2022). As computational analysis opens up new opportunities for researching complex topics using social media data, models are being developed to automatically detect demographic information such as users' age (Klein et al., 2021; Tonja et al., 2022), language (Sarkar et al., 2016) (Aroyehun and Gelbukh, 2020), gender (Markov et al., 2017) (Gómez-Adorno et al., 2019), medical history (Lee et al., 2021), and so on.

More people are using social media in various ways to interact with others, share information, and express their own thoughts. Social media platforms like Twitter and Reddit have cutting-edge technology and are rich with raw, unprocessed data that can be analyzed and transformed into meaningful information. Social media research in different fields, including health (Aroyehun and Gelbukh, 2019), politics (McKeon and Gitomer, 2019), economics (Ojo et al., 2021), have included demographic variables. We participated in the social media mining Task 4 of the SMM4H 2022 shared task which centered on automatically identifying tweets that self-report the user's exact age from those that do not. In Task 9 of the same challenge, we also attempted to automatically classify Reddit posts that self-report the actual age of the online user at the time of posting from those that do not. A detailed overview of the shared tasks in the 7th edition of the workshop can be found in (Weissenbacher et al., 2022).

We applied two (BERT-Large and RoBERTa-Large) transformer models using Hugging Face [1] library. The paper is organized as follows: section 2 describes Task 4 objective, system description, experiment and result. Section 3 describes Task 9 objective, system description, experiment and result. Finally, section 4 concludes the paper and sheds some light on possible future work.

2 Task 4: Classification of Tweets Self-Reporting Exact Age

The objective of this task is automatically distinguish tweets that self-report the user's exact age from those that do not.

2.1 Data description

For Task 4: SMM4H organizers provided us with a dataset which include the Tweet ID, the text of the Tweet Object, and the annotated binary class containing 0 and 1, Tweets were annotated as "1" if the user's exact age could be determined from the tweet or annotated as "0" if the user's exact age could not be determined from the tweet. The training set consists of 8,800 tweets with 5,966 examples labeled as "0" and 2,834 examples labeled as "1". The validation dataset has 2,200 tweets with

[1]https://huggingface.co/

Proceedings of the 29th International Conference on Computational Linguistic, pages 58–61
October 12–17, 2022.

1,491 examples labeled as "0" and 709 examples labeled as "1". Thus, the dataset has a huge class imbalance, 67.7% of the texts were labeled as "0" and 32.3% were labelled as "1". To solve the class imbalance problem in the given dataset we applied the random oversampling method called the Synthetic Minority Oversampling Technique (SMOTE) (Chawla et al., 2002). Before the data record is entered into the text classification model, we cleaned up and pre-processed tweets. We performed the following pre-processing steps to remove unnecessary data from the dataset, this includes removal of urls, removal of tweeter usernames and stop word removal.

2.2 System Description and Experiment

We used BERT-Large-uncased (Devlin et al., 2018) and RoBERTa-large (Liu et al., 2019) implemented using Huggingface toolkit (Wolf et al., 2020) to extract the exact age from social media posts. After pre-processing the textual data as described in section 2.1, the text sequence was tokenized using the subword tokenizer by using both BERT-large and RoBERTa-large modes with maximum text length of 180. To optimize the model, we used a relu optimizer with a batch size of 64 and a learning rate of 0.0001. We used the maximum number of epochs of 10 with early stopping based on the performance of the validation set. We also used dropout of 0.1 to regularize the model. To run our experiment we used Google colab pro + with Python programming language.

2.3 Result

We evaluated the performance of our models on the validation set and the test set, we compared our validation set performance result to check the model performance before evaluating the model in the test set for submission. We evaluated the result of the models with the F1-score for the "positive" class (i.e., tweets that self-report the user's exact age).

The results on the validation set for BERT-Large and RoBERTa-Large models are reported in Table 1. As shown in Table 1, the best performing model on the validation set for Task 4 is RoBERTa-Large model. When we observed the model performance on validation set, both the models showed less result in precision, recall and F1-score for class '1'(positive class) than class '0'. When comparing the performance of the models in test set, RoBERTa-Large model was able to achieve 0.846 F1-score on the test set as seen in Table 2.

Model	Class	P	R	F1
BERT-Large	0	0.67	0.81	0.73
	1	0.27	0.14	0.18
Accuracy				0.60
RoBERTa-Large	0	0.96	0.79	0.87
	1	0.68	0.94	0.79
Accuracy				**0.84**

Table 1: Performance of our models on Task 4 validation set (unofficial results)

Model	P	R	F1-score
RoBERTa-Large	0.804	0.891	**0.846**
BERT-Large	0.737	0.886	0.805

Table 2: Performance of our models in Task 4 in test set (official results)

3 Task 9: Classification of Reddit Posts Self-Reporting Exact Age

The objective of this task is similar with the objective described in section 2, the only difference is that this task used dataset from Reddit social media.

3.1 Data description

The datasets for this task were collected from Reddit posts, the labels were annotated in similar manner as described in subsection 2.1. The dataset is disease-specific and consists of posts collected via a series of keywords associated with dry eye disease. The training set consists of 9000 posts with 6,079 examples labeled as "0" and 2,921 examples labeled as "1". The validation dataset has 1000 posts with 686 examples labeled as "0" and 314 examples labeled as "1". As described in subsection 2.1 this dataset also has class imbalance, we used the same methods to solve the class imbalance issue as described in subsection 2.1. We also followed the same procedure for data pre-processing as described in subsection 2.1

3.2 System Description and Experiment

We used the same system description and experimental setup as in Task 4 described in section 2.2 because the objective of both tasks are the same while the difference is only in the dataset source.

3.3 Result

Similarly we evaluated the performance of selected models on Task 9 validation set before evaluating and submitting the prediction file on test set. As used in section 2.2 for Task 9 we evaluated the performance of the models with F1-score for the positive class (i.e. posts annotated as "1").

The results on the validation set for BERT-Large and RoBERTa-Large models are reported in Table 3. As shown in Table 3 the best performing model on the validation set for Task 9 is RoBERTa-Large model. RoBERTa-Large model showed better result on predicting positive class(1) than BERT-Large. When evaluating the performance of the models on the test set, BERT-Large model was able to achieve an F1-score 0.865 on the test set as seen in Table 4.

Model	Class	P	R	F1
BERT-Large	0	0.97	0.88	0.92
	1	0.78	0.94	0.85
Accuracy				0.90
RoBERTa-Large	0	0.95	0.91	0.93
	1	0.82	0.90	0.86
Accuracy				**0.91**

Table 3: Performance of our models on Task 9 validation set (unofficial results)

Model	Precision	Recall	F1-score
BERT-Large	0.797	0.946	**0.865**

Table 4: Performance of our models on Task 9 test set (official results)

4 Conclusion

In this work, we describe our team submission for Social Media Mining for Health Applications shared task 2022. We have explored an application of RoBERTa and BERT language models to the task of classification of tweets self-reporting exact age and classification of Reddit posts self-reporting exact age. Our experiments have shown that RoBERTa-Large outperforms BERT-Large in classification of tweets self-reporting exact age (Task 4) in both validation and test set. For classification of Reddit posts self-reporting exact age (Task 9), we found that RoBERTa-Large outperforms BERT-Large in validation set.

In the future, we will explore the effect of class imbalance on the performance of classification models, apply different methods to solve class imbalance and their effect on model performance.

Acknowledgements

The work was done with partial support from the Mexican Government through the grant A1S-47854 of CONACYT, Mexico, grants 20220852, 20220859, and 20221627 of the Secretaría de Investigación y Posgrado of the Instituto Politécnico Nacional, Mexico. The authors thank the CONACYT for the computing resources brought to them through the Plataforma de Aprendizaje Profundo para Tecnologías del Lenguaje of the Laboratorio de Supercómputo of the INAOE, Mexico and acknowledge the support of Microsoft through the Microsoft Latin America PhD Award.

References

Segun Taofeek Aroyehun and Alexander Gelbukh. 2019. Detection of adverse drug reaction in tweets using a combination of heterogeneous word embeddings. In *Proceedings of the Fourth Social Media Mining for Health Applications (#SMM4H) Workshop & Shared Task*, pages 133–135, Florence, Italy. Association for Computational Linguistics.

Segun Taofeek Aroyehun and Alexander Gelbukh. 2020. Nlp-cic at hasoc 2020: Multilingual offensive language detection using all-in-one model. In *FIRE*.

Nitesh V Chawla, Kevin W Bowyer, Lawrence O Hall, and W Philip Kegelmeyer. 2002. Smote: synthetic minority over-sampling technique. *Journal of artificial intelligence research*, 16:321–357.

Jacob Devlin, Ming-Wei Chang, Kenton Lee, and Kristina Toutanova. 2018. Bert: Pre-training of deep bidirectional transformers for language understanding. *arXiv preprint arXiv:1810.04805*.

Luis Gasco, Darryl Estrada-Zavala, Eulàlia Farré-Maduell, Salvador Lima-López, Antonio Miranda-Escalada, and Martin Krallinger. 2022. Overview of the SocialDisNER shared task on detection of diseases mentions from healthcare related and patient generated social media content: methods, evaluation and corpora. In *Proceedings of the Seventh Social Media Mining for Health (#SMM4H) Workshop and Shared Task*.

Helena Gómez-Adorno, Roddy Fuentes-Alba, Ilia Markov, Grigori Sidorov, and Alexander Gelbukh. 2019. A convolutional neural network approach for gender and language variety identification. *Journal of Intelligent and Fuzzy Systems*, 36(5):4845–4855.

Ari Z. Klein, Arjun Magge, and Graciela Gonzalez-Hernandez. 2021. Reportage: Automatically extracting the exact age of twitter users based on self-reports in tweets. *CoRR*, abs/2103.06357.

Lung-Hao Lee, Man-Chen Hung, Chien-Huan Lu, Chang-Hao Chen, Po-Lei Lee, and Kuo-Kai Shyu. 2021. Classification of tweets self-reporting adverse pregnancy outcomes and potential covid-19 cases using roberta transformers. In *SMM4H*.

Yinhan Liu, Myle Ott, Naman Goyal, Jingfei Du, Mandar Joshi, Danqi Chen, Omer Levy, Mike Lewis, Luke Zettlemoyer, and Veselin Stoyanov. 2019. Roberta: A robustly optimized BERT pretraining approach. *CoRR*, abs/1907.11692.

Ilia Markov, Helena Gómez-Adorno, Grigori Sidorov, and Alexander Gelbukh. 2017. The winning approach to cross-genre gender identification in russian at rusprofiling 2017. In *FIRE*.

Robin Tamarelli McKeon and Drew H. Gitomer. 2019. Social media, political mobilization, and high-stakes testing. *Frontiers in Education*, 4.

O. E. Ojo, A. Gelbukh, H. Calvo, and O. O. Adebanji. 2021. Performance study of n-grams in the analysis of sentiments. *Journal of the Nigerian Society of Physical Sciences*, 3(4):477–483.

Sandip Sarkar, Saurav Saha, Jereemi Bentham, Partha Pakray, Dipankar Das, and Alexander Gelbukh. 2016. Nlp-nitmz@dpil-fire2016: Language independent paraphrases detection. In *FIRE*.

Atnafu Lambebo Tonja, Muhammad Arif, Olga Kolesnikova, Alexander Gelbukh, and Grigori Sidorov. 2022. Detection of aggressive and violent incidents from social media in spanish using pretrained language model.

Davy Weissenbacher, Ari Z. Klein, Luis Gascó, Darryl Estrada-Zavala, Martin Krallinger, Yuting Guo, Yao Ge, Abeed Sarker, Ana Lucia Schmidt, Raul Rodriguez-Esteban, Mathias Leddin, Arjun Magge, Juan M. Banda, Vera Davydova, Elena Tutubalina, and Graciela Gonzalez-Hernandez. 2022. Overview of the seventh social media mining for health applications smm4h shared tasks at coling 2022. In *In Proceedings of the Seventh Social Media Mining for Health (SMM4H) Workshop and Shared Task*.

Thomas Wolf, Lysandre Debut, Victor Sanh, Julien Chaumond, Clement Delangue, Anthony Moi, Pierric Cistac, Tim Rault, Remi Louf, Morgan Funtowicz, Joe Davison, Sam Shleifer, Patrick von Platen, Clara Ma, Yacine Jernite, Julien Plu, Canwen Xu, Teven Le Scao, Sylvain Gugger, Mariama Drame, Quentin Lhoest, and Alexander Rush. 2020. Transformers: State-of-the-art natural language processing. In *Proceedings of the 2020 Conference on Empirical Methods in Natural Language Processing: System Demonstrations*, pages 38–45, Online. Association for Computational Linguistics.

NCUEE-NLP@SMM4H'22: Classification of Self-reported Chronic Stress on Twitter Using Ensemble Pre-trained Transformer Models

Tzu-Mi Lin, Chao-Yi Chen, Yu-Wen Tzeng and Lung-Hao Lee
Department of Electrical Engineering, National Central University, Taiwan

Abstract

This study describes our proposed system design for the SMM4H 2022 Task 8. We fine-tune the BERT, RoBERTa, ALBERT, XLNet and ELECTRA transformers and their connecting classifiers. Each transformer model is regarded as a standalone method to detect tweets that self-reported chronic stress. The final output classification result is then combined using the majority voting ensemble mechanism. Experimental results indicate that our approach achieved a best F1-score of 0.73 over the positive class.

1 Introduction

The Social Media Mining for Health Application (SMM4H) shared tasks involve NLP challenges on social media data for health research. We participated in the SMM4H 2022 Task 8 (Weissenbacher et al., 2022), focusing on automatically differentiating tweets that exhibit self-reported chronic stress (annotated as "1") from those that do not (annotated as "0"). Chronic stress is a kind of long-term physiological or psychological response which may damage mental health. Deep sparse neural networks were used to detect stress on social media data (Lin et al., 2014). Factor graph models were combined with convolutional neural networks to detect user stress (Lin et al., 2017). An emotion-infused language model was proposed to facilitate stress detection (Turcan et al., 2021). Emotion recognition was investigated as an auxiliary task to improve stress detection (Yao et al., 2021). Recently, novel transformer-based neural networks have achieved highly promising results in many NLP tasks. This trend motivates us to explore the use of pre-trained transformer models to detect chronic stress.

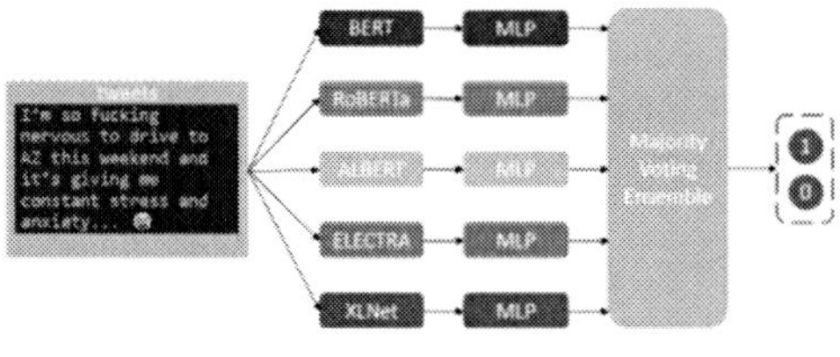

Figure 1: NCUEE-NLP system architecture.

This paper describes a system proposed by the **NCUEE-NLP** (**N**ational **C**entral **U**niversity, Dept. of **E**lectrical **E**ngineering, **N**atural **L**anguage **P**rocessing Lab) for the SMM4H 2022 Task 8 using transformer models including BERT (Devlin et al., 2019), RoBERTa (Liu et al., 2019), ALBERT (Lan et al., 2020), XLNet (Yang et al., 2019), and ELECTRA (Clark et al., 2020), followed by fine-tuning for the classification task. In addition, a majority voting ensemble mechanism is used to enhance system performance. The evaluation metric is F1-score for the positive class (i.e., tweets annotated as "1"). Our best F1-score is 0.73 over the positive class.

2 The NCUEE-NLP System

Figure 1 shows our NCUEE-NLP system architecture for the SMM4H 2022 shared task 8. Our system is composed of two parts: 1) pre-trained transformer models; and 2) ensemble mechanism.

We approach the task using five pre-trained transformer models, including BERT (Devlin et al., 2019), RoBERTa (Liu et al., 2019), ALBERT (Lan et al., 2020), XLNet (Yang et al., 2019), and ELECTRA (Clark et al., 2020). We use training, validation and test datasets to fine-tune the

Proceedings of the 29th International Conference on Computational Linguistic, pages 62–64
October 12–17, 2022.

Pre-trained Transformer Models		Validation set		
		precision	recall	F1-score
Standalone	BERT (B)	0.6567	0.8461	0.7395
	RoBERTa (RB)	0.6823	0.8397	0.7529
	ALBERT (AB)	0.7172	0.6667	0.6910
	XLNet (XL)	0.6683	0.8654	0.7542
	ELECTRA (ET)	0.7076	0.7756	0.7400
Ensemble	B + RB + AB + XL + ET	0.6984	0.8462	0.7652
	B + RB + AB	0.6919	0.8205	0.7507
	B + RB + XL	0.6750	0.8654	0.7584
	B + RB + ET	0.7027	0.8333	0.7625
	B + AB + XL	0.6915	0.8333	0.7558
	B + AB + ET	**0.7151**	0.7885	0.7500
	B + XL + ET	0.6939	**0.8718**	**0.7727**
	RB + AB + XL	0.6825	0.8269	0.7478
	RB + AB + ET	0.7118	0.7756	0.7423
	RB + XL + ET	0.6963	0.8526	0.7666
	AB + XL + ET	0.7022	0.8013	0.7485

Table 1: Submission results on the SMM4H 2022 Task 8 validation dataset.

language model of each transformer thereby improving the embedding representation. Individual language models are then connected to the Multi-Layer Perceptron as a classifier.

The ensemble mechanism uses multiple learning models to obtain better classification performance. We use majority voting ensemble (Lee et al., 2021), in which each transformer model makes an independent classification (i.e., a vote 0 or 1) for each testing instance. The final system prediction output is the one that receives a majority of votes.

3 Evaluation

The experimental datasets were mainly provided by task organizers (Yang et al., 2022), with a total of 2,936 tweets in the training set, including 1,092 positive and 1,844 negative tweets. The validation set contains 534 tweets (156 positive/264 negative).

All tweets were pre-processed using the following steps: 1) Converting username into @USER; 2) Converting emojis into textual representation; 3) Removing all URLs; 4) Removing '#' from hashtags; 5) Segmenting hashtags which contained two or more words into their component words; and 6) Removing multi-lingual characters to retain English letters and numbers only.

The pre-trained transformer models were downloaded from HuggingFace (Wolf et al., 2019). We selected an odd number (i.e., three of five) of models for majority voting ensemble. A total of 12 groups was combined for the ensemble mechanism. The hyper-parameters used were as follows: training batch size 100, learning rate 1e-5, and maximum sequence length 256.

Table 1 shows the results on the SMM4H 2022 Task 8 validation set. Among individual transformer models, ALBERT clearly underperformed the other four models which had similar F1-score results. The five ensemble transformer models enhanced performance compared with the best standalone model, RoBERTa. Among all three ensemble models, those excluding ALBERT resulted in some performance improvements, with the best combination of BERT, XLNet and ELECTRA, achieving an F1-score of 0.7727.

We selected the top three ensemble models that performed well on the validation set as our final testing models. The combination of RoBERTa, XLNET and ELECTRA performed achieved the best F1-score of 0.73 on the test set.

4 Conclusions

This study describes the NCUEE-NLP system submission in SMM4H 2022 Task 8 for self-reported chronic stress, including system design, implementation and evaluation. Our submitted ensemble pre-trained transformer models achieved a high F1-score of 0.73.

Acknowledgments

This study was partially supported by the National Science and Technology Council, Taiwan under the grant MOST 111-2628-E-008-005-MY3.

References

Kevin Clark, Minh-Thang Luong, Quoc V. Le, and Christopher D. Manning. 2020. ELECTRA: Pre-training Text Encoders as Discriminators Rather Than Generators. In *Proceedings of the 8th International Conference on Learning Representations.* https://doi.org/10.48550/arXiv.2003.10555

Jacob Devlin, Ming-Wei Chang, Kenton Lee, and Kristina Toutanova. 2019. BERT: Pre-training of Deep Bidirectional Transformers for Language Understanding. In *Proceedings of the 2019 Conference of the North American Chapter of the Association for Computational Linguistics: Human Language Technologies, Volume 1 (Long and Short Papers)*, pages 4171-4186. https://doi.org/10.48550/arXiv.1810.04805

Zhenzhong Lan, Mingda Chen, Sebastian Goodman, Kevin Gimpel, Piyush Sharma, and Radu Soricut. 2020. Albert: A lite bert for self-supervised learning of language representations. In *Proceedings of the 8th International Conference on Learning Representations.* https://doi.org/10.48550/arXiv.1909.11942

Lung-Hao Lee, Yuh-Shyang Wang, Chao-Yi Chen, and Liang-Chih Yu. 2021. Ensemble Multi-Channel Neural Networks for Scientific Language Editing Evaluation. *IEEE Access,* 9:158540 – 158547. https://doi.org/10.1109/ACCESS.2021.3130042.

Huijie Lin, Jia Jia, Quan Guo, Yuanyuan Xue, Jie Huang, Lianhong Cai, and Ling Feng. 2014. Psychological stress detection from cross-media microblog data using Deep Sparse Neural Network. In *Proceedings of the 2014 IEEE International Conference on Multimedia and Expo*, pages 1-6. https://doi.org/10.1109/ICME.2014.6890213

Huijie Lin, Jia Jia, Jiezhong Qiu, Yongfeng Zhang, Guangyao Shen, Lexing Xie, Jie Tang, Ling Feng, and Tat-Seng Chua. 2017. Detecting Stress Based on Social Interactions in Social Networks. *IEEE Transactions on Knowledge and Data Engineering,* 29(9):1820-1833. https://doi.org/10.1109/tkde.2017.2686382

Yinhan Liu, Myle Ott, Naman Goyal, Jingfei Du, Mandar Joshi, Danqi Chen, Omer Levy, Mike Lewis, Luke Zettlemoyer, and Veselin Stoyanov. 2019. RoBERTa: a robustly optimized BERT pretraining approach. *arXiv preprint,* https://arxiv.org/abs/1907.11692

Elsbeth Turcan, Smaranda Muresan, and Kathleen McKeown. 2021. Emotion-Infused Models for Explainable Psychological Stress Detection. In *Proceedings of the 2021 Conference of the North American Chapter of the Association for Computational Linguistics: Human Language Technologies*, pages 2895–2909. https://doi.org/10.18653/v1/2021.naacl-main.230

Davy Weissenbacher, Ari Z. Klein, Luis Gascó, Darryl Estrada-Zavala, Martin Krallinger, Yuting Guo, Yao Ge, Abeed Sarker, Ana Lucia Schmidt, Raul Rodriguez-Esteban, Mathias Leddin, Arjun Magge, Juan M. Banda, Vera Davydova, Elena Tutubalina, Graciela Gonzalez-Hernandez. 2022. Overview of the Seventh Social Media Mining for Health Applications #SMM4H Shared Tasks at COLING 2022. In *Proceedings of the Seventh Social Media Mining for Health (#SMM4H) Workshop and Shared Task.*

Thomas Wolf, Lysandre Debut, Victor Sanh, Julien Chaumond, Clement Delangue, Anthony Moi, Pierric Cistac, Tim Rault, Remi Louf, Morgan Funtowicz, Joe Davison, Sam Shleifer, Patrick von Platen, Clara Ma, Yacine Jernite, Julien Plu, Canwen Xu, Teven Le Scao, Sylvain Gugger, Mariama Drame, Quentin Lhoest, and Alexander M. Rush. 2019. HuggingFace's transformers: state: of-the-art natural language processing. *arXiv preprint.* https://doi.org/10.48550/arXiv.1910.03771

Yuan-Chi Yang, Angle Xie, Sangmi Kim, Jessica Hair, Mohammed Ali Al-Garadi, and Abeed Sarker. 2022. Automatic detection of twitter users who express chronic stress experiences via supervised machine learning and natural language processing. *Computers, Informatics, Nursing.* article in press.

Yiqun Yao, Michalis Papakostas, Mihai Burzo, Mohamed Abouelenien, and Rada Mihalcea. 2021. MUSER: MUltimodal Stress detection using Emotion Recognition as an Auxiliary Task. In *Proceedings of the 2021 Conference of the North American Chapter of the Association for Computational Linguistics: Human Language Technologies*, pages 2714–2725. https://doi.org/10.18653/v1/2021.naacl-main.216

Zhilin Yang, Zihang Dai, Yiming Yang, Jaime Carbonell, Ruslan Salakhutdinov, and Quoc V. Le. 2019. XLNet: Generalized Autoregressive Pretraining for Language Understanding. In *Proceedings of the 33rd Conference on Neural Information Processing Systems.* https://doi.org/10.48550/arXiv.1906.08237

BioInfo@UAVR@SMM4H'22: Classification and Extraction of Adverse Event mentions in Tweets using Transformer Models

Edgar Morais, José Luís Oliveira, Alina Trifan and **Olga Fajarda**
Department of Electronics, Telecommunications and Informatics (DETI/IEETA),
University of Aveiro, Aveiro, Portugal
{edgarmorais, jlo, alina.trifan, olga.oliveira}@ua.pt

Abstract

This paper describes BioInfo@UAVR team's approach for adressing subtasks 1a and 1b of the Social Media Mining for Health Applications 2022 shared task. These sub-tasks deal with the classification of tweets that contain an Adverse Drug Event mentions and the detection of spans that correspond to those mentions. Our approach relies on transformer-based models, data augmentation, and an external dataset.

1 Introduction

The Social Media Mining for Health Applications shared task (Weissenbacher et al., 2022) addresses challenges relating to the use of social media data for health research. Our team participated in subtasks 1a and 1b. Subtask 1a focuses on the classification of tweets regarding the presence of Adverse Drug Events (ADEs) and subtask 1b focuses on the detection of ADE spans in tweets. In this submission, we performed some experiments that explore the use of transformer-based models to solve the tackled tasks.

2 Datasets

In the experiments executed for these subtasks, we used the datasets provided by the organizers (Magge et al., 2021), which were annotated for each of the subtasks. The training set had 17385 labeled tweets (1239 labeled ADE), the validation set had 915 labeled tweets (65 labeled ADE) and the test set had 10984 unlabeled tweets. Furthermore, an additional dataset was used in the training phase for subtask 1b, along with different balancing techniques.

2.1 Text augmentation

To overcome the class imbalance problem observed in the datasets we used text augmentation to increase the number of positive examples. For each tweet in the minority class, 5 augmented versions of that tweet were created. For text augmentation, we used the TextAttack framework (Morris et al., 2020) to generate new examples, by executing transformations on examples present on the available datasets. The augmenting was done through 4 transformations: replacing characters with random characters, swapping characters with QWERTY adjacent keys, replacing words with synonyms provided by WordNet (Fellbaum, 2010; Miller, 1995), and performing contractions on recognized combinations.

2.2 WEBRADR Benchmark Reference Dataset

The WEBRADR Benchmark Reference Dataset (Dietrich et al., 2020) is a labeled dataset with tweets relating to the presence of ADEs as well as the spans of the identified ADEs in positive examples. Through this dataset, we could extract 31122 tweets (588 labeled ADE). Even though this dataset was not intended for the training of models we explored its use for that purpose.

2.3 Pre-processing

The pre-processing was slightly different for each of the subtasks. For subtask 1a we replaced the special instance "&", with the character "&" and tokenized the tweets with the tokenizer corresponding to the model used. For subtask 1b, besides the steps mentioned for the previous subtask, we also lowercased the tweets.

3 Experiments

3.1 Subtask 1a

In this subtask, we evaluated different transformer-based models, by training them on the training dataset and testing them on the validation dataset. For this task, we evaluated 3 different transformer-based models available on Hugging Face[1]. These

[1]https://huggingface.co/models

Proceedings of the 29th International Conference on Computational Linguistic, pages 65–67
October 12–17, 2022.

Model	Precision	Recall	F1
BERT-large	**0.797**	0.723	0.758
RoBERTa-large	0.778	**0.862**	0.818
BERTweet-large	**0.797**	0.846	**0.821**

Table 1: Comparative results for different transformer modes for subtask 1a.

Training set	Precision	Recall	F1
Base train set	0.797	0.846	0.821
Over-sampling	0.809	0.846	0.827
Under-sampling	0.778	**0.862**	0.818
Augmented set	**0.909**	0.769	**0.833**

Table 2: Results when training a BERTweet-large model classifier with different data for subtask 1a.

models were Bert-large-uncased (Devlin et al., 2019), RoBERTa-large (Liu et al., 2019), and BERTweet-large (Quoc Nguyen et al., 2020). We performed further testing, by training the best-performing model with re-sampled training data or augmented training data. The models were trained with the same hyperparameters and implementation, except for the Bert-large-uncased, which used a batch size of 16 instead of 32 due to memory constraints. All models were trained for 3 epochs with a learning rate of 2e-5.

3.2 Subtask 1b

In this subtask, we used the best performing model from subtask 1a on a Named Entity Recognition (NER) task, and trained it on the training data. In this subtask the experiments focused on evaluating the effect of adding positive examples from the WEBRADR Benchmark Reference Dataset to the training data.

4 Results and discussion

Table 1 presents results obtained from training different models for subtask 1a and testing them with the task validation set. We can observe that the model BERTweet-large presents the best performance. Table 2 presents results obtained from training a BERTweet-large model with different training data obtained through transformations from the original challenge training data. We can observe that the use of text augmentation on the training set yields the best performance. Table 3 presents the results of training a BERTweet-large model for subtask 1b with different data. We can observe that the inclusion of positive examples from the WEBRADR reference dataset in the training set slightly improves the performance of the model.

Training set	Strict Precision	Strict Recall	Strict F1
Base training set	0.598	0.598	0.598
Base set with pos tweets from WEBRADR	**0.609**	**0.609**	**0.609**

Table 3: Results when training a BERTweet-large model NER pipeline with different data.

	Precision	Recall	F1
Submission	0.839	0.598	0.698
Average	0.646	0.497	0.562

Table 4: Results of submission for subtask 1a.

	Precision	Recall	F1
Overlapping results			
Submission	0.828	0.341	0.484
Average	0.539	0.517	0.527
Strict results			
Submission	0.560	0.235	0.331
Average	0.344	0.339	0.341

Table 5: Results of submission for subtask 1b.

The submission for subtask 1a was obtained through the training of a BERTweet-large classification model with the augmented training and validation data. For subtask 1b we used a BERTweet-large NER model trained on the training data with positive examples from the WEBRADR dataset. The submission for subtask 1b was obtained by using the NER system on the predictions submitted for subtask 1a. The results from the final submission are in Tables 4 and 5. In subtask 1a, we obtained results above the average in every metric, however in subtask 1b the only metrics above the average values were the precision metrics.

5 Conclusion

We proposed a text classification and NER pipeline which address the challenges posed by subtask 1a and 1b. Our system was able to achieve a F1 score of 0.698 on subtask 1a and a strict F1 score of 0.331 on subtask 1b.

Acknowledgment

This work was supported by FCT – Fundação para a Ciência e Tecnologia within project DSAIPA/AI/0088/2020.

References

Jacob Devlin, Ming Wei Chang, Kenton Lee, and Kristina Toutanova. 2019. BERT: Pre-training of deep bidirectional transformers for language understanding. In *NAACL HLT 2019 - 2019 Conference of the North American Chapter of the Association for Computational Linguistics: Human Language Technologies - Proceedings of the Conference*, volume 1, pages 4171–4186. Association for Computational Linguistics (ACL).

Juergen Dietrich, Lucie M. Gattepaille, Britta Anne Grum, Letitia Jiri, Magnus Lerch, Daniele Sartori, and Antoni Wisniewski. 2020. Adverse Events in Twitter-Development of a Benchmark Reference Dataset: Results from IMI WEB-RADR. *Drug Safety*, 43(5):467–478.

Christiane Fellbaum. 2010. WordNet: An Electronic Lexical Database. In *Theory and Applications of Ontology: Computer Applications*, pages 231–243. Springer Netherlands.

Yinhan Liu, Myle Ott, Naman Goyal, Jingfei Du, Mandar Joshi, Danqi Chen, Omer Levy, Mike Lewis, Luke Zettlemoyer, and Veselin Stoyanov. 2019. RoBERTa: A Robustly Optimized BERT Pretraining Approach. *CoRR*, abs/1907.1.

Arjun Magge, Elena Tutubalina, Zulfat Miftahutdinov, Ilseyar Alimova, Anne Dirkson, Suzan Verberne, Davy Weissenbacher, and Graciela Gonzalez-Hernandez. 2021. DeepADEMiner: a deep learning pharmacovigilance pipeline for extraction and normalization of adverse drug event mentions on Twitter. *Journal of the American Medical Informatics Association : JAMIA*, 28(10):2184–2192.

George A. Miller. 1995. WordNet: A Lexical Database for English. *Communications of the ACM*, 38(11):39–41.

John X. Morris, Eli Lifland, Jin Yong Yoo, Jake Grigsby, Di Jin, and Yanjun Qi. 2020. TextAttack: A Framework for Adversarial Attacks, Data Augmentation, and Adversarial Training in NLP. pages 119–126.

Dat Quoc Nguyen, Thanh Vu, Anh Tuan Nguyen, and VinAI Research. 2020. BERTweet: A pre-trained language model for English Tweets. pages 9–14.

Davy Weissenbacher, Ari Z. Klein, Luis Gascó, Darryl Estrada-Zavala, Martin Krallinger, Yuting Guo, Yao Ge, Abeed Sarker, Ana Lucia Schmidt, Raul Rodriguez-Esteban, Mathias Leddin, Arjun Magge, Juan M. Banda, Vera Davydova, Elena Tutubalina, and Graciela Gonzalez-Hernandez. 2022. Overview of the Seventh Social Media Mining for Health Applications #SMM4H Shared Tasks at COLING 2022. In *Proceedings of the Seventh Social Media Mining for Health (#SMM4H) Workshop and Shared Task.*

FRE at SocialDisNER: Joint Learning of Language Models for Named Entity Recognition

Kendrick Cetina and **Nuria García-Santa**
Fujitsu Research of Europe (FRE)
Camino Cerro de los Gamos 1, 28224. Pozuelo de Alarcón (Madrid), Spain.
{kendrick.cetina, nuria.garcia.uk} @fujitsu.com

Abstract

This paper describes our followed methodology for the automatic extraction of disease mentions from tweets in Spanish as part of the SocialDisNER challenge within the 2022 Social Media Mining for Health Applications (SMM4H) Shared Task. We followed a Joint Learning ensemble architecture for the fine-tuning of top performing pre-trained language models in biomedical domain for Named Entity Recognition tasks. We used text generation techniques to augment training data. During practice phase of the challenge our approach showed results of 0.87 F1-Score.

1 Introduction

Twitter is one of the main channels for communication at present. The high number of users and the daily activity worldwide on this platform, in addition to the nature of tweets, i.e., short informal text messages, make it as an ideal source of information for Natural Language Processing (NLP) tasks (Yang et al., 2020). The combination of such amount of information with pre-trained transformer language models has changed the way we do NLP nowadays (Qiu et al., 2020). These pre-trained language models have brought significant breakthroughs in deep learning, creating a revolutionary ecosystem for NLP techniques, such as Named Entity Recognition, Text Classification, etc. In the 2022 SocialDisNER challenge, we focused on fine-tuning several pre-trained language models architectures to detect disease mentions in tweets.

1.1 Task Description

Social Media Mining for Health Applications (SMM4H) workshop offers the opportunity to develop NLP systems for automatic extraction of relevant knowledge from social media data. On this regard, the purpose of the 2022 subtask, Mining Social Media Content for Disease Mentions (SocialDisNER) (Gasco et al., 2022), is to recognise disease mentions in tweets written in Spanish, including informal and professional language style. The objective is to extract the beginning and end positions of disease mentions [1].

2 System Description

The summary of our system is the following: first, we search for the two best performing pre-trained language models for our task. Then, we used data augmentation techniques to extend the initial training data and fine-tune the selected language models with the Joint Learning ensemble approach described in (García-Santa and Cetina, 2021). Finally, by qualitative analysis of results, we cleaned the model predictions with post-processing techniques.

2.1 Model Selection

We used Hugging Face[2] models hub to search for Fill-Mask and Token Classification models suitable for Spanish Named Entity Recognition of Diseases. Models we tested include BioBERT, several BERT-Based models, RoBERTa, Multi-lingual BERT and DistilBERT. We tested more than 15 different language models fine-tuning with training and testing data provided by the challenge organizers. After performance analysis of all pre-trained language models tested we selected the two highest performing models to be used in our Joint Learning approach: SINAI: a RoBERTa-based model[3] from SINAI team (Chizhikova et al.) at DISease TExt Mining Shared Task 2022(Miranda-Escalada et al., 2022) and bsc-bio-es model[4] from Barcelona Supercomputing Center that is a RoBERTa-based model trained on a biomedical corpus in Spanish collected from several sources (Carrino et al., 2022).

[1] https://temu.bsc.es/socialdisner/
[2] https://huggingface.co/models
[3] https://huggingface.co/chizhikchi/Spanish_disease_finder
[4] https://huggingface.co/PlanTL-GOB-ES/bsc-bio-es

Proceedings of the 29th International Conference on Computational Linguistic, pages 68–70
October 12–17, 2022.

2.2 Data Augmentation

We used GPT-2 (Radford et al., 2019) for text generation (Li et al., 2022). GPT-2 is a transformers model pre-trained on a very large corpus of English data in a self-supervised fashion trained with the goal of generating texts from a prompt.

For our system we used the challenge training set as input to fine-tune GPT-2[5] for 10 epochs. Then, we used 1000 random noun-chunks from the provided twitter data to generate sentences of 150 words with fine-tuned GPT-2. Next, we used pattern matching with known entities to tag the diseases in generated text.

Figure 1 shows the F1 score in each training epoch of the selected highest performance models: SINAI and bsc-bio-es. In green we can see the performance of augmenting the training set with GPT-2 and fine-tuning bsc-bio-es.

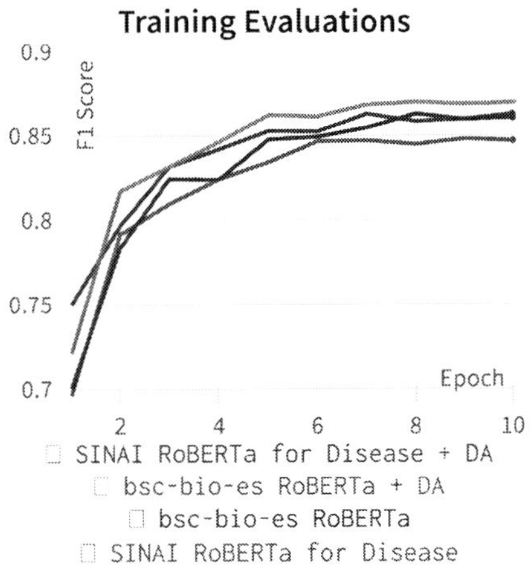

Figure 1: Training performance

2.3 Learning & Post-processing

We used the approach described in (García-Santa and Cetina, 2021) for the joint learning of our selected best-performance models (SINAI & bsc-bio-es). The workflow of our joint learning architecture comprises language models as sub-networks concatenated to form a larger network. The sub-networks learn together the best combined classification, retrieving an ensemble model. This architecture allows weight update of language model sub-networks jointly.

Finally, a qualitative assessment of model prediction lead us to implement a post-processing common to all predictions made. This post-processing includes cleaning of predicted entities, removal of emojis and hashtags and concatenation of consecutive entities predicted as two different entities.

3 Evaluation

During the practice phase of the challenge, participants were given the opportunity to test their approach against the evaluation system from the organizers to assess the quality in an intermediate stage. During this phase we obtained strict F1-Scores of 0.84 with our joint-embedding model fine-tuned with the augmented data and 0.87 with our post-processing techniques.

Two sets of predicted data were submitted for final evaluation. Submission 1 consisted of the Joint Learning Model predictions with our post-processing cleaning and Submission 2 consisted of the Joint Learning Model predictions without any kind of post-processing.

Performances of each submission are presented in Table 1. We can see that model predictions with post-processing resulted in the highest performance.

	Strict P	Strict R	Strict F
Submission 1	0.680	0.805	0.738
Submission 2	0.667	0.805	0.729

Table 1: Final evaluation results of our submitted predictions.

4 Conclusion

In this paper we presented our system focused on fine-tuning two RoBERTa-based language models together with our Joint Learning approach for disease mentions detection task in Spanish tweets. We extended the training data by generating text with a GPT-based model fine-tuned with biomedical and twitter data. Combining the Joint Learning and augmented data, our system achieved 0.87 F1-Score in the practice phase of the challenge. Our final results in evaluation phase of the challenge achieved 0.73 and 0.74 F1-Scores. In the future, we plan to apply transfer learning techniques to extend our system to the analysis of social media messages in other languages.

[5]https://huggingface.co/gpt2

References

Casimiro Pio Carrino, Joan Llop, Marc Pàmies, Asier Gutiérrez-Fandiño, Jordi Armengol-Estapé, Joaquín Silveira-Ocampo, Alfonso Valencia, Aitor Gonzalez-Agirre, and Marta Villegas. 2022. Pre-trained biomedical language models for clinical NLP in Spanish. In *Proceedings of the 21st Workshop on Biomedical Language Processing*, pages 193–199, Dublin, Ireland. Association for Computational Linguistics.

Mariia Chizhikova, Jaime Collado-Montañéz, Pilar López-Úbeda, Manuel C. Díaz-Galiano, L. Alfonso Ureña-López, and M. Teresa Martín-Valdivia. Sinai at clef 2022: Leveraging biomedical transformers to detect and normalize disease mentions. pages 265–273.

Nuria García-Santa and Kendrick Cetina. 2021. Extracting multilingual relations with joint learning of language models. In *Joint European Conference on Machine Learning and Practice Knowledge Discovery in Databases*, pages 401–407. Springer.

Luis Gasco, Darryl Estrada-Zavala, Eulàlia Farré-Maduell, Salvador Lima-López, Antonio Miranda-Escalada, and Martin Krallinger. 2022. Overview of the SocialDisNER shared task on detection of diseases mentions from healthcare related and patient generated social media content: methods, evaluation and corpora. In *Proceedings of the Seventh Social Media Mining for Health (#SMM4H) Workshop and Shared Task*.

Bohan Li, Yutai Hou, and Wanxiang Che. 2022. Data augmentation approaches in natural language processing: A survey. *AI Open*.

Antonio Miranda-Escalada, Luis Gascó, Salvador Lima-López, Eulàlia Farré-Maduell, Darryl Estrada, Anastasios Nentidis, Anastasia Krithara, Georgios Katsimpras, Georgios Paliouras, and Martin Krallinger. 2022. Overview of distemist at bioasq: Automatic detection and normalization of diseases from clinical texts: results, methods, evaluation and multilingual resources.

Xipeng Qiu, Tianxiang Sun, Yige Xu, Yunfan Shao, Ning Dai, and Xuanjing Huang. 2020. Pre-trained models for natural language processing: A survey. *Science China Technological Sciences*, 63(10):1872–1897.

Alec Radford, Jeff Wu, Rewon Child, David Luan, Dario Amodei, and Ilya Sutskever. 2019. Language models are unsupervised multitask learners.

Sihui Yang, Zifan Ning, and Yaping Wu. 2020. Nlp based on twitter information: A survey report. In *2020 2nd International Conference on Information Technology and Computer Application (ITCA)*, pages 620–625. IEEE.

ITAINNOVA at SocialDisNER: A Transformers cocktail for disease identification in social media in Spanish

Rosa M. Montañés-Salas, Irene López-Bosque
Luis García-Garcés and **Rafael del-Hoyo-Alonso**
Technological Institute of Aragón (ITAINNOVA), Zaragoza (Spain)
{rmontanes, ilopez, lggarcia, rdelhoyo}@itainnova.es

Abstract

The Social Media Mining for Health Applications (#SMM4H) Shared Task aims to promote the state of the art of health informatics challenges for social media. The 10th track of the task, SocialDisNER, focuses on the identification of disease mentions in tweets written in Spanish. ITAINNOVA presents a hybrid system based on Transformer Language Models in combination with Natural Language Processing techniques such as the development and use of a diseases gazetteer for approximate string matching. A comprehensive exploration of the components contributions is presented as well as the final test results obtained, which outperform the mean overall performance in the task.

1 Motivation

The detection of disease mentions in tweets (SocialDisNER, Gasco et al., 2022b) is part of the SMM4H Shared Task (Weissenbacher et al., 2022). SocialDisNER proposes a challenging task: NER (Named-Entity Recognition) offset detection of diseases by finding the span of its mentions in tweets published in the Spanish language. Using social media data for health research involves facing multiple Natural Language Processing (NLP) challenges: multilingualism, usage of formal and informal expressions, misspellings, ambiguity and so on, which may be better tackled unifying state of the art approaches and more conventional methods.

In this context, ITAINNOVA participates with a hybrid system which combines Transformer-based Language Models (LMs) with a custom-built gazetteer for Approximate String Matching (ASM) and dedicated text processing techniques for the social media domain. Additionally zero-shot classification capabilities (Pushp and Srivastava, 2017) have been explored in order to support different parts of the system. An extensive analysis on the interactions of these components has been accomplished, making the system stand out above the mean performance of all the participating teams.

2 System description

The overall architecture is illustrated in figure 1.

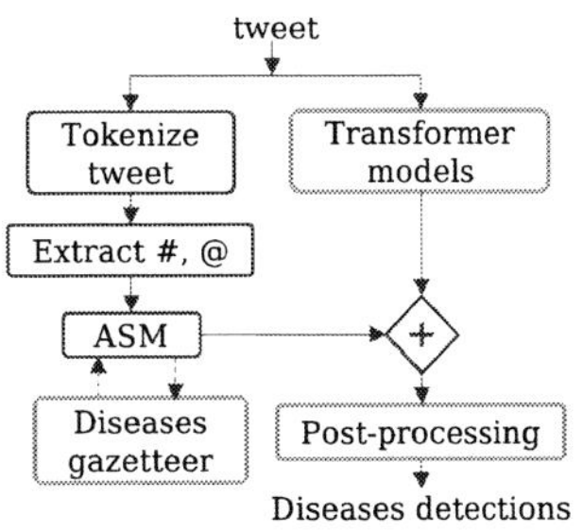

Figure 1: System architecture.

2.1 Transformer-based Language models

The core of the system is a parallel aggregation ensemble of fine-tuned Transformer-based LMs. Nine publicly available pretrained models from the HuggingFace hub, were selected following these requirements: the model has support for Spanish or multiple languages, it has been pretrained with a health related or social media corpus, and its architecture may be fine-tuned for token classification.

Each model is subjected to a hyper-parameter tuning using both gold and silver standard corpus (Gasco et al., 2022a), which allows ranking the models by performance: (1)*wikineural-multilingual-ner* (Tedeschi et al., 2021)[1], (2)*bsc-bio-ehr-es-cantemist* (Miranda-Escalada et al., 2020a), (3)*bertin-base-ner-conll2002-es* (de la Rosa et al., 2022), (4)*bsc-bio-es* (Carrino et al., 2022), (5)*twitter-xlm-roberta-base* (Barbieri et al., 2022)[2], (6)*bsc-bio-ehr-es* and (7)*bsc-bio-ehr-es* (Carrino et al., 2022), (8)*roberta-large-bne* and (9) *roberta-large-bne-capitel-ner* (Gutiérrez-Fandiño et al., 2022)[3]. Tables 3 and 4 in the appendix show the values and metrics of fine-tuned models.

[1]Based on mBERT(Devlin et al., 2018) + Bi-LSTM + CRF
[2]XLM-Roberta (Conneau et al., 2019)
[3]Rest of models are built on RoBERTa (Liu et al., 2019)

Proceedings of the 29th International Conference on Computational Linguistic, pages 71–74
October 12–17, 2022.

2.2 Diseases gazetteer string matching

In conjunction with the neural models, an approximate string matching is performed with an in-domain gazetteer. The gazetteer has been built with diseases-related concepts from publicly available corpora: DisTEMIST (Gasco et al., 2022c), AbreMES-DB (Intxaurrondo, 2018), CodiEsp (Miranda-Escalada et al., 2020b), SNOMED-CT (International Health Terminology Standards Development Organisation - IHTSDO, 2014) and ICD-10-CM (CodeBooks, 2016).

An iterative curating process has been performed over the 122620 entries gathered. Firstly, normalization and duplicates removal is needed. Thereafter a filter on the number of tokens is applied: 5 and 3 tokens-length are considered relating to the average length of tweets. An analysis of n-gram frequencies enables to extract sets of general common-used terms and "stop-terms", which are included if not previously present or removed, respectively. Then a zero-shot classifier built on BETO (Cañete et al., 2020)[4] is used to filter no disease-related terms. Finally, two versions of the gazetteer are consolidated: *Final*, which compile up to 5-token length entries, containing 69655 health-related terms; a 3-token length version of the former called *Reduced* having 32852 terms.

2.3 Text processing and filtering

Various text processing steps are performed within the prediction flow. After ignoring breaklines, hashtags (#) and mentions (@) are extracted as plain tokens, and analyzed applying morpho-lexical rules to gather meaningful words (i.e. *#cancerdemama* gets transformed into "cancer", "de", "mama"). Then, once the disease mentions are extracted, punctuation marks, special characters and emojis at the beginning and the end of each one are removed and offsets are adjusted. After that, the zero-shot model is applied to filter generic mentions. Finally, duplicates and overlapped entities are excluded.

The code of the system is available in GitHub[5].

3 Results and discussion

A thorough study on different state of the art Transformer Language Models for NER in Spanish, their aggregation and integration with other NLP techniques was conducted using the datasets of SocialDisNER. The top three configurations are shown in Table 1, while the whole set of results can be reached at table 5 of the appendix.

Conf.	Gaz.	E.	st.F	st.P	st.R
B2	X	F	**0,817**	**0,840**	0,795
B5	X	T	**0,817**	0,833	**0,802**
B5	X	F	0,815	0,831	0,800

Table 1: Validation results ranked by strict F1.
"B"s refer to the ensemble of the best N top models.

The obtained results demonstrate that ensembling approaches provide the best performance, since standalone models enriched with specific rules work reasonably well on the task. The mBERT with LSTM-CRF model outperforms other architectures, but in terms of model sizes no significant differences have been found. Despite the overall strong performance, many false positives are extracted, so that further research on the effect of the pretraining corpus would be needed.

Using gazetteers alongside LMs, when their size and quality are extensive enough, have a positive impact on the recall. However, large gazetteers are time-consuming in building and predicting phases comparing with their performance contribution. In regard to zero-shot classification, it has contributed to build gazetteers by filtering out of domain terms. Nevertheless, when applied to the prediction pipeline it filters true positives and thus worsens performance, so that it has not been used on test. Domain specific text processing, such as hashtag segmentation and filtering rules to remove false positives, is needed due to the linguistic complexity and orthographic diversity of social media.

Final results in the test set obtained with two different configurations focusing on comparativeness over gazetteer usage, are depicted in table 2. In comparison to the official results of SocialDisNER, the developed system outperforms average values of the participants' submissions.

Model	Gaz./E.	st.F	st.P	st.R
B5	X/T	**0,774**	**0,779**	0,769
B5	Reduce/T	0,752	0,691	**0,826**
Mean task	-	0.675	0.680	0.677
Median task	-	0.761	0.758	0.780

Table 2: Test results.

[4] https://huggingface.co/Recognai/bert-base-spanish-wwm-cased-xnli

[5] https://github.com/ITAINNOVA/SocialDisNER

Acknowledgements

This work has been partially funded by the Department of Big Data and Cognitive Systems at the Technological Institute of Aragon, by IODIDE group of the Government of Aragon, grant number T1720R and by the European Regional Development Fund (ERDF).

References

Francesco Barbieri, Luis Espinosa-Anke, and Jose Camacho-Collados. 2022. XLM-T: Multilingual Language Models in Twitter for Sentiment Analysis and Beyond. In *Proceedings of LREC*.

Casimiro Pio Carrino, Joan Llop, Marc Pàmies, Asier Gutiérrez-Fandiño, Jordi Armengol-Estapé, Joaquín Silveira-Ocampo, Alfonso Valencia, Aitor Gonzalez-Agirre, and Marta Villegas. 2022. Pretrained biomedical language models for clinical NLP in Spanish. In *Proceedings of the 21st Workshop on Biomedical Language Processing*, pages 193–199, Dublin, Ireland. Association for Computational Linguistics.

José Cañete, Gabriel Chaperon, Rodrigo Fuentes, Jou-Hui Ho, Hojin Kang, and Jorge Pérez. 2020. Spanish pre-trained bert model and evaluation data. In *PML4DC at ICLR 2020*.

Medical CodeBooks. 2016. *ICD-10-CM Complete Code Set 2016*, volume 1. Medical Code Books.

Alexis Conneau, Kartikay Khandelwal, Naman Goyal, Vishrav Chaudhary, Guillaume Wenzek, Francisco Guzmán, Edouard Grave, Myle Ott, Luke Zettlemoyer, and Veselin Stoyanov. 2019. Unsupervised cross-lingual representation learning at scale. *arXiv preprint arXiv:1911.02116*.

Javier de la Rosa, Eduardo G. Ponferrada, Paulo Villegas, Pablo Gonzalez de Prado Salas, Manu Romero, and Maria Grandury. 2022. Bertin: Efficient pre-training of a spanish language model using perplexity sampling.

Jacob Devlin, Ming-Wei Chang, Kenton Lee, and Kristina Toutanova. 2018. Bert: Pre-training of deep bidirectional transformers for language understanding.

Luis Gasco, Darryl Estrada, Eulàlia Farré, and Martin Krallinger. 2022a. SocialDisNER corpus: gold standard annotations for detection of disease mentions in Spanish tweets. Funded by the Plan de Impulso de las Tecnologías del Lenguaje (Plan TL).

Luis Gasco, Darryl Estrada-Zavala, Eulàlia Farré-Maduell, Salvador Lima-López, Antonio Miranda-Escalada, and Martin Krallinger. 2022b. Overview of the SocialDisNER shared task on detection of diseases mentions from healthcare related and patient generated social media content: methods, evaluation and corpora. In *Proceedings of the Seventh Social Media Mining for Health (#SMM4H) Workshop and Shared Task*.

Luis Gasco, Eulàlia Farré, Antonio Miranda-Escalada, Salvador Lima, and Martin Krallinger. 2022c. DisTEMIST corpus: detection and normalization of disease mentions in spanish clinical cases. Funded by the Plan de Impulso de las Tecnologías del Lenguaje (Plan TL).

Asier Gutiérrez-Fandiño, Jordi Armengol-Estapé, Marc Pàmies, Joan Llop-Palao, Joaquin Silveira-Ocampo, Casimiro Pio Carrino, Carme Armentano-Oller, Carlos Rodriguez-Penagos, Aitor Gonzalez-Agirre, and Marta Villegas. 2022. Maria: Spanish language models. *Procesamiento del Lenguaje Natural*, 68(0):39–60.

International Health Terminology Standards Development Organisation - IHTSDO. 2014. SNOMED CT. Http://www.ihtsdo.org/snomed-ct/.

Ander Intxaurrondo. 2018. Abremes-db. Funded by the Plan de Impulso de las Tecnologías del Lenguaje (Plan TL).

Yinhan Liu, Myle Ott, Naman Goyal, Jingfei Du, Mandar Joshi, Danqi Chen, Omer Levy, Mike Lewis, Luke Zettlemoyer, and Veselin Stoyanov. 2019. Roberta: A robustly optimized bert pretraining approach.

A Miranda-Escalada, E Farré, and M Krallinger. 2020a. Named entity recognition, concept normalization and clinical coding: Overview of the cantemist track for cancer text mining in spanish, corpus, guidelines, methods and results. In *Proceedings of the Iberian Languages Evaluation Forum (IberLEF 2020), CEUR Workshop Proceedings*.

Antonio Miranda-Escalada, Aitor Gonzalez-Agirre, and Martin Krallinger. 2020b. CodiEsp corpus: gold standard Spanish clinical cases coded in ICD10 (CIE10) - eHealth CLEF2020. Funded by the Plan de Impulso de las Tecnologías del Lenguaje (Plan TL).

Pushpankar Kumar Pushp and Muktabh Mayank Srivastava. 2017. Train once, test anywhere: Zero-shot learning for text classification. *CoRR*, abs/1712.05972.

Simone Tedeschi, Valentino Maiorca, Niccolò Campolungo, Francesco Cecconi, and Roberto Navigli. 2021. WikiNEuRal: Combined neural and knowledge-based silver data creation for multilingual NER. In *Findings of the Association for Computational Linguistics: EMNLP 2021*, pages 2521–2533, Punta Cana, Dominican Republic. Association for Computational Linguistics.

Davy Weissenbacher, Ari Z. Klein, Luis Gascó, Darryl Estrada-Zavala, Martin Krallinger, Yuting Guo, Yao Ge, Abeed Sarker, Ana Lucia Schmidt, Raul Rodriguez-Esteban, Mathias Leddin, Arjun Magge, Juan M. Banda, Vera Davydova, Elena Tutubalina, and Graciela Gonzalez-Hernandez. 2022. Overview of the seventh social media mining for health applications #smm4h shared tasks at coling 2022. In *Proceedings of the Seventh Social Media Mining for Health (#SMM4H) Workshop and Shared Task.*

A Complete result tables

Model	Learning rate	Epochs	Batch size	Warmup ratio
M1	1e-05	6	4	0.05
M2	2e-06	5	4	0.1
M3	1e-05	10	6	0.2
M4	2e-06	5	4	0.1
M5	1e-05	8	4	0.01
M6	1e-05	6	4	0.05
M7	5e-06	10	16	0.005
M8	3.4e-05	10	8	0.1
M9	5e-05	10	10	0.01

Table 3: Hyperparameter tuning.

R.	Model	ov.F	ov.P	ov.R	st.F	st.P	st.R
1	Babelscape/wikineural-multilingual-ner	0,882	0,937	0,833	0,769	0,813	0,729
2	PlanTL-GOB-ES/bsc-bio-ehr-es-cantemist	0,775	0,891	0,686	0,715	0,822	0,632
3	bertin-project/bertin-base-ner-conll2002-es	0,713	0,835	0,622	0,674	0,758	0,565
4	PlanTL-GOB-ES/bsc-bio-es	0,708	0,832	0,616	0,655	0,769	0,570
5	cardiffnlp/twitter-xlm-roberta-base	0,704	0,836	0,607	0,650	0,772	0,562
6	PlanTL-GOB-ES/bsc-bio-ehr-es	0,706	0,841	0,608	0,648	0,773	0,557
7	PlanTL-GOB-ES/roberta-base-biomedical-clinical-es	0,699	0,814	0,612	0,630	0,733	0,552
8	PlanTL-GOB-ES/roberta-large-bne	0,694	0,822	0,601	0,626	0,741	0,541
9	PlanTL-GOB-ES/roberta-large-bne-capitel-ner	0,675	0,811	0,578	0,596	0,716	0,511

Table 4: Ranking of fine-tuned pretrained models according to strict F1 metric. Columns "R.", "ov." and "st." refer to *Ranking*, *overlap* and *strict* respectively, and *F*, *P*, *R* to F1, Precision and Recall.

Config	Gaz.	Gaz v.	Zero-shot	Emoji	ov.F	ov.P	ov.R	st.F	st.P	st.R
Best 2	F	X	F	F	0,899	0,928	0,872	0,817	0,840	0,795
Best 5	F	X	F	T	0,899	0,919	0,879	0,817	0,833	0,802
Best 5	F	X	F	F	0,899	0,919	0,879	0,815	0,831	0,800
Model 1	T	Final	F	F	0,894	0,859	0,932	0,807	0,768	0,649
Best 2	T	Final	F	F	0,898	0,854	0,947	0,807	0,761	0,859
Best 5	T	Final	F	F	0,897	0,849	0,952	0,805	0,754	0,862
Model 2	T	Final	F	F	0,887	0,873	0,901	0,802	0,784	0,820
All 9	F	X	F	F	0,887	0,889	0,885	0,796	0,794	0,798
Best 5	T	Reduced	F	T	0,886	0,827	0,954	0,793	0,732	0,864
Best 2	T	Reduced	F	F	0,886	0,831	0,950	0,792	0,735	0,858
Best 5	T	Reduced	F	F	0,886	0,827	0,954	0,791	0,731	0,861
Model 4	T	Reduced	F	F	0,877	0,848	0,907	0,790	0,758	0,825
Model 2	T	Reduced	F	F	0,879	0,847	0,914	0,788	0,754	0,825
Model 1	F	X	F	F	0,882	0,937	0,833	0,769	0,813	0,729
All 9	F	X	T	T	0,850	0,857	0,842	0,760	0,763	0,757
Model 1	T	Reduced	F	F	0,882	0,833	0,938	0,759	0,709	0,815
Best 5	T	Reduced	T	T	0,853	0,802	0,911	0,758	0,706	0,819
All 9	T	Reduced	T	T	0,845	0,784	0,916	0,744	0,682	0,817
Model 2	F	X	F	F	0,775	0,891	0,686	0,715	0,822	0,632
Model 1	T	Reduced	T	T	0,837	0,798	0,881	0,713	0,673	0,757

Table 5: System configurations ranked by strict F1.
Gaz and *Gaz v.* refers to the use of the gazetteer and its corresponding version, *Zero-shot* and *Emoji* indicates whether zero-shot filter and emoji cleaning is performed.

mattica@SMM4H'22: Leveraging sentiment for stance & premise joint learning

Oscar Lithgow-Serrano[1] †, Joseph Cornelius[1], Fabio Rinaldi[1], Ljiljana Dolamic[2]
[1]IDSIA, Lugano, Switzerland
[2]armasuisse S+T, Thun, Switzerland
†Corresponding author: `oscar@idsia.ch`

Abstract

This paper describes our submissions to the Social Media Mining for Health Applications (SMM4H) shared task 2022. Our team (mattica) participated in detecting stances and premises in tweets about health mandates related to COVID-19 (Task 2). Our approach was based on using an in-domain Pretrained Language Model, which we fine-tuned by combining different strategies such as leveraging an additional stance detection dataset through two-stage fine-tuning, joint-learning Stance and Premise detection objectives; and ensembling the sentiment-polarity given by an off-the-shelf fine-tuned model.

1 Introduction

The Social Media Mining for Health Applications (SMM4H) shared task 2022 (Weissenbacher et al., 2022) is aimed to apply Natural Language Processing to address different challenges of using social media for Health research. Specifically, we participated in Task-2, which consisted of detecting the stance of a tweet towards a given topic (subtask 2a), and detecting if a tweet contained or not a premise (subtask 2b). The organizers provided a manually labeled dataset (Davydova and Tutubalina, 2022) split into training (3,669 tweets) and validation (600 tweets). Each record in the dataset was labeled on both axis; the *Stance* classification included the labels FAVOR, AGAINST, and NONE; and the *Premise* axis with values 1 and 0 indicating the presence or absence of a premise respectively. For the evaluation phase, an unlabeled test set of 2,000 tweets was provided.

Our approach consisted of leveraging an in-domain Pretrained Language Model **(PLM)** and adapting it to the tasks through two consecutive fine-tuning steps. In the second, we used the predictions of an off-the-shelf model for sentiment polarity as additional features and applied joint learning of the stance and premise detection.

Related work (Sun et al., 2019; Li and Caragea, 2019; Fang et al., 2019) use sentiment analysis as an auxiliary task in an MTL setting. Whereas (Mohammad et al., 2017, 2016), like us, use the sentiment as an additional input feature for the stance classification. To our knowledge, no other approaches jointly-learned stance and premise detection leveraging sentiment polarity.[1]

2 Extra data and Pre-processing

We used COVIDLies (Hossain et al., 2020) as an extra training dataset. The dataset consists of 6,761[2] COVID-19-related tweets, each paired with a misconception and annotated with the concerning tweet's stance. In COVIDLies, the misconceptions are misinformation statements that were manually rephrased as short and positive expressions (e.g., "Coronavirus is caused by 5G"). Thus, different from a topic stance, in this dataset, the annotated standpoint is with respect to a specific statement. To combine both datasets, we migrated the provided data from topic-stance to the COVIDLies statement-stance format by manually formulating a positive statement from each related topic (see table 1). We also pre-processed the tweets by delexicalizing user mentions and URLs and replacing them with *@user* and *URL* correspondingly; striping out the hash character from hashtags, removing extra spaces; and replacing emojis with their corresponding short-code aliases (i.e., demojize[3]).

Topic	Statement
face masks	Face masks help to protect us.
stay at home orders	Stay at home is a needed measure.
school closures	Schools need to remain closed.

Table 1: Map from *topic* to *statement*.

[1]Code available at https://github.com/OWLmx/ws_ssm4h22
[2]Only 3,256 could be recovered.
[3]We used the package https://pypi.org/project/emoji

Proceedings of the 29th International Conference on Computational Linguistic, pages 75–77
October 12–17, 2022.

3 System description

The base of our approach is a transformer-based language model pretrained on a corpus of Twitter messages on the topic of COVID-19 (CT-BERT V2) (Müller et al., 2020).

We leveraged the COVIDLies dataset by an initial fine-tuning **(ft)** for stance-detection. Next, we performed a second fine-tuning on the data provided for this task. The second fine-tuning included a multitask-learning **(MTL)** setup where the objectives were Stance and Premise classification. Both losses were computed by cross entropy and combined using homoscedastic uncertainty to weight each task loss (Kendall et al., 2018). Also, in the second fine-tuning, the resulting logits of sentiment classification were appended directly to the pooled output of the encoder just before passing it to the classification heads[4]. The sentiment classification was obtained with an already fine-tuned roBERTa-base model for sentiment-analysis in tweets (Loureiro et al., 2022).

We trained the models with AdamW (Loshchilov and Hutter, 2019) optimizer without a warm-up period. A weighted random sampling with replacement was used in both fine-tuning steps (except in the multi-task setups), with learning rates of $\alpha = 10^{-4}$ and $\alpha = 10^{-5}$ respectively. For single task configurations, we used a training batch size of $bs = 8$ and a $bs = 4$ for MTL configurations; in both cases, the maximum sequence length was set to 160. An early stop with patience of 3 using losses scores of the held-out validation split ($bs = 16$) was applied during training. [5]

4 Results and Discussion

We used the COVIDLies dataset for the first fine-tuning. For the second **ft** we used the provided data as follows: we created a validation split with 5% of the training and 12% of the validation data, and used the rest of the training data as a training split and the rest of the validation data as a test split.

All our submissions were based on the two-stage fine-tuning **(2st-ft)** and what differed was the strategies combination in the 2nd ft. For subtask-2a, we submitted a run with a base 2st-ft, another including the logits from sentiment classification **(sent)**, another with the 2st-ft and MTL, and finally, one

Setting	Accuracy	F1-macro
2st-ft	0.840	0.836
2st-ft + sent	0.848	0.841
2st-ft + MTL	0.853	0.849
2st-ft + sent + MTL	0.857	0.853

Table 2: Stance detection results in our held-out test set.

that combined all the strategies (2st-ft + sent + MTL). For subtask-2b, we used the two setups that included MTL. This is, our two submissions corresponded to the premise inference from the (2st-ft + MTL) and the (2st-ft + sent + MTL) runs.

Evaluating our test-set, we observed that the cumulative combination of the different strategies resulted in small but consistent gains for the stance detection task performance (see Table 2). We analyzed each strategy's impact on identifying the different stances (see Fig. 1). We observed that integrating the sentiment polarity gives an important boost to detecting the FAVOR stance, whereas jointly learning to predict the presence of Premises is more beneficial to recognizing the AGAINST stance. Combining all the strategies resulted in the best balance for both -stances.

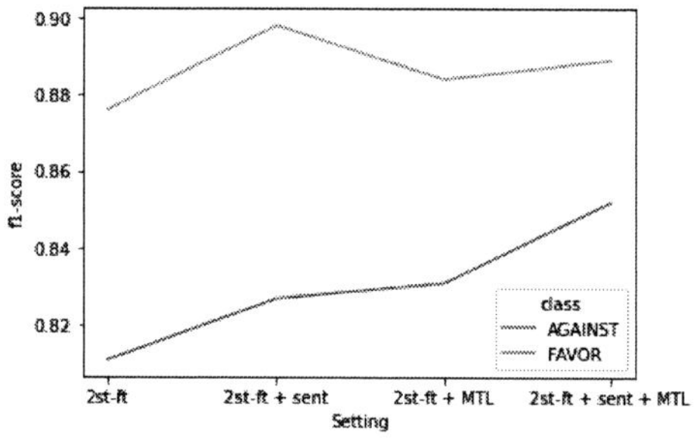

Figure 1: Stances' F1 score with the different strategies.

5 Conclusions

Our approach involved leveraging a related dataset by a preliminary fine-tuning, and combining sentiment analysis along with multi-task learning of premise and stance.

In the official test set, our best result for stance detection (subtask 2a) was 0.633, which is 14 percentage points (p.p.) above the mean and 8 p.p. above the median of all participants' submissions. For the premise detection (subtask 2b), our best score was 0.647, which is precisely the median but 7 p.p. above the mean of all submissions.

The results show that jointly learning to detect premise and stance is beneficial for both tasks. Combined with the tweet's sentiment polarity, the two-stage fine-tuned model gave the best results.

[4]Linear -> ReLU -> Dropout (0.1) -> Softmax

[5]The models were implemented using Pytorch (Paszke et al., 2019), Pytorch Lightning (Falcon and The PyTorch Lightning team, 2019) and Huggingface's Transformers (Wolf et al., 2020) library.

Acknowledgements

This research was partly supported by the project MisInfoFastCatch, funded by the armasuisse Science and Technology Swiss Federal Office (contract nr. 8203004722).

References

Vera Davydova and Elena Tutubalina. 2022. Smm4h 2022 task 2: Dataset for stance and premise detection in tweets about health mandates related to covid-19. In *Proceedings of the Seventh Social Media Mining for Health Applications (#SMM4H) Workshop Shared Task*, pages –.

William Falcon and The PyTorch Lightning team. 2019. PyTorch Lightning.

Wei Fang, Moin Nadeem, Mitra Mohtarami, and James Glass. 2019. Neural Multi-Task Learning for Stance Prediction. pages 13–19.

Tamanna Hossain, Robert L. Logan IV, Arjuna Ugarte, Yoshitomo Matsubara, Sean Young, and Sameer Singh. 2020. COVIDLies: Detecting COVID-19 Misinformation on Social Media.

Alex Kendall, Yarin Gal, and Roberto Cipolla. 2018. Multi-task Learning Using Uncertainty to Weigh Losses for Scene Geometry and Semantics. *Proceedings of the IEEE Computer Society Conference on Computer Vision and Pattern Recognition*, pages 7482–7491.

Yingjie Li and Cornelia Caragea. 2019. Multi-Task Stance Detection with Sentiment and Stance Lexicons. In *Proceedings of the 2019 Conference on Empirical Methods in Natural Language Processing and the 9th International Joint Conference on Natural Language Processing (EMNLP-IJCNLP)*, pages 6299–6305, Hong Kong, China. Association for Computational Linguistics.

Ilya Loshchilov and Frank Hutter. 2019. Decoupled weight decay regularization. *7th International Conference on Learning Representations, ICLR 2019*.

Daniel Loureiro, Francesco Barbieri, Leonardo Neves, Luis Espinosa Anke, and Jose Camacho-Collados. 2022. TimeLMs: Diachronic Language Models from Twitter.

Saif M. Mohammad, Svetlana Kiritchenko, Parinaz Sobhani, Xiaodan Zhu, and Colin Cherry. 2016. SemEval-2016 task 6: Detecting stance in tweets. *SemEval 2016 - 10th International Workshop on Semantic Evaluation, Proceedings*, pages 31–41.

Saif M. Mohammad, Parinaz Sobhani, and Svetlana Kiritchenko. 2017. Stance and Sentiment in Tweets. *ACM Transactions on Internet Technology*, 17(3):1–23.

Martin Müller, Marcel Salathé, and Per E Kummervold. 2020. COVID-Twitter-BERT: A Natural Language Processing Model to Analyse COVID-19 Content on Twitter.

Adam Paszke, Sam Gross, Francisco Massa, Adam Lerer, James Bradbury, Gregory Chanan, Trevor Killeen, Zeming Lin, Natalia Gimelshein, Luca Antiga, Alban Desmaison, Andreas Kopf, Edward Yang, Zachary DeVito, Martin Raison, Alykhan Tejani, Sasank Chilamkurthy, Benoit Steiner, Lu Fang, Junjie Bai, and Soumith Chintala. 2019. PyTorch: An Imperative Style, High-Performance Deep Learning Library. In *Advances in Neural Information Processing Systems 32*, pages 8024–8035. Curran Associates, Inc.

Qingying Sun, Zhongqing Wang, Shoushan Li, Qiaoming Zhu, and Guodong Zhou. 2019. Stance detection via sentiment information and neural network model. *Frontiers of Computer Science*, 13(1):127–138.

Davy Weissenbacher, Ari Z. Klein, Luis Gascó, Darryl Estrada-Zavala, Martin Krallinger, Yuting Guo, Yao Ge, Abeed Sarker, Ana Lucia Schmidt, Raul Rodriguez-Esteban, Mathias Leddin, Arjun Magge, Juan M. Banda, Vera Davydova, Elena Tutubalina, and Graciela Gonzalez-Hernandez. 2022. Overview of the seventh social media mining for health applications smm4h shared tasks at coling 2022. In *Proceedings of the Seventh Social Media Mining for Health Applications (#SMM4H) Workshop Shared Task*, pages –.

Thomas Wolf, Lysandre Debut, Victor Sanh, Julien Chaumond, Clement Delangue, Anthony Moi, Perric Cistac, Clara Ma, Yacine Jernite, Julien Plu, Canwen Xu, Teven Le Scao, Sylvain Gugger, Mariama Drame, Quentin Lhoest, and Alexander M Rush. 2020. Transformers: State-of-the-Art Natural Language Processing. pages 38–45. Association for Computational Linguistics.

KU_ED at SocialDisNER: Extracting Disease Mentions in Tweets Written in Spanish

Antoine Lain[1] **Wonjin Yoon**[2] **Hyunjae Kim**[2] **Jaewoo Kang**[2,3] **T Ian Simpson**[1]
[1] Institute for Adaptive and Neural Computation, The University of Edinburgh,
10 Crichton Street, Edinburgh EH8 9AB, UK [2]Korea University [2]AIGEN Sciences
{Antoine.Lain,ian.simpson}@ed.ac.uk
{wjyoon,hyunjae-kim,kangj}@korea.ac.kr

Abstract

This paper describes our system developed for the Social Media Mining for Health (SMM4H) 2022 SocialDisNER task. We used several types of pre-trained language models, which are trained on Spanish biomedical literature or Spanish Tweets. We showed the difference in performance depending on the quality of the tokenization as well as introducing silver standard annotations when training the model. Our model obtained a strict F1 of 80.3% on the test set, which is an improvement of +12.8% F1 (24.6 std) over the average results across all submissions to the SocialDisNER challenge.

1 Introduction

In this system description paper, we aim to detect disease mentions from tweets written in Spanish as part of the Social Media Mining for Health (SMM4H) 2022 SocialDisNER task (Gasco et al., 2022). The organizers provided the participants with various sets of data, either labelled by healthcare experts or machine-generated by the organizers themselves. All the sets contain anonymous tweets written in Spanish, each saved as a txt file, with a corresponding tsv file that contains the label for each tweet. Since the data is user-generated text limited to 280 characters as per the limitation from Twitter, it contains misspelled words, abbreviations, emojis, links, hashtags, and mentions to other users of the platform, making the task challenging.

Acknowledging the outstanding performance for named entity recognition presented in Devlin et al. (2019) and from its adaptation to the biomedical domain presented in Lee et al. (2020), we decided to use pre-trained biomedical language models for detection of disease mentions in the SocialDisNER task. We selected four publicly available pre-trained language models on Spanish corpora (Carrino et al., 2022; Cañete et al., 2020; Chizhikova et al.; Huertas-Tato et al., 2022), one pre-trained on English corpora (Sanh et al., 2019) to retrain one of the models for named entity recognition. We reported the results of our experiments that range from 62.4% to 82.9% strict F1 score on the validation set.

Model	Precision	Recall	Tag-F1
Carrino et al. (2022)	0.94	0.95	0.94
Chizhikova et al.	0.95	0.94	0.94
Cañete et al. (2020)	0.91	0.92	0.92
Sanh et al. (2019)	0.88	0.90	0.89
Huertas-Tato et al. (2022)	0.92	0.92	0.92

Table 1: Token-level performance of pre-trained language models on the SocialDisNER validation set.

2 Data Description

The first set released by the organizers was the gold standard (GS), 7,500 tweets, it was accompanied by a tab-separated file with healthcare experts' annotations containing the unique tweet ID, beginning position, end position, type and extraction. This set was divided into 2 subsets, a training set of 5,000 tweets and a development set of 2,500 tweets. The second set, 85,077 tweets, is similar as it offers the same amount of information but is generated by a machine at the discretion of the organizers, making this set a silver standard ultimately less reliable than human expert annotations. Finally, the third set is the test set. The organizers used 2,000 of 23,430 tweets from the test set to evaluate the performance of each team but the labels were not disclosed by the time of submission. For our final system, we only used the training dataset annotated from healthcare professionals (GS). There is an average of 3,3 labels per tweet in the training set when the average is 1,7 labels per tweet in the development set supplied by the organizers.

3 System Description

BERT type models have demonstrated improvement compared to previous state of the art meth-

Proceedings of the 29th International Conference on Computational Linguistic, pages 78–80
October 12–17, 2022.

ods in NER (Devlin et al., 2019; Lee et al., 2020). We decided to select a few pre-trained language models from Hugging Face[1] to train each of these models for the NER task using the transformer architecture (Vaswani et al., 2017) and the seqeval library implemented in Python (Nakayama, 2018). We first selected four models trained on Spanish corpora (Carrino et al., 2022; Cañete et al., 2020; Chizhikova et al.; Huertas-Tato et al., 2022) and one trained on English corpora (Sanh et al., 2019). We also needed to convert the data to the correct format. We decided to use the BIO tagging format (Begin, Interior, Outside) (Ramshaw and Marcus, 1999), where each token within a tweet was coupled with one of the three tags.

3.1 Experiments and System Selection

First, we ran all five models using the same input parameters in order to select the best performing model for the task, the results are reported in Table 1. The model from Carrino et al. (2022) had the best F1-score.

Due to the challenging style of writing that tweets present, we worked on improving the quality of the tokenization as it has shown improvement in Kim et al. (2021). We split each tweet by space, then we split every token to separate the punctuation from the rest. Hashtags (#) and at (@) were separated from the original token. We also cut the words where only parts of the word were annotated. We ran three experiments: simple tokenization, improved tokenization and simple tokenization + silver standard data. Each experiment uses the same pre-trained language model 'PlanTL-GOB-ES/bsc-bio-es'[2], with 3 epochs, learning rate = $1e^{-4}$ and weight decay = $1e^{-5}$. Since we needed to give the span for each extraction, we replaced every emoji with '@' since it was changing the position of the text. As a post-processing rule, if the length of the prediction of the model was shorter than 3 characters the prediction was ignored.

4 Results and analysis of the error

We reported the results of our experiments in Table 2. The best performing model on the validation set was when we improved the quality of the tokenization. We gained 6.1% F1-score compared to a tokenization where the punctuation was kept. The difference in performance between the overlap and strict measures showed that the model was able to identify a part of the mentions but still resulted in around 10% difference between both measures. Looking at the mentions from the model compared to the mentions from the healthcare expert it seems that extra rules in the post-processing step could have been implemented to reduce the gap between both performances. For example when the identified token is part of a word, which is not a hashtag, extract the word instead of the token.

Model	Set	Measure	Precision	Recall	F1
Simple	Validation	Strict	71.9	82.4	76.8
Improved	Validation	Strict	83.1	82.7	82.9
Simple + Silver	Validation	Strict	66.0	59.2	62.4
Simple	Validation	Overlap	84.5	95.1	89.5
Improved	Validation	Overlap	94.6	92.7	93.7
Simple + Silver	Validation	Overlap	89.0	78.6	83.5
All participants	Test	Strict Mean	68.0	67.7	67.5
All participants	Test	Strict Median	75.8	78.0	76.1
Our Model	Test	Strict	80.9	79.8	80.3

Table 2: Summary of our results based on the official overlap and strict evaluation.

5 Conclusion

In this work, we studied the difference in performance of pre-trained language models depending on the quality of the labels and tokenization for detection of disease mentions in tweets. At the end, the model achieved 80.3% F1 score on the test set. For future work, one direction would be to use ensemble models for token classification, where we combine a Twitter pre-trained language model with a biomedical pre-trained language model.

Acknowledgments

This research was supported by (1) the MSIT (Ministry of Science and ICT), Korea, under the ICT Creative Consilience program (IITP-2022-2020-0-01819) supervised by the IITP (Institute for Information & communications Technology Planning & Evaluation), (2) a grant of the Korea Health Technology R&D Project through the Korea Health Industry Development Institute (KHIDI), funded by the Ministry of Health & Welfare, Republic of Korea (grant number: HR20C0021), (3) National Research Foundation of Korea (NRF-2014M3C9A3063541, NRF-2020R1A2C3010638) and (4) School of Informatics at UoE.

[1]https://huggingface.co/

[2]PlanTL-GOB-ES/bsc-bio-es/

References

Casimiro Pio Carrino, Joan Llop, Marc Pàmies, Asier Gutiérrez-Fandiño, Jordi Armengol-Estap'e, Joaqu'in Silveira-Ocampo, Alfonso Valencia, Aitor Gonzalez-Agirre, and Marta Villegas. 2022. Pre-trained biomedical language models for clinical nlp in spanish. In *BIONLP*.

José Cañete, Gabriel Chaperon, Rodrigo Fuentes, Jou-Hui Ho, Hojin Kang, and Jorge Pérez. 2020. Spanish pre-trained bert model and evaluation data. In *PML4DC at ICLR 2020*.

Mariia Chizhikova, Jaime Collado-Montañéz, Pilar López-Úbeda, Manuel C. Díaz-Galiano, L. Alfonso Ureña-López, and M. Teresa Martín-Valdivia. Sinai at clef 2022: Leveraging biomedical transformers to detect and normalize disease mentions. pages 265–273.

Jacob Devlin, Ming-Wei Chang, Kenton Lee, and Kristina Toutanova. 2019. BERT: Pre-training of deep bidirectional transformers for language understanding. In *Proceedings of the 2019 Conference of the North American Chapter of the Association for Computational Linguistics: Human Language Technologies, Volume 1 (Long and Short Papers)*, pages 4171–4186.

Luis Gasco, Darryl Estrada-Zavala, Eulàlia Farré-Maduell, Salvador Lima-López, Antonio Miranda-Escalada, and Martin Krallinger. 2022. Overview of the SocialDisNER shared task on detection of diseases mentions from healthcare related and patient generated social media content: methods, evaluation and corpora. In *Proceedings of the Seventh Social Media Mining for Health (#SMM4H) Workshop and Shared Task*.

Javier Huertas-Tato, Alejandro Martín, and David Camacho. 2022. Bertuit: Understanding spanish language in twitter through a native transformer. volume abs/2204.03465.

Hyunjae Kim, Mujeen Sung, Wonjin Yoon, Sungjoon Park, and Jaewoo Kang. 2021. Improving tagging consistency and entity coverage for chemical identification in full-text articles. In *Proceedings of the seventh BioCreative challenge evaluation workshop*.

Jinhyuk Lee, Wonjin Yoon, Sungdong Kim, Donghyeon Kim, Sunkyu Kim, Chan Ho So, and Jaewoo Kang. 2020. Biobert: a pre-trained biomedical language representation model for biomedical text mining. *Bioinformatics*, 36:1234 – 1240.

Hiroki Nakayama. 2018. seqeval: A python framework for sequence labeling evaluation. Software available from https://github.com/chakki-works/seqeval.

Lance A Ramshaw and Mitchell P Marcus. 1999. Text chunking using transformation-based learning. In *Natural language processing using very large corpora*, pages 157–176. Springer.

Victor Sanh, Lysandre Debut, Julien Chaumond, and Thomas Wolf. 2019. Distilbert, a distilled version of bert: smaller, faster, cheaper and lighter. *ArXiv*, abs/1910.01108.

Ashish Vaswani, Noam Shazeer, Niki Parmar, Jakob Uszkoreit, Llion Jones, Aidan N Gomez, Łukasz Kaiser, and Illia Polosukhin. 2017. Attention is all you need. In *Advances in neural information processing systems*, pages 5998–6008.

CHAAI@SMM4H'22: RoBERTa, GPT-2 and Sampling - An interesting concoction

Christopher Palmer, Sedigheh Khademi, Muhammad Javed,
Gerardo Luis Dimaguila, Jim Buttery

Murdoch Children's Research Institute

{chris.palmer, sedigh.khademi, muhammad.javed, gerardoluis.dimaguil, jim.buttery}@mcri.edu.au

Abstract

This paper describes the approaches to the SMM4H 2022 Shared Tasks that were taken by our team for tasks 1 and 6. Task 6 was the "Classification of tweets which indicate self-reported COVID-19 vaccination status (in English)". The best test F1 score was 0.82 using a CT-BERT model, which exceeded the median test F1 score of 0.77, and was close to the 0.83 F1 score of the SMM4H baseline model. Task 1 was described as the "Classification, detection and normalization of Adverse Events (AE) mentions in tweets (in English)". We undertook task 1a, and with a RoBERTa-base model achieved an F1 Score of 0.61 on test data, which exceeded the mean test F1 for the task of 0.56.

1 Introduction

The shared tasks of the Social Media Mining for Health (SMM4H) are focused on overcoming difficult challenges in utilizing natural language processing techniques for deriving health-related information from social media data (Weissenbacher et al., 2022).

Task 6 of the seventh SMM4H (in 2022) was the "Classification of tweets which indicate self-reported COVID-19 vaccination status (in English)". Our team, "Champions of Health and Artificial Intelligence" (CHAAI), found task 6 particularly interesting, as the challenge is similar to a key research area we are conducting in SAEFVIC (SAEFVIC, 2022) to leverage social media monitoring to improve vaccine safety surveillance. Detecting self-reporting in health-related online posts helps to identify genuine descriptions of health events. It is as important to know *who* is speaking as it is to know *what* is being spoken about; as in detecting adverse events following immunization (AEFI) (Habibabadi et al., 2022; Wang et al., 2019; Lian et al., 2022); experiences of COVID-19 disease (Valdes et al., 2021); and drug effects (Aji et al., 2021).

Task 1 was the "Classification, detection and normalization of Adverse Events (AE) mentions in tweets (in English)". The adverse events mentioned in the task description are Adverse Drug Events (ADE), which are unexpected side effects following consumption of medications. Task 1a was for the classification of ADE, task 1b was to identify the spans of the relevant text, and task 1c required matching the colloquially expressed reactions to MedDRA standard concept IDs. Whereas task 6 was akin to obtaining an online data stream of likely candidates for AEFI detection, task 1a's challenge of precisely identifying an ADE was akin to differentiating specific AEFI in that data stream.

In both tasks, we had to deal with massive imbalances between the under-represented positive class and the negative class, which matches real-world conditions. Dealing with class imbalances was a major focus of our response to the tasks.

2 Task 6: Classification of tweets which indicate self-reported COVID-19 vaccination status

Identifying tweets of users reporting on their COVID-19 vaccination status.

2.1 Task 6 Data and Pre-processing

There were 13,692 records in the supplied training data (SMM4H, 2022), with 1,495 records with the positive label of "Self_reports", and 12,197 with the negative "Vaccine_chatter" label. There were 2,783 unlabeled records in the validation dataset. However, 2,478 records in the validation dataset were also found in the training data.

Text preparation consisted of replacing user mentions and URLs with placeholder expressions, converting emoji into text equivalents, and splitting words on apostrophes.

The dataset specification described the positive class as "unambiguous tweets of users clearly stating that they have been vaccinated." However, the training data did not align with this description.

Proceedings of the 29th International Conference on Computational Linguistic, pages 81–84
October 12–17, 2022.

We found that almost half of the positive labels (705/1495) included general descriptions of non-personal vaccination, personal statements of *not* being vaccinated, and even declarations and sentiment opposing vaccination. There were also personal reports of recent vaccination in the negative labels (392/12197).

We responded this labelling inconsistency by added additional clearly positive (and some clarifying negative) labelled texts to help train a model more consistently on the characteristics of a text discussing personal vaccination status. These were obtained from our previous work on COVID-19 vaccination reactions data (Khademi Habibabadi et al., 2022), which had been based on prior research that combined topic modelling and classification for filtering tweets to obtain *vaccine adverse event mentions* (VAEM) (Khademi Habibabadi et al., 2019), (Khademi et al., 2022).

After evaluation of classification using the added data, we also bolstered the existing shared task positive labelled data (despite its lack of accuracy), by generating additional similar examples. We used a GPT-2 (Alec Radford, Jeffrey Wu, Rewon Child, David Luan, Dario Amodei, 2020) model to learn from the positively labelled texts, then generated new texts by using a seed phrase of the initial few words of each existing positive record. Various combinations of the separately obtained texts and generated texts were used and assessed with F1 scoring. We found that adding the generated texts improved the score. Eventually, an enlarged and balanced dataset of 29,470 records was obtained. This was split 90/10 to create training and validation datasets of 26,524 and 2,946 records respectively.

During the evaluation phase we obtained scores from the CodaLab system, which used the separate task 6 validation dataset. Labels were only supplied to competitors *after* evaluation completion.

The task 6 hold-out test dataset was used to obtain the final scores of the models, which are used by the task examiners to rank the entries.

2.2 Task 6 System Description

Models' results are presented in Table 1. We have previously used the BERTweet-Large (Nguyen et al., 2020) model for similar tasks, so we evaluated fine tuning of the original model for Model 1, abbreviated as BERTweet-Lg in the table. A checkpoint of a BERTweet-Large model, previously fine-tuned by our team (Khademi Habibabadi et al., 2022) was used for Model 2 (abbreviated as BTL-PrevFT), and likewise a checkpoint of a previously fine-tuned (Khademi et al., 2022) RoBERTa-Large model (Liu et al., 2019) was used for Model 3 (RL-PrevFT), resulting in two new viable checkpoints (Model 3 and Model 3a). Additionally, we evaluated fine-tuning of the COVID-Twitter-BERT (CT-BERT) model (Müller et al., 2020) for Model 4.

We used the HuggingFace Trainer with an AdamW optimizer, for 3 to 5 epochs, and hyper-parameters of learning rate of 2e-5 and weight decay of 0.01. Other settings were Trainer defaults, including batch size of 8. We retained the best checkpoints of each training run, some of which favored recall, and others precision, as prior experience indicated models for classifier ensembles should emphasize recall and precision differently. We created a high-scoring ensemble that included the two versions of Model 3.

2.3 Task 6 Results

Model #, Descrip	Precision	Recall	F1
1 – BERTweet-Lg	0.82	0.75	0.78
2 – BTL-PrevFT	0.75	0.74	0.75
3 – RL-PrevFT	0.79	0.84	0.82
3a – RL-PrevFT	0.88	0.74	0.81
4 – CT-BERT	0.80	0.64	0.71
Ensemble (1,3,3a)	0.86	0.79	0.83

Table 1: Task 6 validation scores

Table 1 shows precision, recall and F1 scores on the positive class in the supplied validation data, for the best (by F1) of each the four models we assessed, and for an ensemble of 3 models using max voting.

Upon receiving the validation labels, we analyzed our incorrect predictions and found many validation records that we considered as incorrectly labelled. Although we had 103/2783 incorrect predictions, by our judgement of the labels, we concluded that the model was incorrect for only 15 of the 103 records.

The false positives tended to have the language of self-reports, but the vaccination mentions were either for someone else such as a parent, or for a future appointment – e.g., "*I just qualified to be eligible for the #COVID19 vaccine*", and "*Today he received his first dose of a COVID-19 vaccine*". The false negatives often had an indirect reference that implied that a previous vaccination had occurred – e.g., "*I am going to get my booster*", and "*I will not get the second shot*".

3 Task 1a: Classification of Adverse Events mentions in tweets

Task 1a was to classify English tweets containing one or more Adverse Drug Events (ADE) or no ADE.

3.1 Task 1a Data and Pre-processing

The training dataset consisted of 17,385 tweets, with 1,235 positive labels (ADE) and 16,150 negative labels (noADE). To overcome the class imbalance, we performed oversampling and under-sampling on the training dataset.

Oversampling: Applied data augmentation (Lemaitre et al., 2017) to increase the size of positive samples in the training dataset. We used contextual word embedding techniques (Ma, 2019) to insert or substitute words randomly in copies of the positively labelled texts, using a pretrained Roberta base model. After insertion and substitution of words in the posts, the positive samples were increased to 2,905.

Under-sampling: 0.5 under-sampling was applied to the majority negative class, which reduced the negative samples to 8,075.

In the pre-processing step, we removed the hashtags, user mentions, and special characters but as these changes did not make a positive contribution to model scores, so we progressed to fine-tune the models without pre-processing.

3.2 Task 1a System Description

We evaluated the RoBERTa-base transformer model (Liu et al., 2019), the BERT-base-uncased and BERT-large-uncased models (Devlin et al., 2018), and the distilBERT-base-uncased (Sanh et al., 2019) model.

3.3 Task 1a Experiments

We fine-tuned and evaluated the various models, dividing the dataset into 80%, 10%, and 10% for training, testing, and validation respectively. After analyzing the results of these trainings, we decided to use the Roberta-base model and trained for 5 epochs with a batch size of 32, a learning rate of 1e-5, a dropout rate of 0.1, and the Adam optimizer.

3.4 Task 1a Results

Model	Precision	Recall	F1
RoBERTa-base	0.72	0.83	0.77
RoBERTa-large	0.75	0.73	0.74
BERT-base-uncased	0.63	0.62	0.62
Distilbert-uncased	0.52	0.82	0.63

Table 2: Task 1a validation scores

The validation scores on the positive label of the best model are in Table 2. The F1 score of the selected model on the validation dataset was 0.77.

One of the major reasons for false negatives in the validation dataset are subtleties in the expressions to explain ADEs – e.g., "*debating on taking a trazodone and literally passing out for the day*". In some tweets people wanted to know the causes of adverse events with an unclear description of an experienced ADE – e.g., "*21y.o. w/ sickle-cell anemia and taking trazodone presents w/ priapism. what's the cause?*".

4 Conclusion

For task 6, although our best model on validation data was an ensemble of a BERTweet-Large and two RoBERTa-Large models, surprisingly, the CT-BERT model (Model 4) was our best model on test data. Its F1 score of 0.82 was close to the 0.83 F1 score of the baseline SMM4H model, and significantly better than the median F1 score of 0.77 for the task. However, the ensemble was only marginally behind – to 3 decimals its F1 score was 0.814, compared to 0.819 for the CT-BERT.

For task 1a, our best result on the test dataset was from the RoBERTa-base model. Notably, thanks to its balance of precision and recall, its F1 score matched last year's winning F1 score of 0.61 (Ramesh et al., 2021). Moreover, it exceeded this year's mean F1 score of 0.56. Scores are presented in Table 3.

Task	Model	Precision	Recall	F1
6	CT-BERT	0.86	0.78	0.82
	Ensemble	0.87	0.77	0.81
	Baseline	0.90	0.77	0.83
	Median	0.90	0.68	0.77
1a	RoBERTa-base	0.61	0.61	0.61
	Mean	0.65	0.50	0.56

Table 3: Test scores for both tasks

References

Alham Fikri Aji, Made Nindyatama Nityasya, Haryo Akbarianto Wibowo, Radityo Eko Prasojo, and Tirana Fatyanosa. 2021. BERT Goes Brrr: A Venture Towards the Lesser Error in Classifying Medical Self-Reporters on Twitter. :58–64.

Ilya Sutskever Alec Radford, Jeffrey Wu, Rewon Child, David Luan, Dario Amodei. 2020. Language Models are Unsupervised Multitask Learners. *OpenAI Blog*, 1(May):1–7.

Jacob Devlin, Ming-Wei Wei Chang, Kenton Lee, and Kristina Toutanova. 2018. BERT: Pre-training of deep bidirectional transformers for language understanding. *NAACL HLT 2019 - 2019 Conference of the North American Chapter of the Association for Computational Linguistics: Human Language Technologies - Proceedings of the Conference*, 1(Mlm):4171–4186.

Sedigheh Khademi Habibabadi, Pari Delir Haghighi, Frada Burstein, and Jim Buttery. 2022. Vaccine Adverse Event Mining of Twitter Conversations: 2-Phase Classification Study. *JMIR Med Inform 2022;10(6):e34305 https://medinform.jmir.org/2022/6/e34305*, 10(6):e34305.

Sedigheh Khademi Habibabadi, Pari Delir Haghighi, Sedigheh Khademi Habibabadi, Pari Delir Haghighi, Sedigheh Khademi, and Pari Delir Haghighi. 2019. Topic Modelling for Identification of Vaccine Reactions in Twitter. *ACM International Conference Proceeding Series*:31.

Sedigheh Khademi Habibabadi, Christopher Palmer, Gerardo Luis Dimaguila, Muhammad Javed, Hazel Clothier, and Jim Buttery. 2022. Automated social media surveillance for detection of vaccine safety signals: a validation study. *Applied Clinical Informatics*, in press.

Sedigheh Khademi, Pari Delir Haghighi, Frada Burstein, and Jim Buttery. 2022. Vaccine adverse event mentions in social media: Mining the language of Twitter conversations. *JMIR Medical Informatics preprint*.

Guillaume Lemaitre, Fernando Nogueira, and Christos K. Aridas. 2017. Imbalanced-learn: A Python Toolbox to Tackle the Curse of Imbalanced Datasets in Machine Learning. *Journal of Machine Learning Research*, 18(17):1–5.

Andrew T Lian, Jingcheng Du, and Lu Tang. 2022. Using a Machine Learning Approach to Monitor COVID-19 Vaccine Adverse Events (VAE) from Twitter Data. *Vaccines*, 10(1):103.

Yinhan Liu, Myle Ott, Naman Goyal, Jingfei Du, Mandar Joshi, Danqi Chen, Omer Levy, Mike Lewis, Luke Zettlemoyer, and Veselin Stoyanov. 2019. RoBERTa: A Robustly Optimized BERT Pretraining Approach. (1).

Edward Ma. 2019. nlpaug: NLP Augmentation.

Martin Müller, Marcel Salathé, and Per E Kummervold. 2020. COVID-Twitter-BERT: A Natural Language Processing Model to Analyse COVID-19 Content on Twitter.

Dat Quoc Nguyen, Thanh Vu, and Anh Tuan Nguyen. 2020. BERTweet: A pre-trained language model for English Tweets. *arXiv preprint arXiv:2005.10200*.

Sidharth Ramesh, Abhiraj Tiwari, Parthivi Choubey, Saisha Kashyap, Sahil Khose, Kumud Lakara, Nishesh Singh, and Ujjwal Verma. 2021. BERT based Transformers lead the way in Extraction of Health Information from Social Media. :33–38.

SAEFVIC. 2022. Surveillance of Adverse Events Following Vaccination in the Community.

Victor Sanh, Lysandre Debut, Julien Chaumond, and Thomas Wolf. 2019. DistilBERT, a distilled version of BERT: smaller, faster, cheaper and lighter. :2–6.

SMM4H. 2022. Social Media Mining for Health 2022 (#SMM4H) | Health Language Processing Lab @ Penn IBI.

Alberto Valdes, Jesus Lopez, and Manuel Montes. 2021. UACH-INAOE at SMM4H: a BERT based approach for classification of COVID-19 Twitter posts. :65–68.

Junxiang Wang, Liang Zhao, and Yanfang Ye. 2019. Semi-supervised Multi-instance Interpretable Models for Flu Shot Adverse Event Detection. *Proceedings - 2018 IEEE International Conference on Big Data, Big Data 2018*(October):851–860.

Davy Weissenbacher, Ari Z. Klein, Luis Gascó, Darryl Estrada-Zavala, Martin Krallinger, Yuting Guo, Yao Ge, Abeed Sarker, Ana Lucia Schmidt, Raul Rodriguez-Esteban, Mathias Leddin, Arjun Magge, Juan M. Banda, Vera Davydova, Elena Tutubalina, Elena Tutubalina, and Graciela Gonzalez-Hernandez. 2022. Overview of the Seventh Social Media Mining for Health Applications #SMM4H Shared Tasks at COLING 2022. In *Proceedings of the Seventh Social Media Mining for Health (#SMM4H) Workshop and Shared Task*.

IAI @ SocialDisNER : Catch me if you can! Capturing complex disease mentions in tweets

Aman Sinha*[1,2], Cristina García Holgado*[3], Marianne Clausel[1], Mathieu Constant[2]
[1]IECL, Université de Lorraine, Nancy, France ; [2]ATILF, Université de Lorraine, Nancy, France
[3]Lattice, École Normale Supérieure, Paris, France
{aman.sinha, marianne.clausel, mathieu.constant}@univ-lorraine.fr
cristina.gholgado@gmail.com

Abstract

Biomedical NER is an active research area today. Despite the availability of state-of-the-art models for standard NER tasks, their performance degrades on biomedical data due to OOV entities and the challenges encountered in specialized domains. We use Flair-NER framework to investigate the effectiveness of various contextual and static embeddings for NER on Spanish tweets, in particular, to capture complex disease mentions.

1 Motivation

In this paper, we present our system which recognizes disease mentions in Spanish tweets as part of the SocialDisNER challenge (Gasco et al., 2022). The motivation of our work is to:

(a) *Investigate the problem of identifying disease mentions in tweets* : Disease mentions occur with a variable length. While NER systems easily recognize simple-word entities, the task becomes exponentially difficult as the length increases (Dai, 2018; Shen et al., 2021), and also, it may complicate the gold annotation process. Moreover, entities can be composed of subgroups of entities (Dai et al., 2020). In this work, we deal with a particular linguistic context: tweet data. Here, we encounter an additional challenge as diseases mentions are specialized domain terms, but they occur in a social media platform characterized by informal language, comprising various noise-inducing elements such as hashtags, emojis, typos, and code-switching.

(b) *Investigate the performance of embedding resources* : Transfer learning (Pan and Yang, 2009) has been found to be very useful for downstream NLP tasks (Ruder et al., 2019). In this work, we explore the "capability" of different language models to use instilled knowledge from similar and different domains (Gururangan et al., 2020) and languages (Pfeiffer et al., 2020) pre-training datasets to learn NER on Spanish Twitter data, to see their effectiveness in addressing the above challenges.

GROUP	CONTEXTUAL	STATIC
domain	**xlrsc** (Lange et al., 2021) **rbbce** (Carrino et al., 2021) **sdf** (Chizhikova et al., 2022)	**es+clinical** **es+en+clinical**
multilingual	**xrl** (Conneau et al., 2019) **bbmcn** (Adelani, 2021) **wmn** (Tedeschi et al., 2021)	
es	**bscfn** (Cañete et al., 2020)	**es**
en	**bbucn** (Rawal, 2021)	

Table 1: Grouping of the different embeddings resources

2 Experimental Setup

2.1 Approach

To investigate the challenges of identifying complex disease mentions in tweets, we tested different embeddings using the Flair-NER (Akbik et al., 2019) framework using two types of models: Flair-S (Simple) and Flair-T (Transformers). Both of these models consist of a CRF-based SequenceTagger which takes input from a WordEmbedding/FlairEmbedding (Akbik et al., 2018) module for the static word embeddings (Flair-S), or from a TransformerWordEmbedding module for the contextual embeddings (Flair-T). We further defined four categories to group the different used embeddings as shown in Table 1.

2.2 Dataset

We used the GOLD SocialDisNER dataset (Gasco et al., 2022) provided by the organizers. It consists of a collection of health-related tweets from general users and a disease mention file annotated by medical experts. The disease mention file contains the annotated diseases mentions along with their begin and end offsets, and their corresponding

* Authors have equal contribution.

Proceedings of the 29th International Conference on Computational Linguistic, pages 85–89
October 12–17, 2022.

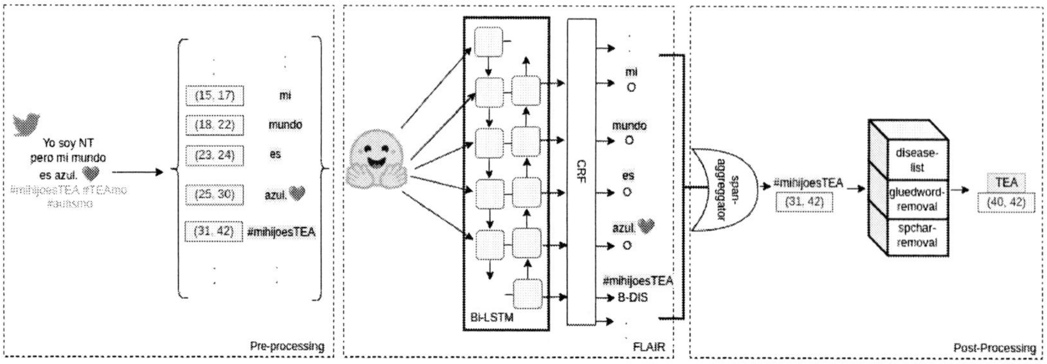

Figure 1: Experiment Pipeline

tweet file id. The dataset has 5000 training tweets (19425 mentions), 2500 validation tweets (4252 mentions), and 2500 test tweets files respectively.

2.3 Text-Processing

We perform a minimal pre-processing to keep all information in the text as we noticed a benefit from keeping the noise. We therefore propose the use of post-processing to obtain the strict spans from the identified disease mentions by removing different types of noise.

Pre-processing Tweets were tokenized on white space and converted into CoNLL format with no further pre-processing. Spans were generated for every token considering their character position in the text. BIO labels were assigned to every token matching the provided disease mentions in each tweet.

Post-processing Predictions were done over the tokenized tweets. The resulting predictions were converted into the mentions' file format where every detected disease mention span was aggregated. In a later post-processing step we fixed the aggregated spans. This fix of the original spans was performed on every detected disease mention when this contained any type of surrounding noise in the begin and/or end (See fig.1). This span-fix step determines the strict-f1 metric and includes three noise-treatment parts:
(a) Disease list: string-matching of disease mentions using an external custom list of disease words from a combination of the training disease mentions and online medical disease glossaries. This steps facilitates the removal of agglutinated words or attached noise (e.g. *#labioRojoContraLaMigraña*⋆ [0-26 → 18-25]).
(b) Glued words removal: removal of outer noise specific to Twitter hashtags when no disease mention is matched. The begin and the end of the mention is checked to remove this specific noise (e.g. *#DíaNacionalDelPárkinson* [0-24 → 16-24], *#TodasContraElCáncerDeMama* [0-26 → 15-26]).
(c) Special characters (Spchar) removal: removal of emojis, punctuation signs and other related characters when no disease mention is matched from the list (e.g. *#autismo*♡ [0-9 → 1-8]).

2.4 Implementation

For the Flair-T and Flair-S experiments, the models were trained for 50 epochs with a decaying learning rate (lr). Our setting were Flair-T (lr=5e-6, batch_size=4) and Flair-S (lr=1e-1, batch_size=2). The code is available at our Github repository.

	LMs	*Tag-F1*	*lenient-f1*	*strict-f1*
domain	**xrlsc**	**0.93**	**0.955**	**0.759**
	rbbce	0.93	0.951	0.757
	sdf	0.93	0.949	0.757
	es+clinical[†]	0.87	0.907	0.729
	es+en+clinical[†]	0.88	0.910	0.731
multilingual	**xrl**	0.93	0.950	0.756
	bbmcn	0.90	0.927	0.739
	wmn	0.89	0.925	0.735
es	**bscfn**	0.90	0.922	0.741
	es[†]	0.80	0.830	0.660
en	**bbucn**	0.87	0.914	0.725

Table 2: Final experiments results on Validation set; (†) correspond to Flair-S setting and rest of the entries correspond to Flair-T setting results.

3 Results

Table 2 shows the results on validation set. The systems were evaluated on a span-level using two metrics: *lenient-f1* and *strict-f1*. We also evaluated the systems on a token-level for BIO-tag classification (*Tag-f1*). Our final test set submission **xrlsc**

	lenient-f1	*strict-f1*
xrlsc	**0.941**	0.647
Mean	0.844 ± 0.168	0.675 ± 0.246
Median	0.898	0.761

Table 3: Comparison of our best submitted system (xrlsc) on Test set with other participants submissions

obtained 0.941 (*lenient-f1*) and 0.647 (*strict-f1*). Although *tag-f1* and *lenient-f1* reported similar scores, we observed a decrease on *strict-f1*. Compared to other submissions, our system performed ~10% better over mean w.r.t. lenient-f1 whereas, it falls ~2.5% short w.r.t. strict-f1 (see Table 3).

4 Discussion

Model	*without PP*	*with PP*
xrlsc	0.318	0.759
rbbce	0.319	0.757
sdf	0.318	0.757
es+clinical	0.308	0.729
es+en+clinical	0.313	0.731
xrl	0.317	0.756
bbmcn	0.311	0.739
wmn	0.306	0.735
bscfn	0.316	0.741
es	0.294	0.660
bbucn	0.298	0.725

Table 4: Effect of post-processing (PP) with metric *strict-f1* on Validation set

Impact of post-processing on span aggregation While the models were efficient at identifying the disease mentions on a token-level (as indicated by lenient-f1 scores in Table 2), the presence of noise in the extracted disease mentions leads to low strict-f1 scores without post-processing as shown in Table 4. The post-processing strategy of noise removal[1] increases significantly the strict-f1 by more than 40% (except for es-model). However, there is still a noticeable difference with the lenient-f1 which shows that this step encountered limitations to capture properly the strict spans. String-matching to remove surrounding noise on target entities is an effective strategy but it is however limited to known diseases from the provided list. When encountering unknown detected mentions, it becomes an arduous task due to the shifting and variation of different agglutinated noise-words (e.g. #ElVPHEsCosaDeTodos / #vacunavph / #DiaInternacionalContraelVPH). We also associate the reason of a lower strict-f1 to a displacement of the tokens' spans on the dataset due to the interference of special characters generated by multi-character emojis and a few encoding conflicts from the tweets extraction process. On further analysis, we found 170 tweet files potentially affected by this issue out of 2500 files in the validation set which comprises 7.92% of the total disease mentions.

Complex and discontinuous named entities Such entities are particularly problematic as they can lead to multiple disease mention identification. Variable context boundaries (e.g. *#dolorneuropatico en #COVID19?* → single or multiple disease mentions?) may lead to an alteration of the computation of the spans. We found that the error for capturing long and discontinuous entities was 26.11% lower for Flair-T models than Flair-S models. For noisy and agglutinated words such as *#HablemosDeVIH* or *#diabetestipo2*, Flair-T are more effective than Flair-S models. Besides the small number of errors on Flair-T in this sense, we noted a negative effect of agglutinated words with emojis (e.g. *#autismo*♡).

Transformation of entities With the character limit of tweets and the length of certain entities (e.g. *Enfermedad pulmonar intersticial difusa*), we encounter an increased use of acronyms (e.g. *#EPID*), which can be challenging for NER systems. We observed Flair-T models were 49.2% less prone to fail in detecting diseases' acronyms. Other transformations include flexions or verbal derivations (e.g. *resfriado→resfriarse*), where Flair-T was found to be consistently more effective.

Multilinguality Domain- and multilingual-specific embeddings have a comparable performance (refer Table 2). Both performed better than es-models as they benefited from the knowledge from the clinical datasets (Lange et al., 2021) that were used to pre-train them. Irrespective of the adaptive fine-tuning, en-specific models performed lower than es-specific models. Their marginal difference can be attributed to common standard disease names used by users on Twitter.

Gold Annotations Domain embeddings helped to identify unknown diseases and possible incom-

[1]We use blue font color to distinguish between target entities and noise

plete spans on the gold annotations. This raises the question of what sequence we can consider a disease and to what extent we determine its span. While our error analysis showed identification of words not strictly corresponding to diseases (e.g. *esquizofrenia cultural*), new ones were identified (e.g. *Alteraciones cutáneas*), in both Spanish and English (e.g. *#epilepsywarrior*), and the complete span was captured (e.g. *esteatosis hepática grado II-III*) when a gold annotation was incomplete. We found this to be an interesting strategy to explore to improve the quality of gold annotations.

5 Conclusion

In this paper, we studied the existing problems for Bio-Medical NER on Twitter data and further shown the effectiveness of contextualized embeddings. In addition, we observed a potential in these systems to improve the quality of gold annotations. Regardless of the high performance of the presented systems, post-processing remains a key step to achieve high quality extractions of target entities. In future work, we would like to investigate more effective post-processing strategies and a finer tagging schema such as nested annotations for a more accurate detection of complex disease mentions.

References

David Adelani. 2021. bert-base-multilingual-cased-ner-hrl.

Alan Akbik, Tanja Bergmann, Duncan Blythe, Kashif Rasul, Stefan Schweter, and Roland Vollgraf. 2019. FLAIR: An easy-to-use framework for state-of-the-art NLP. In *NAACL 2019, 2019 Annual Conference of the North American Chapter of the Association for Computational Linguistics (Demonstrations)*, pages 54–59.

Alan Akbik, Duncan Blythe, and Roland Vollgraf. 2018. Contextual string embeddings for sequence labeling. In *COLING 2018, 27th International Conference on Computational Linguistics*, pages 1638–1649.

Casimiro Pio Carrino, Jordi Armengol-Estapé, Asier Gutiérrez-Fandiño, Joan Llop-Palao, Marc Pàmies, Aitor Gonzalez-Agirre, and Marta Villegas. 2021. Biomedical and clinical language models for spanish: On the benefits of domain-specific pretraining in a mid-resource scenario.

José Cañete, Gabriel Chaperon, Rodrigo Fuentes, Jou-Hui Ho, Hojin Kang, and Jorge Pérez. 2020. Spanish pre-trained bert model and evaluation data. In *PML4DC at ICLR 2020*.

Mariia Chizhikova, Jaime Collado-Montañéz, Pilar López-Úbeda, Manuel C. Díaz-Galiano, L. Alfonso Ureña-López, and M. Teresa Martín-Valdivia. 2022. Sinai at clef 2022: Leveraging biomedical transformers to detect and normalize disease mentions. pages 265–273.

Alexis Conneau, Kartikay Khandelwal, Naman Goyal, Vishrav Chaudhary, Guillaume Wenzek, Francisco Guzmán, Edouard Grave, Myle Ott, Luke Zettlemoyer, and Veselin Stoyanov. 2019. Unsupervised cross-lingual representation learning at scale. *CoRR*, abs/1911.02116.

Xiang Dai. 2018. Recognizing complex entity mentions: A review and future directions. In *Proceedings of ACL 2018, Student Research Workshop*, pages 37–44, Melbourne, Australia. Association for Computational Linguistics.

Xiang Dai, Sarvnaz Karimi, Ben Hachey, and Cecile Paris. 2020. An effective transition-based model for discontinuous NER. In *Proceedings of the 58th Annual Meeting of the Association for Computational Linguistics*, pages 5860–5870, Online. Association for Computational Linguistics.

Luis Gasco, Darryl Estrada-Zavala, Eulàlia Farré-Maduell, Salvador Lima-López, Antonio Miranda-Escalada, and Martin Krallinger. 2022. Overview of the SocialDisNER shared task on detection of diseases mentions from healthcare related and patient generated social media content: methods, evaluation and corpora. In *Proceedings of the Seventh Social Media Mining for Health (#SMM4H) Workshop and Shared Task*.

Suchin Gururangan, Ana Marasović, Swabha Swayamdipta, Kyle Lo, Iz Beltagy, Doug Downey, and Noah A Smith. 2020. Don't stop pretraining: adapt language models to domains and tasks. *arXiv preprint arXiv:2004.10964*.

Lukas Lange, Heike Adel, Jannik Strötgen, and Dietrich Klakow. 2021. Clin-x: pre-trained language models and a study on cross-task transfer for concept extraction in the clinical domain.

Sinno Jialin Pan and Qiang Yang. 2009. A survey on transfer learning. *IEEE Transactions on knowledge and data engineering*, 22(10):1345–1359.

Jonas Pfeiffer, Ivan Vulić, Iryna Gurevych, and Sebastian Ruder. 2020. Mad-x: An adapter-based framework for multi-task cross-lingual transfer. *arXiv preprint arXiv:2005.00052*.

Sam Rawal. 2021. bert-base-uncased-clinical-ner.

Sebastian Ruder, Matthew E Peters, Swabha Swayamdipta, and Thomas Wolf. 2019. Transfer learning in natural language processing. In *Proceedings of the 2019 conference of the North American chapter of the association for computational linguistics: Tutorials*, pages 15–18.

Yongliang Shen, Xinyin Ma, Zeqi Tan, Shuai Zhang, Wen Wang, and Weiming Lu. 2021. Locate and label: A two-stage identifier for nested named entity recognition. In *Proceedings of the 59th Annual Meeting of the Association for Computational Linguistics and the 11th International Joint Conference on Natural Language Processing (Volume 1: Long Papers)*, pages 2782–2794, Online. Association for Computational Linguistics.

Simone Tedeschi, Valentino Maiorca, Niccolò Campolungo, Francesco Cecconi, and Roberto Navigli. 2021. WikiNEuRal: Combined neural and knowledge-based silver data creation for multilingual NER. In *Findings of the Association for Computational Linguistics: EMNLP 2021*, pages 2521–2533, Punta Cana, Dominican Republic. Association for Computational Linguistics.

UCCNLP@SMM4H'22:Label distribution aware long-tailed learning with post-hoc posterior calibration applied to text classification

Paul Trust
University College Cork
Cork, Ireland

Provia Kadusabe
Worldquant University
Louisiana, USA

Rosane Minghim
University College Cork
Cork, Ireland

Ahmed Zahran
University College Cork
Cork, Ireland

Kizito Omala
Makerere University
Kampala, Uganda

Abstract

This paper describes our submissions for the Social Media Mining for Health (SMM4H) workshop 2022 shared tasks. We have participated in 2 tasks: (1) classification of adverse drug events (ADE) mentions in English tweets (Task-1a) and (2) classification of self-reported intimate partner violence (IPV) on Twitter (Task 7). We propose an approach that uses RoBERTa (Robustly Optimized BERT Pretraining Approach) fine-tuned with a label distribution-aware margin loss function and post-hoc posterior calibration for robust inference against class imbalance. We achieved a 4% and 1% increase in performance on IPV and ADE respectively, when compared with the traditional fine-tuning strategy with unweighted cross-entropy loss.

1 Introduction

Social media platforms are becoming part of our every day life with an estimation of about 4.2 billion people using some sort of social media (Hsieh-Yee, 2021). The 7th Social Media Mining for Health Applications Workshop (SMM4H) 2022 organized tasks involving automatic methods for collection, extraction, representation, analysis and validation of social media data for health informatics. The tasks we have participated in utilized data from Twitter, which is one of the largest social media platforms with approximately 192 million daily active users (Conger, 2021).

Our team UCCNLP participated in 2 tasks: (1) classification of adverse drugs events (ADE) mentions in english tweets (Task-1a) and (2) classification of self-reported intimate partner violence (IPV) on twitter (Task 7). Adverse drug events (ADEs) are negative effects related to the use of drugs. ADEs mentions on social media are a useful indicator of the efficacy of medications and also can potentially reveal previously unknown side effects of drugs (Ramesh et al., 2021).

Intimate partner violence (IPV) on the other hand refers to the abuse or aggression that occurs in a romantic relationship. IPV is a serious health problem and can have lifelong impact on health and well-being of individuals. Victims of IPV sometimes share their stories on twitter making it an important tool in early identification for timely intervention and support (Ramesh et al., 2021).

Exploration of the provided training datasets of both tasks revealed that the datasets were heavily imbalanced with long tailed distributions for example the IPV data contains only 11% of the provided tweets identified as self-reported IPV.

The skewed nature of these datasets complicates efficient training even for state of the art models like BERT (Bidirectional Embeddings from Transformers) (Devlin et al., 2019) and RoBERTa (A Robustly Optimized BERT Pretraining Approach) (Liu et al., 2019). The difficulty in model training arises from the fact that naive empirical risk minimization for imbalanced datasets tends to be biased towards learning the distribution on these head classes which deteriorates its performance on tail classes.

In this work, we proposed to solve the class imbalance in classification on both tasks using RoBERTa with label distribution-aware loss function (LDAM) combined and posterior calibration to alleviate the problem of class imbalance during training.

2 Related work

2.1 Twitter data for Intimate partner violence and Adverse drug events Classification

The task of ADE detection from English tweets featured as a shared task at SMM4H in 2021 and the top systems on the task used RoBERTa combined with under sampling and oversampling (Ramesh et al., 2021), a combination of BERT and drug embeddings obtained by chemical struc-

Proceedings of the 29th International Conference on Computational Linguistic, pages 90–94
October 12–17, 2022.

ture BERT-based encoder (Sakhovskiy et al., 2021). Other methods used BERT with naive class weights (Yaseen and Langer, 2021) and BERT combined with data augmentation (Ji et al., 2021).

The task of classification of self-reported partner violence on twitter had not appeared before as a shared task, but previous work on IPV classification fine-tuned RoBERTa (Al-Garadi et al., 2021) using a standard training procedure without consideration of class imbalance.

2.2 Long-tailed learning and class imbalance

Most of the existing algorithms for handling class imbalance can be divided in the following categories: re-sampling, re-weighting and confidence calibration and regularization.

2.2.1 Re-sampling

There are two groups of re-sampling strategies: over-sampling the the minority classes (Buda et al., 2018; Byrd and Lipton, 2019; Shen et al., 2016) and under-sampling the frequent classes (He and Garcia, 2009; Haixiang et al., 2017). The limitation of under-sampling is that it discards a large portion of the data making it infeasible when data imbalance is extreme. Over-sampling may also lead to over-fitting the minority classes.

2.2.2 Re-weighting

Cost-sensitive re-weighting assigns weights for different classes or for different samples. The traditional re-weighting scheme assigns weights to classes proportional to the inverse of their frequency (class balanced loss (CB)) (Huang et al., 2016, 2019). However, under extreme data imbalance and large-scale scenarios, re-weighting makes deep learning models difficult to optimize during training (Zhong et al., 2021). In this work, we explore efficient loss functions adapted for class imbalance mostly used in computer vision for our text classification tasks. Focal loss (FS), which a weighted version of cross-entropy loss with sample-specific weight $(1-p_i)^\gamma$, where p_i is the model output probability for class i (Lin et al., 2017). Label-aware distribution loss (LDAM) derives a generalization error bound for imbalanced training and proposes a margin-aware weighted cross-entropy loss (Cao et al., 2019) and Maximum Margin LDAM loss (MM-LDAM) minimizes a margin-based generalization bound by shifting decision bound (Kang et al., 2021).

2.2.3 Thresholding and calibration

These are methods that are applied at test time to calibrate confidence scores generated by the machine learning model to be robust to class imbalance. Representative methods in this category are post-hoc logit adjustment (Menon et al., 2020) and posterior calibration (PC) (Tian et al., 2020).

3 Methodology

3.1 Training phase

We fine-tuned transformer language models (RoBERTa and BERTweet) with a label distribution-aware margin loss function (LDAM). More formally, consider our classifier $f = g \odot (RoBERTa \vee BERTweet)$, which consists of two parts: a pre-trained language model that outputs hidden representations of input samples and g, which is the classification head on our imbalanced dataset, that outputs k-dimensional vector, where each dimension corresponds to the prediction confidence of a specific class.

Instead of using a standard cross entropy loss, we fine-tuned our model f with (LDAM) (Cao et al., 2019):

$$\mathcal{L}_{LDAM}((x,y);f) = -\log \frac{e^{zy-\Delta y}}{e^{z_y-\Delta_y} + \sum_{j\neq y} e^{z_j}} \tag{1}$$

where $\Delta_j = \frac{C}{n_j^{1/4}}$ for $j \in \{1,..,k\}$, Δ_y is chosen to be a label independent constant C, n_j is the number of instance per class j.

We adopted a deferred re-weighting procedure as in (Cao et al., 2019), since naive re-weighting and re-sampling were found to be inferior to the vanilla empirical risk minimization (ERM).

3.2 Prediction phase

In the prediction phase, our aim is to use the learned parameters in the training phase $\{w, \theta\}$ to make predictive inference on the test data with unknown labels. This is done by taking the most likely label on the test data defined by $h^*(x) = \text{argmax}_{y\in K} P_{test}(y|x)$ under the model's posterior distribution. By using parameters learned on the training data to make inference on the test data, we are assuming that test and train data are drawn from the same distribution.

This was not the case, since our training data was imbalanced and we had no information about the distribution on the test data. We therefore calibrated the posterior distribution on the test data to be robust to class imbalance through posterior re-balancing (PC) proposed in (Tian et al., 2020).

More formally, the optimal Bayes classifier $h_{test}(x)$ on the test set is an approximation based on the training posterior distribution defined as:

$$h_{test}(x) = \underset{y \in K}{\operatorname{argmax}} \frac{p_{train}(y|x) p_{test}(y)}{p_{train}(y)} \quad (2)$$

We can view $h_{test}(x)$ as a posterior distribution from the training dataset weighted by $p_{test}(y)/p_{train}(y)$ which we denoted as a re-balanced posterior ($p_{rbal}(y|x)$). This weighting $p_{test}(y)/p_{train}(y)$ introduced at testing phase can be too big or too small depending on the label distribution on test and train datasets.

To compensate for imperfect learning due to diference in train and test distribution, we solve an approximation through Kullback-Leibler (KL) divergence by treating the re-balanced posterior and original posterior as an approximation to the true posterior proposed by (Tian et al., 2020) as follows:

$$\begin{aligned} p^*(y|x) &= \operatorname{argmin}_p (1-\lambda) KL(p, p_{train}(y|x) \\ &\quad + \lambda KL(p, p_{rbal}(y|x)) \end{aligned}$$

A closed form solution to the optimization exists and is defined as follows:

$$p^*(y|x) = \frac{1}{Z(x)} (p_{train}(y|x))^{(1-\lambda)} (p_{rbal}(y|x))^{\lambda} \quad (3)$$

where $Z(x)$ is the normalization constant

We can thus balance the posterior distribution on the test data against class imbalance through a hyper parameter λ without any need to retrain the model.

4 Experiments and Results

The datasets for both tasks: (IPV and ADE) were released by SMM4H 2022 shared tasks. IPV self report classification (Task 7) consisted of 4523 training tweets, 534 validation tweets and 1291 test tweets (Al-Garadi et al., 2022). ADE detection task contained 17173 train tweets , 908 validation tweets and 10968 test tweets.

We conducted experiments with pre-trained transformer language models; RoBERTa (Liu et al., 2019) and BERTweet (Nguyen et al., 2020) with different loss functions and also with post-hoc posterior calibration (PC). Experiments were done for 10 epochs, max length of 128, batch size of 10 and the learning rate was set at 0.0005. The final submission were evaluated using a median $f1$-score over 5 runs. The system was written in PyTorch using hugging-face transformer library (Wolf et al., 2019).

Table 1 demonstrates that cost sensitive learning under class imbalance is superior to naive training with cross entropy loss with both RoBERTa and BERTweet models. Another observation is that doing posterior calibration on the test performance also improves performance. All these observations and conclusions are consistent with other works in computer vision (Tian et al., 2020; Cao et al., 2019; Lin et al., 2017) Maximum Margin LDAM (MM-LDAM-PC) with posterior calibration achieved the best performance in both datasets and with both models. RoBERTa achieves the best perfomance on the IPV dataset (69.37% versus 69.15%) while BERTweet achieves the best performance on the ADE dataset (47.81% versus 47.10%)

Model	IPV	ADE
BERTweet-CEL	62.66	46.90
BERTweet-FL	60.48	46.20
BERTweet-LDAM	67.06	47.62
BERTweet-MM-LDAM	68.74	47.77
BERTweet-MM-LDAM-PC	**69.15**	**47.81**
RoBERTa-CEL	65.34	46.29
RoBERTa-FL	66.05	46.02
RoBERTa-LDAM	67.25	47.03
RoBERTa-MM-LDAM	68.51	47.09
RoBERTa-LDAM-PC (Ours)	**69.37**	**47.10**

Table 1: The table shows results of the median f-scores on the validation sets for 5 runs on Intimate Partner violence (IPV) dataset and adverse drug effects (ADE) dataset. CEL represents cross entropy loss, FL represents focal length, LDAM represents label distribution-aware loss function, MM represents maximum margin and PC represents posterior calibration

5 Conclusion

In this work, we developed two systems based on pre-trained transformer based language models fine-tuned with label distribution aware loss functions and posterior calibration on the test set. We experimented with various loss functions as well as different transformer models. The results on the validation set revealed that incorporating class prior probabilities at both training and testing helps to boost performance under class imbalance. Further more our system achieved 62.25% $f1$-score on IPV (Task 7) and 8% for ADE task (Task $1a$) on our codalab submission.

6 Acknowledgements

We thank Science Foundation Ireland (SFI) Center for Research Training in Advanced Networks and Future communications at University College Cork for funding this research and Irish Centre for High-End Computing (ICHEC) for providing access to computing power for running some of our experiments.

References

Mohammed Ali Al-Garadi, Sangmi Kim, Yuting Guo, Elise Warren, Yuan-Chi Yang, Sahithi Lakamana, and Abeed Sarker. 2021. Natural language model for automatic identification of intimate partner violence reports from twitter. *medRxiv*.

Mohammed Ali Al-Garadi, Sangmi Kim, Yuting Guo, Elise Warren, Yuan-Chi Yang, Sahithi Lakamana, and Abeed Sarker. 2022. Natural language model for automatic identification of intimate partner violence reports from twitter. *Array*, page 100217.

Mateusz Buda, Atsuto Maki, and Maciej A Mazurowski. 2018. A systematic study of the class imbalance problem in convolutional neural networks. *Neural networks*, 106:249–259.

Jonathon Byrd and Zachary Lipton. 2019. What is the effect of importance weighting in deep learning? In *International Conference on Machine Learning*, pages 872–881. PMLR.

Kaidi Cao, Colin Wei, Adrien Gaidon, Nikos Arechiga, and Tengyu Ma. 2019. Learning imbalanced datasets with label-distribution-aware margin loss. *Advances in neural information processing systems*, 32.

Kate Conger. 2021. Twitter shakes off the cobwebs with new product plans. *The New York Times*.

Jacob Devlin, Ming-Wei Chang, Kenton Lee, and Kristina Toutanova. 2019. BERT: Pre-training of deep bidirectional transformers for language understanding. In *Proceedings of the 2019 Conference of the North American Chapter of the Association for Computational Linguistics: Human Language Technologies, Volume 1 (Long and Short Papers)*, pages 4171–4186, Minneapolis, Minnesota. Association for Computational Linguistics.

Guo Haixiang, Li Yijing, Jennifer Shang, Gu Mingyun, Huang Yuanyue, and Gong Bing. 2017. Learning from class-imbalanced data: Review of methods and applications. *Expert systems with applications*, 73:220–239.

Haibo He and Edwardo A Garcia. 2009. Learning from imbalanced data. *IEEE Transactions on knowledge and data engineering*, 21(9):1263–1284.

Ingrid Hsieh-Yee. 2021. Can we trust social media? *Internet Reference Services Quarterly*, 25(1-2):9–23.

Chen Huang, Yining Li, Chen Change Loy, and Xiaoou Tang. 2016. Learning deep representation for imbalanced classification. In *Proceedings of the IEEE conference on computer vision and pattern recognition*, pages 5375–5384.

Chen Huang, Yining Li, Chen Change Loy, and Xiaoou Tang. 2019. Deep imbalanced learning for face recognition and attribute prediction. *IEEE transactions on pattern analysis and machine intelligence*, 42(11):2781–2794.

Zongcheng Ji, Tian Xia, and Mei Han. 2021. PAII-NLP at SMM4H 2021: Joint extraction and normalization of adverse drug effect mentions in tweets. In *Proceedings of the Sixth Social Media Mining for Health (#SMM4H) Workshop and Shared Task*, pages 126–127, Mexico City, Mexico. Association for Computational Linguistics.

Haeyong Kang, Thang Vu, and Chang D Yoo. 2021. Learning imbalanced datasets with maximum margin loss. In *2021 IEEE International Conference on Image Processing (ICIP)*, pages 1269–1273. IEEE.

Tsung-Yi Lin, Priya Goyal, Ross Girshick, Kaiming He, and Piotr Dollár. 2017. Focal loss for dense object detection. In *Proceedings of the IEEE international conference on computer vision*, pages 2980–2988.

Yinhan Liu, Myle Ott, Naman Goyal, Jingfei Du, Mandar Joshi, Danqi Chen, Omer Levy, Mike Lewis, Luke Zettlemoyer, and Veselin Stoyanov. 2019. Roberta: A robustly optimized bert pretraining approach. *arXiv preprint arXiv:1907.11692*.

Aditya Krishna Menon, Sadeep Jayasumana, Ankit Singh Rawat, Himanshu Jain, Andreas Veit, and Sanjiv Kumar. 2020. Long-tail learning via logit adjustment. *arXiv preprint arXiv:2007.07314*.

Dat Quoc Nguyen, Thanh Vu, and Anh Tuan Nguyen. 2020. BERTweet: A pre-trained language model for English tweets. In *Proceedings of the 2020 Conference on Empirical Methods in Natural Language Processing: System Demonstrations*, pages 9–14, Online. Association for Computational Linguistics.

Sidharth Ramesh, Abhiraj Tiwari, Parthivi Choubey, Saisha Kashyap, Sahil Khose, Kumud Lakara, Nishesh Singh, and Ujjwal Verma. 2021. BERT based transformers lead the way in extraction of health information from social media. In *Proceedings of the Sixth Social Media Mining for Health (#SMM4H) Workshop and Shared Task*, pages 33–38, Mexico City, Mexico. Association for Computational Linguistics.

Andrey Sakhovskiy, Zulfat Miftahutdinov, and Elena Tutubalina. 2021. KFU NLP team at SMM4H 2021 tasks: Cross-lingual and cross-modal BERT-based models for adverse drug effects. In *Proceedings of the Sixth Social Media Mining for Health (#SMM4H) Workshop and Shared Task*, pages 39–43, Mexico City, Mexico. Association for Computational Linguistics.

Li Shen, Zhouchen Lin, and Qingming Huang. 2016. Relay backpropagation for effective learning of deep convolutional neural networks. In *European conference on computer vision*, pages 467–482. Springer.

Junjiao Tian, Yen-Cheng Liu, Nathaniel Glaser, Yen-Chang Hsu, and Zsolt Kira. 2020. Posterior recalibration for imbalanced datasets. *Advances in Neural Information Processing Systems*, 33:8101–8113.

Thomas Wolf, Lysandre Debut, Victor Sanh, Julien Chaumond, Clement Delangue, Anthony Moi, Pierric Cistac, Tim Rault, Rémi Louf, Morgan Funtowicz, et al. 2019. Huggingface's transformers: State-of-the-art natural language processing. *arXiv preprint arXiv:1910.03771*.

Usama Yaseen and Stefan Langer. 2021. Neural text classification and stacked heterogeneous embeddings for named entity recognition in SMM4H 2021. In *Proceedings of the Sixth Social Media Mining for Health (#SMM4H) Workshop and Shared Task*, pages 83–87, Mexico City, Mexico. Association for Computational Linguistics.

Zhisheng Zhong, Jiequan Cui, Shu Liu, and Jiaya Jia. 2021. Improving calibration for long-tailed recognition. In *Proceedings of the IEEE/CVF conference on computer vision and pattern recognition*, pages 16489–16498.

HaleLab_NITK@SMM4H'22: Adaptive Learning Model for Effective Detection, Extraction and Normalization of Adverse Drug Events from Social Media Data

Reshma Unnikrishnan, Sowmya Kamath S, Ananthanarayana V S
Department of Information Technology
National Institute of Technology Karnataka, Surathkal, Mangalore 575025, India
{reshmau.197it008, sowmyakamath, anvs} @nitk.edu.in

Abstract

This paper describes the techniques designed for detecting, extracting and normalizing adverse events from social data as part of the submission for the Shared task, Task 1-SMM4H'22. We present an adaptive learner mechanism for the foundation model to identify Adverse Drug Event (ADE) tweets. For the detected ADE tweets, a pipeline consisting of a pre-trained question-answering model followed by a fuzzy matching algorithm was leveraged for the span extraction and normalization tasks. The proposed method performed well at detecting ADE tweets, scoring an above-average F1 of 0.567 and 0.172 overlapping F1 for ADE normalization. The model's performance for the ADE extraction task was lower, with an overlapping F1 of 0.435.

1 Introduction

Social media data is extensively used for a wide variety of informed decision-making applications in varied domains like e-business, healthcare, policy making etc (Aichner et al., 2021; Unnikrishnan et al., 2021; Saini et al., 2022). Analyzing drug post-marketing management policies is a major aspect of pharmacovigilance (Nikfarjam et al., 2015) and has received significant research attention in recent years. As part of the Shared task, participants were provided with Twitter data (Magge et al., 2021b; Weissenbacher et al., 2022) that includes ADE mentions for detection, extraction and normalization purposes to aid pharmacovigilance studies. While detecting adverse reactions from social data having informal language is a complex task, the availability of valid ADE samples is scarce, making the data highly imbalanced (Klein et al., 2020; Magge et al., 2021a; Weissenbacher et al., 2022). Each of the tasks are correlated, thus, it is necessary to ensure that authentic ADE tweet samples are detected before utilizing them for ADE extraction and normalization processes. With these objectives, our work encompasses the design of an adaptive learning mechanism for the foundation model for effective ADE identification, extraction and normalization.

2 Methodology

Extracting ADEs from informal social data is challenging due to the short text length and style-variant nature (Weissenbacher et al., 2022). As stated earlier, we focused on developing an efficient adaptive learning method for foundation models due to the association between the three subtasks. We adopted RoBERTa (Liu et al., 2019) as the foundation model for applying adaptive learning techniques. The empirical and theoretical aspects of the data-specific features learnt by the higher-order layers were explored, for understanding the contribution of different layers in a foundation model and relearning task-specific requirements.

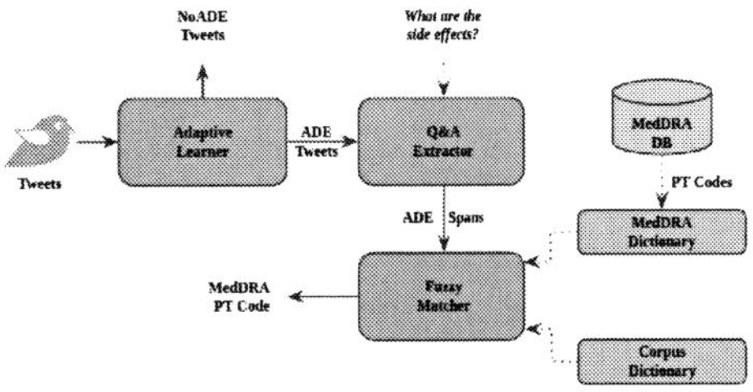

Figure 1: Proposed Workflow

During knowledge distillation, it is essential to ensure that the pre-trained weights are not affected by learning divergence, due to absence of previously learned weights. With this in mind, we employed mixout (Lee et al., 2019) with the weight reinitialization technique (Li et al., 2020), which helps adaptively retrain the last few layers. While mixout helps retain the weights from the base foundation model without losing the pre-trained features, weight reinitialization helps relearn these specific weights for the higher order layers to pre-

Proceedings of the 29th International Conference on Computational Linguistic, pages 95–97
October 12–17, 2022.

Table 1: Results on Test set

Task		Precision	Recall	F1
Mean	1a	0.646	0.497	0.562
Ours	1a	**0.674**	0.489	**0.567**

Approach	Task	Overlapping			Strict		
		Precision	*Recall*	*F1*	*Precision*	*Recall*	*F1*
Mean	1b	0.539	0.517	0.527	0.344	0.339	0.341
	1c	0.120	0.112	0.116	0.085	0.082	0.083
Ours	1b	**0.562**	0.354	0.435	0.194	0.114	0.144
	1c	**0.232**	**0.137**	**0.172**	**0.088**	0.052	0.065

dict ADE tweets.

The span extraction task was approached as a Question Answering (QA) task, rather than as a conventional named entity recognition (NER) task (Dima et al., 2021; Balumuri et al., 2021). To extract the side effects or adverse medical events from the Adaptive Learner output, Roberta-base (Liu et al., 2019) fine-tuned on the SQuAD2.0 dataset (Rajpurkar et al., 2018) was trained for a batch size of 96 for two epochs, a learning rate of 3e-5, handling a maximum sequence length of 386 and query length of 96. Since all the spans here represent ADEs on individuals, we considered a fixed query ("*What are the side effects?*") to extract spans from ADE tweets.

Due to fewer number of ADE tweets with varied MedDRA codes[1](Mozzicato, 2009) (no duplicates), formulating it as a supervised classification problem is practically unfeasible. In view of this, the conventional fuzzy matching algorithm built on Levenshtein distance (Yujian and Bo, 2007) based similarity measure was used for mapping the extracted spans to MedDRA codes. We created two dictionaries: MedDRA terms and their codes (Preferred Terms - PT) from the MedDRA database and the dictionary from the gold standard train data with spans and MedDRA codes. The spans extracted from the QA model were matched against the two dictionary terms, and the MedDRA code for one that weighted more was used as the label for the ADE spans.

3 Experiments and Observation

The SMM4H'22 task organizers provided data (Magge et al., 2021b; Weissenbacher et al., 2022) for this experimentation. We observed that the data was highly imbalanced along with a unique set of ADE events, thus increasing the learning complexity of a system. For ADE detection, experimentation on weight retraining varied between 1 to 4 layers and on mixout from 0.5 to 0.7 were performed. Empirically, we chose to reinitialize the top 4 layers by fixing the mixout to 0.7. While approaching the entity recognition task as QA, we found that it returned context chunks comprising adverse events rather than short entities. Since the MedDRA database holds proper clinical expressions, we believe that including the gold standard data covering the colloquial mentions of adverse effects as another dictionary helps improve the span normalization.

Table 2 presents the observed results for all three subtasks on the validation set, which achieved promising performance in terms of precision, recall and F1 score for tasks 1a and 1b. Task 1c falls into the 0.2 range for all, which is still better than all system submissions' baseline mean average overlapping score on test data. While observing the test data scores in Table 1, task 1a attains above average F1 score of 0.567 and an improved overlapping precision, recall and F1 score for task 1c. In contrast, task 1b exhibits degraded performance across all measures in the test set compared to the validation data performance. As stated earlier, the measurement fails to give a better score in test data due to the return of context chunks rather than word entities from the QA model. For ensuring reproducibility of the results, the source code of the proposed approach is made available publicly[2].

4 Conclusion

For addressing the SMM4H'22 Shared task 1, we concentrated on developing an adaptive learning technique for the foundation model using Roberta

[1]https://www.meddra.org/

[2]https://github.com/Reshma-U/SMM4H-22

Table 2: Results on Validation set

Shared Task	Precision	Recall	F1
Task 1a	0.71	0.72	0.72
Task 1b	1.00	0.63	0.77
Task 1c	0.24	0.22	0.21

base for ADE detection. While achieving a promising F1 measure of 0.72 on the validation set, the same model on the test set produced an above-average F1 of 0.567. We approached ADE span extraction as a QA task that showed degraded results due to the return of contextual chunks rather than word entities. Fuzzy matching for span normalization gave above-average overlapping P, R and F1 scores. In future, we plan to fine-tune the QA model to return adverse word level entities for improved performance in the span prediction task.

References

Thomas Aichner, Matthias Grünfelder, Oswin Maurer, and Deni Jegeni. 2021. Twenty-five years of social media: a review of social media applications and definitions from 1994 to 2019. *Cyberpsychology, behavior, and social networking*, 24(4):215–222.

Spandana Balumuri, Sony Bachina, and Sowmya Kamath. 2021. Sb_nitk at mediqa 2021: Leveraging transfer learning for question summarization in medical domain. In *Proceedings of the 20th workshop on biomedical language processing*, pages 273–279.

George-Andrei Dima, Dumitru-Clementin Cercel, and Mihai Dascalu. 2021. Transformer-based multi-task learning for adverse effect mention analysis in tweets. In *Sixth Social Media Mining for Health (# SMM4H) Workshop and Shared Task*, pages 44–51.

Ari Klein, Ilseyar Alimova, Ivan Flores, Arjun Magge, Zulfat Miftahutdinov, Anne-Lyse Minard, Karen O'connor, Abeed Sarker, Elena Tutubalina, Davy Weissenbacher, et al. 2020. Overview of the fifth social media mining for health applications (# smm4h) shared tasks at coling 2020. In *Fifth Social Media Mining for Health Applications Workshop & Shared Task*, pages 27–36.

Cheolhyoung Lee, Kyunghyun Cho, and Wanmo Kang. 2019. Mixout: Effective regularization to finetune large-scale pretrained language models. *arXiv preprint arXiv:1909.11299*.

Xingjian Li, Haoyi Xiong, Haozhe An, Cheng-Zhong Xu, and Dejing Dou. 2020. Rifle: Backpropagation in depth for deep transfer learning through reinitializing the fully-connected layer. In *International Conference on Machine Learning*, pages 6010–6019. PMLR.

Yinhan Liu, Myle Ott, Naman Goyal, Jingfei Du, Mandar Joshi, Danqi Chen, Omer Levy, Mike Lewis, Luke Zettlemoyer, and Veselin Stoyanov. 2019. Roberta: A robustly optimized bert pretraining approach. *arXiv preprint arXiv:1907.11692*.

Arjun Magge, Ari Klein, Antonio Miranda-Escalada, Eulalia Al-Garadi, Farre, Salvador Lima-López, et al. 2021a. Overview of the sixth social media mining for health applications (# smm4h) shared tasks at naacl 2021. In *Sixth Social Media Mining for Health (# SMM4H) Workshop and Shared Task*, pages 21–32.

Arjun Magge, Elena Tutubalina, Zulfat Miftahutdinov, Ilseyar Alimova, Anne Dirkson, Suzan Verberne, Davy Weissenbacher, and Graciela Gonzalez-Hernandez. 2021b. Deepademiner: a deep learning pharmacovigilance pipeline for extraction and normalization of adverse drug event mentions on twitter. *Journal of the American Medical Informatics Association*, 28(10):2184–2192.

Patricia Mozzicato. 2009. Meddra. *Pharmaceutical Medicine*, 23(2):65–75.

Azadeh Nikfarjam, Abeed Sarker, Karen O'connor, Rachel Ginn, and Graciela Gonzalez. 2015. Pharmacovigilance from social media: mining adverse drug reaction mentions using sequence labeling with word embedding cluster features. *Journal of the American Medical Informatics Association*, 22(3):671–681.

Pranav Rajpurkar, Robin Jia, and Percy Liang. 2018. Know what you don't know: Unanswerable questions for squad. *arXiv preprint arXiv:1806.03822*.

Gurdeep Saini, Naveen Yadav, and Sowmya Kamath S. 2022. Ensemble neural models for depressive tendency prediction based on social media activity of twitter users. In *Security, Privacy and Data Analytics*, pages 211–226. Springer.

Reshma Unnikrishnan, S Sowmya Kamath, and VS Ananthanarayana. 2021. Benchmarking shallow and deep neural networks for contextual representation of social data. In *2021 IEEE 18th India Council International Conference (INDICON)*, pages 1–8. IEEE.

Davy Weissenbacher, Ari Z. Klein, Luis Gascó, Darryl Estrada-Zavala, Raul Rodriguez-Esteban, Mathias Leddin, Arjun Magge, Juan M. Banda, Vera Davydova, Elena Tutubalina, and Graciela Gonzalez-Hernandez. 2022. Overview of the seventh social media mining for health applications #smm4h shared tasks at coling 2022. In *Seventh Social Media Mining for Health (#SMM4H) Workshop and Shared Task*.

Li Yujian and Liu Bo. 2007. A normalized levenshtein distance metric. *IEEE transactions on pattern analysis and machine intelligence*, 29(6):1091–1095.

Yet@SMM4H'22: Improved BERT-based classification models with Rdrop and PolyLoss

Yan Zhuang
University Of Electronic Science And Technology Of China
delecisz@gmail.com

Yanru Zhang*
University Of Electronic Science And Technology Of China
Shenzhen Institute for Advanced Study, UESTC
yanruzhang@uestc.edu.cn

Abstract

This paper describes our approach for 11 classification tasks (Task1a, Task2a, Task2b, Task3a, Task3b, Task4, Task5, Task6, Task7, Task8 and Task9) from Social Media Mining for Health (SMM4H) 2022 Shared Tasks. We developed a classification model that incorporated rdrop to augment data and avoid overfitting, poly loss and focal loss to alleviate sample imbalance, and pseudo labels to improve model performance. The results of our submissions are over or equal to the median scores in almost all tasks. In addition, our model achieved the highest score in Task4, with a higher 7.8% and 5.3% F1-score than the median scores in Task2b and Task3a respectively.

1 Introduction

Nowadays, people are more and more willing to share their life status on social networks, and analyze and discuss their illness and medication. These texts can be useful for analyzing an individual's health, but they are rare among the vast number of social tweets and particularly difficult to collect. The Social Media Mining for Health Applications (SMM4H) 2022 workshop aims to bring together researchers interested in automatic methods for the collection, extraction, representation, analysis, and validation of social media data (e.g., Twitter, Facebook) for health informatics (Weissenbacher et al., 2022). There are total 14 tasks in SMM4H 2022 shared tasks, and we participated in all classification tasks, including Task1a: Classification of Adverse Drug Events (ADEs) in English tweets (Magge et al., 2021); Task2a & Task2b: Classification of stance and premise in tweets about health mandates (COVID-19) (Davydova and Tutubalina, 2022); Task3a & Task3b: Classification of changes in medication treatments in tweets and WebMD reviews; Task4: Classification of Tweets Self-Reporting Exact Age (in English)(Klein et al., 2022); Task5: Classification of tweets of self-reported COVID-19 symptoms in Spanish; Task6: Classification of tweets indicating self-reported COVID-19 vaccination status; Task7: Classification of self-reported intimate partner violence on twitter(Al-Garadi et al., 2022); Task8: Classification of self-reported chronic stress on Twitter(Yuan-Chi et al., 2022); Task9: Classification of social media forum posts self-reporting exact age.

To solve the above classification challenges, we developed an improved BERT-based classification model. It incorporates rdrop for data augmentation, poly loss and focal loss for balancing label distributions, and pseudo label for boosting the model performance. Our model scored above the median score in all tasks, and finally, our model achieved the highest score in Task4, with a higher 7.8% and 5.3% F1-score than the median scores in Task2b and Task3a respectively.

2 Data Analysis

Of all the 11 tasks we participated, only Task2a and Task5 are three-way classification tasks, the others are all binary-classification tasks, and all tasks use F1-score of certain categories as the evaluation metric. Each of the eleven classification tasks has a different classification objective and the data takes different forms. For example, Task2a is a sentence pair matching task, while Task3b contains several meta-information items such as 'Effectiveness', 'Satisfaction' and so on. However, all of these tasks can be transformed into either a single text classification or a sentence pair matching task, both of which can be trained and tested with the same model structure. Therefore, we focus more on designing generic models to solve as many problems as possible.

After analysing the training and test data, we found that the distribution of labels for each task was consistent between the training and valid sets, but the ratio between the categories within the task

*Corresponding author

Proceedings of the 29th International Conference on Computational Linguistic, pages 98–102
October 12–17, 2022.

Task	Task2a	Task2b	Task3b	Task4	Task8	Task9
Training	38:37:25	63:37	55:45	68:32	63:37	69:31
Validation	41:33:26	63:37	57:43	68:32	63:37	69:31
Task	**Task1a**	**Task3a**	**Task5**	**Task6**	**Task7**	
Training	93:7	91:9	60:24:16	89:11	89:11	
Validation	90:10	91:9	60:24:16	89:11	90:10	

Table 1: Label distribution in each task. Among all tasks, only task2a and task5 are three-way classification tasks, the others are all binary classification tasks. The ratios in the table are sorted in descending order by the number of samples in each category in each task.

Task	Task2a	Task2b	Task3b	Task8	Task9
BioBERT	98.73	**98.62**	86.43	80.48	94.20
pubmedBERT	97.41	98.57	85.12	81.19	93.8
DeBERTa	97.55	98.09	**88.67**	84.05	94.5
BERTweet	98.85	98.57	87.51	84.52	94.5
Best+RD+PLoss	**98.88**	98.45	87.58	**85.03**	**94.80**

Table 2: Model performance in Task2a, Task2b, Task3b, Task8 and Task9 valid sets. To simplify the model, we here have uniformly used micro F1 as the evaluation metric rather than the metrics set for each task. So there may be deviations in the values. RD denotes the models using rdrop. PLoss represents the model with Taylor expansion of focal loss as the new loss function. Best means the model with the best performance in each task. The best performance in each task is highlighted in boldface, and the second highest F1 score is highlighted by underline.

was not consistent, as shown in the Table 1. The distribution of labels in these tasks, especially in Task1a, Task3a, Task6 and Task7, is quite unbalanced, reaching 93:7 in the worst cases, which requires a method to alleviate such situation.

3 Method

The whole procedure of our system model follows the BERT-style training and inference frame. The difference between the different models lies in the process of data pre-processing, encoders and tricks.

3.1 Data Pre-Processing

Since the data format of each task is different, we convert them all into the single text and text pairs for classification. Firstly, we normalized the data by changing the split words start with @ into @USER and removing the urls, @USER strings, non-English, non-numeric, and non-punctuation characters. Then we modified the task data for inputting into the classification model. For Task2a and Task2b, we selected the 'Tweet' and 'Claim' columns as the text pairs, and took 'Stance' and 'Premise' as the label respectively. While in Task3b, we transformed the number in "Ease of Use, Effectiveness, Satisfaction" columns into text, taking "Ease of Use" as an example, from 1 to 5 were transformed into "not easy, hardly easy, easy, quite easy, and extremely easy". Then the transformed data was appended together with the columns "DRUG" before the report text, and the whole data were then put into the model for training and inference. In other tasks, we just took the text and label columns for training and testing.

3.2 Encoders

Different tasks may involve different languages. For example, Task5 aims to classify the self-reported COVID-19 symptoms in Spanish tweets while the others using English tweets. So we chose various encoders, including the ones pretrained on general corpus and domain-specific corpus, to accommodate this situation.

BERTweet(Nguyen et al., 2020) is the fisrt large-scale pretrained model for English tweets. It was trained on 850M English tweets using RoBERTa(Liu et al., 2019) pre-training procedure, and showed good performance on tweet-related NLP tasks.

BioBERT(Lee et al., 2020) is short for "Bidirectional Encoder Representations from Transformers for Biomedical Text Mining". It is a domain specific large-scale pret-rained model, and tries to handle word distribution shift from general domain corpora to biomedical corpora.

pubmedBERT (Gu et al., 2021) is pre-trained

Task	Task1a	Task3a	Task5	Task6	Task7	Task4
BioBERT	93.00	92.88	86.40	95.59	94.94	99.52
pubmedBERT	93.00	93.13	84.89	95.11	94.01	99.01
DeBERTa	92.90	94.02	86.35	95.55	94.76	<u>99.73</u>
BERTIN	-	-	84.93	-	-	-
BERTweet	94.97	93.51	-	94.78	94.76	<u>99.73</u>
Best_performing+RD	96.00	93.51	86.35	<u>95.77</u>	95.13	99.57
Best_performing+PLoss	<u>96.28</u>	<u>94.21</u>	<u>86.50</u>	**95.78**	<u>95.32</u>	99.57
Best_performing+RD+PLoss	**96.61**	**94.51**	**86.54**	<u>95.77</u>	**95.73**	99.58
Best_performing+RD+PLoss+PL	-	-	-	-	-	**99.90**

Table 3: Model performance in Task1a, Task3a, Task5, Task6, Task7 and Task4 validation sets. To simplify the model, we also here uniformly used micro F1 as the evaluation metric rather than the metrics set for each task. RD and PL denote the models using rdrop and pseudo label respectively. PLoss represents the model with Taylor expansion of focal loss as the new loss function. Best_performing means the model used the encoder with the best performance in each task. '-' means we did not implement the model on that task. The best performance in each task is highlighted in boldface, and the second highest F1 score is highlighted by underline.

on the PubMed abstracts, including 14 million abstracts and 3.2 billion words.

DeBERTa(He et al., 2020) is short for Decoding-enhanced BERT with disentangled attention. It improves the BERT(Kenton and Toutanova, 2019) and RoBERTa by introducing disentangled attention mechanism and an enhanced mask decoder. Given the same pretraining corpus, DeBERTa shows better performance in many downstream tasks.

BERTIN (De la Rosa et al., 2022) is pre-trained on Spanish corpus following the roberta-style pretraining procedure. It uses a data-centric technique, which is called 'perplexity sampling', to train the model with roughly half the amount of the steps and one fifth of the data.

3.3 Tricks

In order to alleviate the unbalanced label distribution and overfitting, we took the following tricks to increase the model performance.

Rdrop(Wu et al., 2021) is a contrastive learning technique. It generates positive samples through putting the input in two sequential dropout layers, and computes the Kullback-Leibler (KL) divergence between the two outputs. It can be used for data augmentation and alleviating overfitting problems.

FocalLoss(Lin et al., 2017) is a new loss function which focuses more on the hard-to-classify samples during training by reducing the weights of the easy-to-classify samples. Thus, it can be used to alleviate the class imbalance through reshaping the standard cross entropy.

PolyLoss(Leng et al., 2022) approximates the loss functions via Taylor expansion, and designs the loss functions as a linear combination of polynomial functions. Thus it can be combined with focal loss and cross entropy functions.

Pseudo Label(Lee et al., 2013) can be seen as the semi-supervised learning for it predicts the labels of the unannotated samples, and re-inputs the samples with the predicted labels into training, thus can improve the model performance in some cases.

4 Experiments and Analysis

All the model we used shared a fixed training config. They were all trained for 5 epochs with learning rate $4e-5$. Besides, the max length of the input was 128, the weight decay rate was 0.01 and the Adam parameter was 1e-8. Table 2 and Table 3 show the performance of different models in each task. We implemented four different encoders to do the classification in each task and the micro F_1-score was chosen as the evaluation metric on our validation process. BioBERT, pubmedBERT, DeBERTa and BERTweet were first applied in the tasks shown in Table 2, then we modified the loss function in the best performing model using rdrop, poly loss tricks to increase the performance.

Results show that the introduction of the tricks help the model get the highest performance in Task2a, Task8 and Task9. While in Task2b, BioBERT achieved the best performance, and the tricks reduced the F_1 by 0.17. The reason for this decline is likely to be the way we processed the data. The purpose of Task2b is to predict whether at least one premise/argument is mentioned in the text, but we transformed this into a text pair match-

Task	Ours			Median		
	Precision	Recall	F_1-Score	Precision	Recall	F_1-Score
Task1a	76.5	58.4	66.2	-	-	-
Task2a	-	-	**57.1(+2.1)**	-	-	55.0
Task2b	-	-	**65.3(+7.9)**	-	-	57.4
Task3a	68.3	60.1	**63.9(+5.3)**	61.7	55.8	58.6
Task3b	85.7	85.4	**85.6(+1.3)**	84.4	86.5	84.3
Task4	93.3	90.8	**92.0(+5.1)**	86.9	88.9	86.9
Task5	85.0	85.0	**85.0(+1.0)**	84.0	84.0	84.0
Task6	90.0	74.0	**81.0(+4.0)**	90.0	68.0	77.0
Task7	90.3	70.5	**79.1(+2.8)**	79.0	71.6	76.3
Task8	79.7	76.9	**78.3(+3.3)**	72.0	76.0	75.0
Task9	95.7	91.9	**93.8(+4.7)**	89.6	91.9	89.1

Table 4: Our model performance and median scores in all tasks we participated. All results and evaluation metrics were provided by the official. '-' denotes officials did not provide that figure. The highest F_1-score in each task is highlighted in boldface. The numbers in brackets indicate how much higher our model's results are than the median score.

ing problem where the claim part may have been introduced as noise, and the tricks increased the noise.

Besides, without tricks, DeBERTa performed the best in Task3b, but with tricks, the scores dropped by 1.09 to 87.58. It may imply that there is a problem with the way we changed meta-information into text information and concatenate it.

Except for Task4, the label distribution of other tasks in Table 3 is very unbalanced. After the introduction of poly loss, the performance of all models has been improved, just as Table 3 shows. In addition, with the addition of rdrop, the model's performance is further improved, which help the model reach the highest F_1 score in Task1a and, Task3a, Task5 and Task7. Besides, the best score of Task1a and Task4 were 0.63 and 0.914 respectively (Magge et al., 2021; Klein et al., 2022), and our model achieved the state-of-the-art performance. In Task6, however, the best encoder using poly loss performed the best, with the additional rdrop dropping slightly.

However, in Task4, these tricks did not seem to improve the performance of the model, but decreased the score a lot, which may be caused by the relatively balanced label distribution of the task. So we took the pseudo label trick to augment the training data.

Specifically, we first used the three best-performing models to predict the samples in the validation set (here is BERTweet, DeBERTa and BioBERT, and we use the test set in the testing phase), and then processed the obtained three logits in the following way: if the two logits are greater than 0.8, it means that the sample is likely to belong to a certain category, and we would add the sample and the predicted label to the training set as a new training sample. Based on the augmented training dataset, we retrained BERTweet encoder with rdrop and poly loss, and got the highest score, which helped us win the first place in this task.

We submitted the highest scoring models incorporating tricks at each task, and the final scores are shown in the Table 4. The official did not provide the median score but provide the mean score of the Task1a, of which the Precison, Recall and F_1-Score are 64.6, 49.7 and 56.2 respectively. It can be seen that in all the tasks we participated in, our results exceeded the median scores. In addition, our model achieved the highest score in Task4, with a higher 7.8% and 5.3% F1-score than the median scores in Task2b and Task3a respectively.

5 Conclusion

In this work, we developed a classification model based on the BERT. It incorporated with rdrop to do the data augmentation, with poly loss to mitigate the label imbalance, and with pseudo label to boost the model performance. We participated 11 classification tasks in SMM4H 2022 shared tasks, and our model scored above the median score in all tasks. Finally, our model achieved the highest score in Task4, with a higher 7.8% and 5.3% F1-score than the median scores in Task2b and Task3a respectively.

References

Mohammed Ali Al-Garadi, Sangmi Kim, Yuting Guo, Elise Warren, Yuan-Chi Yang, Sahithi Lakamana, and Abeed Sarker. 2022. Natural language model for automatic identification of intimate partner violence reports from twitter. *Array*, page 100217.

Vera Davydova and Elena Tutubalina. 2022. Smm4h 2022 task 2: Dataset for stance and premise detection in tweets about health mandates related to covid-19. In *Proceedings of the Seventh Social Media Mining for Health Applications (#SMM4H) Workshop Shared Task*, pages –.

Javier De la Rosa, Eduardo G Ponferrada, Manu Romero, Paulo Villegas, Pablo González de Prado Salas, and María Grandury. 2022. Bertin: Efficient pre-training of a spanish language model using perplexity sampling. *Procesamiento del Lenguaje Natural*, 68:13–23.

Yu Gu, Robert Tinn, Hao Cheng, Michael Lucas, Naoto Usuyama, Xiaodong Liu, Tristan Naumann, Jianfeng Gao, and Hoifung Poon. 2021. Domain-specific language model pretraining for biomedical natural language processing. *ACM Transactions on Computing for Healthcare (HEALTH)*, 3(1):1–23.

Pengcheng He, Xiaodong Liu, Jianfeng Gao, and Weizhu Chen. 2020. Deberta: Decoding-enhanced bert with disentangled attention. *arXiv preprint arXiv:2006.03654*.

Jacob Devlin Ming-Wei Chang Kenton and Lee Kristina Toutanova. 2019. Bert: Pre-training of deep bidirectional transformers for language understanding. In *Proceedings of NAACL-HLT*, pages 4171–4186.

Ari Z Klein, Arjun Magge, and Graciela Gonzalez-Hernandez. 2022. Reportage: Automatically extracting the exact age of twitter users based on self-reports in tweets. *PloS one*, 17(1):e0262087.

Dong-Hyun Lee et al. 2013. Pseudo-label: The simple and efficient semi-supervised learning method for deep neural networks. In *Workshop on challenges in representation learning, ICML*, volume 3, page 896.

Jinhyuk Lee, Wonjin Yoon, Sungdong Kim, Donghyeon Kim, Sunkyu Kim, Chan Ho So, and Jaewoo Kang. 2020. Biobert: a pre-trained biomedical language representation model for biomedical text mining. *Bioinformatics*, 36(4):1234–1240.

Zhaoqi Leng, Mingxing Tan, Chenxi Liu, Ekin Dogus Cubuk, Xiaojie Shi, Shuyang Cheng, and Dragomir Anguelov. 2022. Polyloss: A polynomial expansion perspective of classification loss functions. *arXiv preprint arXiv:2204.12511*.

Tsung-Yi Lin, Priya Goyal, Ross Girshick, Kaiming He, and Piotr Dollár. 2017. Focal loss for dense object detection. In *Proceedings of the IEEE international conference on computer vision*, pages 2980–2988.

Yinhan Liu, Myle Ott, Naman Goyal, Jingfei Du, Mandar Joshi, Danqi Chen, Omer Levy, Mike Lewis, Luke Zettlemoyer, and Veselin Stoyanov. 2019. Roberta: A robustly optimized bert pretraining approach. *arXiv preprint arXiv:1907.11692*.

Arjun Magge, Elena Tutubalina, Zulfat Miftahutdinov, Ilseyar Alimova, Anne Dirkson, Suzan Verberne, Davy Weissenbacher, and Graciela Gonzalez-Hernandez. 2021. Deepademiner: a deep learning pharmacovigilance pipeline for extraction and normalization of adverse drug event mentions on twitter. *Journal of the American Medical Informatics Association*, 28(10):2184–2192.

Dat Quoc Nguyen, Thanh Vu, and Anh Tuan Nguyen. 2020. Bertweet: A pre-trained language model for english tweets. *arXiv preprint arXiv:2005.10200*.

Davy Weissenbacher, Ari Z. Klein, Luis Gascó, Darryl Estrada-Zavala, Martin Krallinger, Yuting Guo, Yao Ge, Abeed Sarker, Ana Lucia Schmidt, Raul Rodriguez-Esteban, Mathias Leddin, Arjun Magge, Juan M. Banda, Vera Davydova, Elena Tutubalina, and Graciela Gonzalez-Hernandez. 2022. Overview of the seventh social media mining for health applications smm4h shared tasks at coling 2022. In *Proceedings of the Seventh Social Media Mining for Health Applications (#SMM4H) Workshop Shared Task*, pages –.

Lijun Wu, Juntao Li, Yue Wang, Qi Meng, Tao Qin, Wei Chen, Min Zhang, Tie-Yan Liu, et al. 2021. R-drop: Regularized dropout for neural networks. *Advances in Neural Information Processing Systems*, 34:10890–10905.

Yang Yuan-Chi, Xie Angel, Kim Sangmi, Hair Jessica, Al-Garadi Mohammed Ali, and Sarker Abeed. 2022. Automatic detection of twitter users who express chronic stress experiences via supervised machine learning and natural language processing. *Computers, Informatics, Nursing*, pages –.

Fraunhofer FKIE @ SMM4H 2022: System Description for Shared Tasks 2, 4 and 9

Daniel Claeser
Fraunhofer FKIE
Fraunhoferstraße 20
53343 Wachtberg
daniel.claeser@fkie.fraunhofer.de

Samantha Kent
Fraunhofer FKIE
Fraunhoferstraße 20
53343 Wachtberg
samantha.kent@fkie.fraunhofer.de

Abstract

We present our results for the shared tasks 2, 4 and 9 of the SMM4H Workshop at COLING 2022 achieved by succesfully fine-tuning pre-trained language models on the downstream tasks. We identify the occurence of code-switching in the test data for task 2 as a possible source of considerable performance degradation on the test set scores. We successfully exploit structural linguistic similarities in the datasets of tasks 4 and 9 for training on joined datasets, scoring first in task 9 and on par with SOTA in task 4.

1 Introduction

This contribution describes the system submissions for thee shared tasks at the Social Media Mining for Health (#SMM4H) Workshop at COLING 2022 (Weissenbacher et al., 2022). We participated in tasks 2, 4 and 9.

All models were developed using the Flair framework (Akbik et al., 2019) and we used the pre-trained models based on PyTorch (Paszke et al., 2019) provided by Huggingface (Wolf et al., 2020). Based on previous experience and the baseline results provided for the stance detection COVID-19 tweets in (Glandt et al., 2021), we focused mainly on the models BERT (Devlin et al., 2019), RoBERTa (Liu et al., 2019), and BERTweet (Nguyen et al., 2020).

2 Task 2

We participated in both subtasks of shared task 2. In task 2a, stance detection, participants were asked to determine an author's stance in relation to three different mandates related to the COVID-19 pandemic (Davydova and Tutubalina, 2022). The following tweets are examples relating to the mandate school closures:

- (FAVOR) - @anonymous NO TO REOPEN SCHOOLS.
- (AGAINST) - Society is bound to fall If Schools fall.
- (NONE) - The UK government tried to reopen schools and the people of Scotland refused.

In task 2b, premise classification, the aim is to determine whether or not a tweet contains an argument that could be used to convince an opponent about one of the given COVID-19 mandates. It is a binary classification task in which the data is annotated as positive (contains a premise) or negative (does not contain a premise). The following two tweets are examples from the mask wearing mandate:

- (1) - If masks work, then why are people working from home?
- (0) - @Anonymous @Anonymous is this about mask wearing?

The training data consists of 3,556 tweets, the validation data of 600 tweets, and the test data of 10,000 tweets.

2.1 System Description

We conducted preliminary experiments using the validation data as a held-out test set. The preliminary models were submitted during the evaluation phase of the shared task. For training data, we split the original training data into a train (3,000 tweets) and development (556 tweets) set. For this task, we used RoBERTa-large and BERTweet-large transformer document embeddings. RoBERTa-large is based on the BERT-large architecture, has 24-layers, 1,024 hidden layer dimension and 16 attention heads with 335M parameters. BERTweet-large is a language model pre-trained specifically on 850M English tweets, 5 million specifically related to the COVID pandemic, and is in turn based on the RoBERTa training procedure and has the same architecture.

Proceedings of the 29th International Conference on Computational Linguistic, pages 103–107
October 12–17, 2022.

Model	Weighted F1
`RoBERTa`	0.7458 - 0.7856
`BERTweet`	0.6889 - 0.7804

Table 1: Eight-fold cross validation for stance detection using RoBERTa and BERTweet.

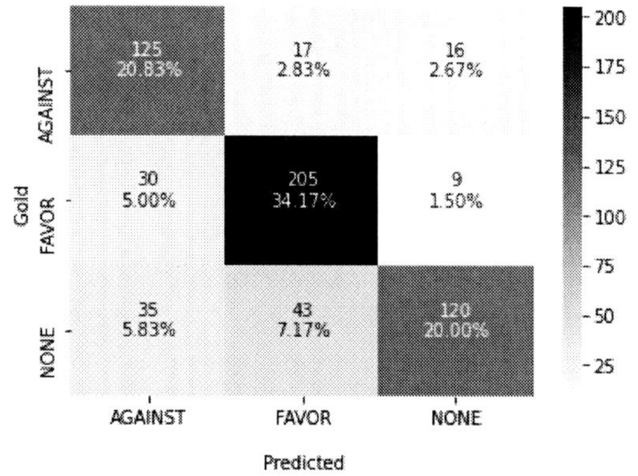

Figure 1: A confusion matrix showing the errors made by the stance detection system.

More specifically after parameter optimization, we fine-tuned both the RoBERTa-large and BERTweet large models and trained the embeddings using a learning rate of 0.005, with a mini-batch size of 32, for a maximum of 50 epochs for the stance detection task. For the premise classification sub-task we found similar parameters to be optimal, except we changed the mini-batch size to 16.

2.1.1 Stance Detection

Table 1 shows the result of performing eight-fold cross-validation at the training stage for the two different pre-trained models. The highest and the lowest weighted F1 scores are shown in the table, with the RoBERTa model slightly outperforming the BERTweet model with an F1 of 0.7856.

We conducted an error analysis on the misclassifications made by the systems. The confusion matrix in Figure 1 shows that the most errors are made in the class 'NONE'.

2.1.2 Premise Classification

We followed a similar procedure for premise classification, the results of which can be found in Table 2. Both models perform similarly, with RoBERTa (weighted F1 0.8224) slightly outperforming BERTweet (weighted F1 0.8138), and the results show that the variance in the eight system runs is 4.55 for RoBERTa and 3.16 for BERTweet.

Model	Weighted F1
`RoBERTa`	0.7769 - 0.8224
`BERTweet`	0.7822 - 0.8138

Table 2: Eight-fold cross validation for premise classification using RoBERTa and BERTweet.

2.2 Final Results

Comparing the results for stance detection to those in a paper by (Glandt et al., 2021), we wanted to maximize the available training data in order to boost performance of our systems. To train the final models, we incorporated the original validation data as training data for the final models. The data was split as follows: train (3,500), development (556) and test (10,000). The test data was pre-processed to match the training data using the Python emoji [1] library.

To gain a better understanding of the differences in scores, we examined the three different datasets provided by the task organizers. Specifically, we observed that the test data contains many more multilingual tweets compared to the training and validation data (see Table 3). In total, 2.93% of test tweets are identified as being non-English, using the FastText language identification model (Joulin et al., 2016). Furthermore, 13.2% of tweets receive a confidence score of less than 0.4 for English, indicating that the language is not so clearly identifiable. An example of a tweet containing Spanish-English code-switching can be found below (Poplack, 1980). Since the pre-trained language models are monolingual and have been fine-tuned on monolingual data, it is imaginable that the models struggle with accurately classifying the test data. It would be interesting to further analyze this issue and other reasons for inaccuracies once the gold standard for the test data is released.

- —échale un vistazo a esto... ... a fair piece on comprehensive contrasting views on the virus so-called 'crisis'...

3 Task 4

Task 4 is a binary classification task in which participants were asked to determine whether a tweet contains a self-report of an exact age or not. For example, in tweet 1 below, the person posting the

[1]https://pypi.org/project/emoji/

#Tweets	Train	Val.	Test
Total	3556	600	10.000
English	3548	599	9707
Other	8	1	293
conf.< 0.4	293	48	1316

Table 3: Distribution of the languages of tweets in the data. Conf. refers to the confidence score given by the language detection algorithm. The table includes all tweets with a confidence score lower than 0.4 out of 1.

tweet reports that they are currently 29. In tweet 0, an age is reported, but it is annotated as negative because the age that is reported is that of the user's dog, and not their own age.

- (1) - My birthday is in 9 days. And just am here to say I am enjoying being still 29 the fullest.
- (0) - My 15yo Lab got into an unopened box of breakfast cereal, which made him sick...: I guess you can't feed an old dog new Trix.

The data consists of 8,800 tweets as training data, 2,200 tweets as validation data, and 10,000 tweets as test data.

3.1 System Description

We trained various configurations of BERT-large and RoBERTa-large (see section 2.1). As a specific contribution, we took advantage of the expected limited inventory available to authors independent of medium or register to express, both explicitely and implicitely, age: A basic n-gram analysis showed a high degree of overlap between the datasets of tasks 4 and 9 despite the differences in medium, register, topic and document length. Trigrams explicitely specifying age like 'n years old', '[the] age of n [years]' as well as trigrams enabling deduction of age in context, such as 'n years ago' are similarly distributed in both datasets to a degree that motivated the idea of merging both training sets.

We found the augmented model to clearly outperform the best configuration trained on the task 4 specific dataset only: Providing the model with an additional 9,000 training instances from the Reddit dataset led to an improved F1 of 0.9526, a statistically significant (8x cross-validated, p=0.000046) margin of 0.0193 over the best-performing model employing only the task-specific

Model	Macro F1	SD
RoBERTa-large	0.9333	0.0069
RoBERTa-large+T9	0.9526	0.0041

Table 4: Eight-fold cross validation for classification of tweets self-reporting exact age with basic and augmented training data.

training set (F1 0.9333). The best results were achieved within 10 epochs at a learning rate of 0.005, AdamW optimizer (Loshchilov and Hutter, 2017) and a mini-batch size of 8. In order to retain good generalization capability in light of unknown label distribution in the blind test set, we evaluated for macro-F1 during fine-tuning.

3.2 Error Analysis

Eight cross-validation runs with models trained in the best-performing parameter configuration produced 913 misclassification instances out of 17,600 samples (551 false positives and 362 false negatives composed of multiple occurences of 148 and 94 unique instances, respectively). The mean error margin was 0.051875 (spread 107-126/2200, σ=6.07), with an average false positive rate of 0.0313 (51-87, σ=11.68) and a false negative rate of an average 0.0206 (36-57, σ=8.2424).

4 Task 9

Task 9 is very similar to task 4, in that it also involves a binary classification task on self-reported ages. The main difference is that the data stems from Reddit, not Twitter, and that the data is disease-specific and was collected using specific keywords related to dry eye disease. For example, in (1) a specific age is provided, and the second post in annotated as (0) because the user only provides an age range.

- (1) - How old are you? I would be surprised if your eyes have stabilized after only 5 years unless you were in your 30s when diagnosed. I was diagnosed at 21, had CXL in left eye that year and been monitoring both eyes since (now 33). Still waiting fir them to stabilize...
- (0) - is a .5D reduction in astigmatism for a developing cataract (age 60's) likely to be due to changes in the lens or cornea ?

The training data consists of 9,000 posts, the validation data of 1,000 posts, and the test data of 2,000 posts.

Model	Macro F1	SD
`RoBERTa-large`	0.9520	0.014043
`RoBERTa-large+T4`	0.9695	0.008629

Table 5: Eight-fold cross validation for classification of Reddit posts self-reporting exact age with basic and augmented training data.

4.1 System Description

In light of the performance gains on the evaluation set of task 4 from training a model based on the combined training sets of tasks 4 and 9, we opted to apply the same strategy for task 9 with its identical binary classification goal: Augmenting the tasks' original training data by the dataset provided for training task 4. Providing the model with an additional 9,000 training instances from the Twitter dataset led to an improved validation set F1 of 0.9695, a statistically significant (8x cross-validated, p=0.0033173) margin of 0.0175 over the best-performing model employing only the task-specific training set (F1 0.9520). Unsurprisingly, this result was accomplished by a configuration identical to that of task 4.

4.2 Error analysis

Eight cross-validation runs with models trained in the best-performing parameter configuration yielded 206 misclassification instances out of 8,000 samples (118 false positives and 88 false negatives composed of multiple occurences of 35 and 22 unique instances, respectively). The mean error margin was 0.02575 (spread 22-30/1000, σ=2.39), with an average false positive rate of 0.01475 (11-22, σ=3.38) and a false negative rate of an average 0.011 (6-16, σ=2.78).

Eleven unique samples accounting for 53 instances of false positive classification should actually be considered true positive since their gold label apparently did not comply to the annotation guidelines. Six of those samples contained an explicit mention of an exact age and five contained implicit, abbreviated or indirect (inferable) age mentions. Unambiguous examples of this phenomenon are e.g.

- (11287) [...] I am now 66 years old, finished high school, college, [...]
- (11239) [...] I was 26 when I was diagnosed too (am now 28) [...]

The complete list of such instances in the conducted validation runs is *11012, 11153, 11173, 11396, 11833, 11294, 11863, 11539* with more examples supposedly identifiable with additional runs.

Considering these samples to be correctly classified as true positive reduces the actual false positive rate to 65 in 8,000 classifications (0.8125%).

There were 88 total instances of false negatives consisting of multiple occurences of 22 unique samples. While 53 of those were genuine cases of false positives, there were six misannotated unique instances accounting for 35 cases where the 'negative' classification conflicting with the gold labels should be considered appropriate.

- (11384) CRVO in a 27 y.o. is a very unusual occurrence. Did your doctors figure out what caused it?

The complete list of samples incorrectly annotated as 'positive' surfacing in our evaluation series is *14485, 11520, 11522, 11194, 11960.*

Given the significant share of debatable gold labels in both false positive and false negative classifications, we suggest revising the dataset based on an investigation of the aforementioned error patterns of repeated evaluation runs.

5 Conclusion

We applied state-of-the art transformer language models to three different shared tasks, succesfully employing dataset fusion to broaden the training base for two closely related tasks. We observed on par results between models in the vicinity of human performance on the gold annotation. The resulting model for task 4 achieved F1 scores of 0.912 (P 0.924, R 0.901) and 0.917 (P 0.891, R 0.904) on the blind test set, outperforming both mean and median F1 scores (0.847, 0.869) of all submissions in the task by a significant margin. These results, despite generalization loss, are also on par with the results of the datataset authors' benchmark F1 on the validation set of 0.914 (Klein et al., 2022). Our best model for task 9 based on the same strategy scored 0.956 F1 on the blind test set (P=0.948, R=0.963), making it the best performing contribution in the competition.

References

Alan Akbik, Tanja Bergmann, Duncan Blythe, Kashif Rasul, Stefan Schweter, and Roland Vollgraf. 2019.

FLAIR: An easy-to-use framework for state-of-the-art NLP. In *NAACL 2019, 2019 Annual Conference of the North American Chapter of the Association for Computational Linguistics (Demonstrations)*, pages 54–59.

Vera Davydova and Elena Tutubalina. 2022. Smm4h 2022 task 2: Dataset for stance and premise detection in tweets about health mandates related to covid-19. In *Proceedings of the Seventh Social Media Mining for Health Applications (#SMM4H) Workshop Shared Task*, pages –.

Jacob Devlin, Ming-Wei Chang, Kenton Lee, and Kristina Toutanova. 2019. BERT: Pre-training of deep bidirectional transformers for language understanding. In *Proceedings of the 2019 Conference of the North American Chapter of the Association for Computational Linguistics: Human Language Technologies, Volume 1 (Long and Short Papers)*, pages 4171–4186, Minneapolis, Minnesota. Association for Computational Linguistics.

Kyle Glandt, Sarthak Khanal, Yingjie Li, Doina Caragea, and Cornelia Caragea. 2021. Stance detection in COVID-19 tweets. In *Proceedings of the 59th Annual Meeting of the Association for Computational Linguistics and the 11th International Joint Conference on Natural Language Processing (Volume 1: Long Papers)*, pages 1596–1611, Online. Association for Computational Linguistics.

Armand Joulin, Edouard Grave, Piotr Bojanowski, and Tomas Mikolov. 2016. Bag of tricks for efficient text classification. *arXiv preprint arXiv:1607.01759*.

Ari Z Klein, Arjun Magge, and Graciela Gonzalez-Hernandez. 2022. Reportage: Automatically extracting the exact age of twitter users based on self-reports in tweets. *PloS one*, 17(1):e0262087.

Yinhan Liu, Myle Ott, Naman Goyal, Jingfei Du, Mandar Joshi, Danqi Chen, Omer Levy, Mike Lewis, Luke Zettlemoyer, and Veselin Stoyanov. 2019. Roberta: A robustly optimized bert pretraining approach. Cite arxiv:1907.11692.

Ilya Loshchilov and Frank Hutter. 2017. Decoupled weight decay regularization. *arXiv preprint arXiv:1711.05101*.

Dat Quoc Nguyen, Thanh Vu, and Anh Tuan Nguyen. 2020. BERTweet: A pre-trained language model for English tweets. In *Proceedings of the 2020 Conference on Empirical Methods in Natural Language Processing: System Demonstrations*, pages 9–14, Online. Association for Computational Linguistics.

Adam Paszke, Sam Gross, Francisco Massa, Adam Lerer, James Bradbury, Gregory Chanan, Trevor Killeen, Zeming Lin, Natalia Gimelshein, Luca Antiga, et al. 2019. Pytorch: An imperative style, high-performance deep learning library. *Advances in neural information processing systems*, 32.

Shana Poplack. 1980. Sometimes i'll start a sentence in spanish y termino en espaÑol: toward a typology of code-switching1. 18(7-8):581–618.

Davy Weissenbacher, Ari Z. Klein, Luis Gascó, Darryl Estrada-Zavala, Martin Krallinger, Yuting Guo, Yao Ge, Abeed Sarker, Ana Lucia Schmidt, Raul Rodriguez-Esteban, Mathias Leddin, Arjun Magge, Juan M. Banda, Vera Davydova, Elena Tutubalina, and Graciela Gonzalez-Hernandez. 2022. Overview of the seventh social media mining for health applications smm4h shared tasks at coling 2022. In *Proceedings of the Seventh Social Media Mining for Health Applications (#SMM4H) Workshop Shared Task*, pages –.

Thomas Wolf, Lysandre Debut, Victor Sanh, Julien Chaumond, Clement Delangue, Anthony Moi, Pierric Cistac, Tim Rault, Remi Louf, Morgan Funtowicz, Joe Davison, Sam Shleifer, Patrick von Platen, Clara Ma, Yacine Jernite, Julien Plu, Canwen Xu, Teven Le Scao, Sylvain Gugger, Mariama Drame, Quentin Lhoest, and Alexander Rush. 2020. Transformers: State-of-the-art natural language processing. In *Proceedings of the 2020 Conference on Empirical Methods in Natural Language Processing: System Demonstrations*, pages 38–45, Online. Association for Computational Linguistics.

5q032e@SMM4H'22: Transformer-based classification of premise in tweets related to COVID-19

Vadim Porvatov
Sberbank / Moscow 117997, Russia
eighonet@gmail.com

Natalia Semenova
AIRI / Moscow 105064, Russia
Sberbank / Moscow 117997, Russia
semenova.bnl@gmail.com

Abstract

Automation of social network data assessment is one of the classic challenges of natural language processing. During the COVID-19 pandemic, mining people's stances from public messages have become crucial regarding understanding attitudes towards health orders. In this paper, the authors propose the predictive model based on transformer architecture to classify the presence of premise in Twitter texts. This work is completed as part of the Social Media Mining for Health (SMM4H) Workshop 2022. We explored modern transformer-based classifiers in order to construct the pipeline efficiently capturing tweets semantics. Our experiments on a Twitter dataset showed that RoBERTa-large is superior to the other transformer models in the case of the premise prediction task. The model achieved competitive performance with respect to ROC AUC value 0.807, and 0.7648 for the F1 score.

1 Introduction

Modern natural language processing methods emerged as tools of outstanding performance in many classic machine learning tasks. Along with their other achievements, the introduction of transformer architecture (Vaswani et al., 2017) allowed to develop of domain-specific models in the interlingual transfer of social media texts (Miftahutdinov et al., 2020), identification of drug similarity (Tutubalina et al., 2017), and detection of adverse drug effects (Sakhovskiy et al., 2021).

One of the prospective domains of language model development is an automatic extraction and further assessment of insights gained from Twitter texts (Pamungkas et al., 2019). In addition, stance and premise classification tasks are frequently interpreted as critical challenges in social media text analysis (Bar-Haim et al., 2017; Go et al., 2009).

Contribution. In this paper, we extensively evaluate premise classification approaches and partially explore dependencies between configurations of the considered models and their performance.

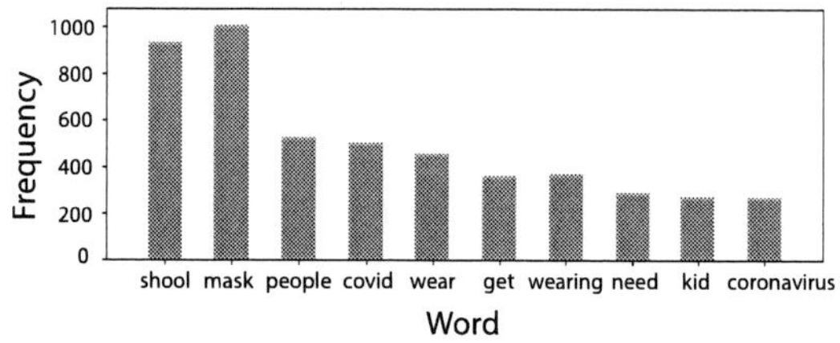

Figure 1: Top-10 frequent words in the dataset.

2 Data

Available labeled data (Davydova and Tutubalina, 2022) includes 4155 tweets divided into train and test samples in a ratio of 17:3. Regarding the premise classification, the dataset contains a subset of 2445 tweets with a positive label and 1710 tweets with a negative label. As an additional metadata, there are 1402 tweets tagged as *stay at home orders*, 1526 related to the *face masks*, and 1227 tweets marked as opinions about *school closures*. Established text data could be assessed from the perspective of trending word frequencies, Figure 1.

3 Method

In order to reach the best score, we performed analysis of the state-of-the-art architectures for text classification and selected the following models: BERT (Kenton and Toutanova, 2019), RoBERTa (Liu et al., 2019), AlBERT (Lan et al., 2019), DeBERTa (He et al., 2020), RemBERT (Chung et al., 2020), and Longformer (Beltagy et al., 2020). Generally, these models encode each tweet to the fixed-size vector and further apply the classifier layers in an end-to-end manner. The output activation function (softmax) converts the hidden representation of the text to the desired class probabilities which are further used during the

Proceedings of the 29th International Conference on Computational Linguistic, pages 108–110
October 12–17, 2022.

Split	Train			Test		
	Accuracy	F1-score	ROC AUC	Accuracy	F1-score	ROC AUC
Random	0.4986	0.5592	0.5014	0.4959	0.4302	0.5016
BERT (base, uncased)	0.9446	0.9268	0.9392	0.7947	0.7185	0.7793
BERT (base, uncased, ml)	0.913	0.889	0.9005	0.7813	0.7385	0.7742
AlBERTv2 (base)	0.9781	0.9708	0.9758	0.7746	0.6853	0.7581
AlBERTv2 (xlarge)	0.9215	0.8992	0.9126	0.7496	0.6681	0.7314
DeBERTa (base)	0.9194	0.8995	0.9095	0.813	0.7607	0.799
Longformer (large)	0.9758	0.9681	0.9721	0.8097	0.7522	0.7956
RemBERT	**0.9885**	**0.9846**	**0.9871**	0.7997	0.7309	0.7842
RoBERTa (base)	0.9213	0.9024	0.9118	0.7997	0.7521	0.7882
RoBERTa (large)	0.9837	0.9761	0.9795	**0.8214**	**0.7648**	**0.807**

Table 1: Evaluation of methods for a premise classification task.

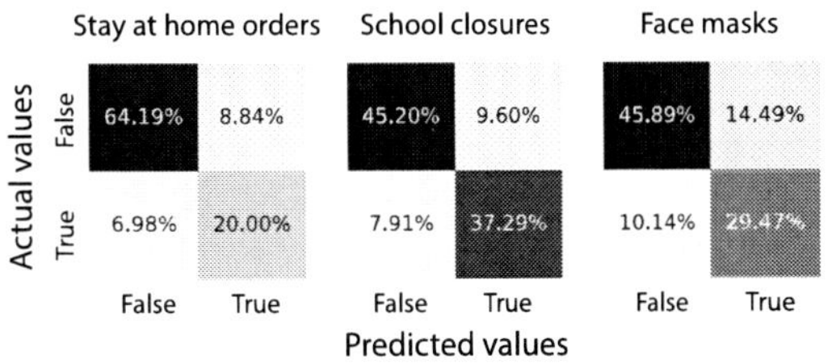

Figure 2: Confusion matrices of RoBERTa-large regarding the different tweets categories of test data.

computation of binary cross entropy loss function:

$$-\frac{1}{n}\sum_{i=1}^{n} y_i \cdot \log \hat{y}_i + (1 - y_i) \cdot \log (1 - \hat{y}_i), \quad (1)$$

where n is the number of samples, y_i is the true label of a tweet, and $\hat{y}_i$ is the predicted one.

Relatively small train data encouraged us to use pre-trained variations of each considered model. As long as tweets comprise of divergent content, it is required to inspect the dataset at the preprocessing stage carefully. To move further with model training, we need to handle the presence of nametags, hashtags, emoticons, and other additional symbols. We perform processing procedures with the help of *tweet-preprocessor*[1] package. First, we remove abundant text pieces (e. g., user mentions) as well as replace web pages links and hashtags with placeholders. Before tokenization, we convert cased words to uncased ones.

4 Experiments

We measure the performance of the models on the main task via three commonly used metrics for the binary classification: ROC AUC, Accuracy, and F1-score.

As the optimizer for the selected models, we used AdamW (Loshchilov and Hutter, 2019). The models were trained along the 20 epochs with learning rates varying from 0.001 to 0.00001 and batch sizes from the range [4, 8, 16, 32, 48]. Experiments were done with 3 Tesla V100 GPUs and 512 Gb of RAM. The training time of the models lies in the interval from 3.5 hours up to 5 hours.

For the premise classification, the best performance was achieved by the large RoBERTa model, Table 1. Obtained metrics are tangibly dissimilar from the other architectures despite RemBERT which suffers from overfitting and thus converges to the better values in the training sample. Detailed analysis of RoBERTa-large performance on different tweets categories is given in Figure 2.

To ensure the statistical significance of applied preprocessing procedures, we leverage the Mann–Whitney U test regarding the null hypothesis that the F1 scores obtained on the initial and preprocessed tweets belong to the same distribution. We rejected the null hypothesis with a p-value ≤ 0.05.

5 Conclusion

In this work, we have explored the application of different transformer models to the task of premise classification. We extensively evaluated BERT variants and obtained the best architecture during the computational experiments. In future work, we intend to focus on the ensembling methods applied to the presented task.

[1] https://pypi.org/project/tweet-preprocessor/

References

Roy Bar-Haim, Indrajit Bhattacharya, Francesco Dinuzzo, Amrita Saha, and Noam Slonim. 2017. Stance classification of context-dependent claims. In *Proceedings of the 15th Conference of the European Chapter of the Association for Computational Linguistics: Volume 1, Long Papers*, pages 251–261, Valencia, Spain. Association for Computational Linguistics.

Iz Beltagy, Matthew E. Peters, and Arman Cohan. 2020. Longformer: The long-document transformer. *arXiv:2004.05150*.

Hyung Won Chung, Thibault Fevry, Henry Tsai, Melvin Johnson, and Sebastian Ruder. 2020. Rethinking embedding coupling in pre-trained language models. In *International Conference on Learning Representations*.

Vera Davydova and Elena Tutubalina. 2022. Smm4h 2022 task 2: Dataset for stance and premise detection in tweets about health mandates related to covid-19. In *Proceedings of the Seventh Social Media Mining for Health Applications (SMM4H) Workshop & Shared Task*, pages –.

Alec Go, Richa Bhayani, and Lei Huang. 2009. Twitter sentiment classification using distant supervision. *Processing*, 150.

Pengcheng He, Xiaodong Liu, Jianfeng Gao, and Weizhu Chen. 2020. Deberta: Decoding-enhanced bert with disentangled attention. In *International Conference on Learning Representations*.

Jacob Devlin Ming-Wei Chang Kenton and Lee Kristina Toutanova. 2019. Bert: Pre-training of deep bidirectional transformers for language understanding. In *Proceedings of NAACL-HLT*, pages 4171–4186.

Zhenzhong Lan, Mingda Chen, Sebastian Goodman, Kevin Gimpel, Piyush Sharma, and Radu Soricut. 2019. Albert: A lite bert for self-supervised learning of language representations. In *International Conference on Learning Representations*.

Yinhan Liu, Myle Ott, Naman Goyal, Jingfei Du, Mandar Joshi, Danqi Chen, Omer Levy, Mike Lewis, Luke Zettlemoyer, and Veselin Stoyanov. 2019. Roberta: A robustly optimized bert pretraining approach.

Ilya Loshchilov and Frank Hutter. 2019. Decoupled weight decay regularization. In *International Conference on Learning Representations*.

Zulfat Miftahutdinov, Ilseyar Alimova, and Elena Tutubalina. 2020. On biomedical named entity recognition: experiments in interlingual transfer for clinical and social media texts. In *European Conference on Information Retrieval*, pages 281–288. Springer.

EW Pamungkas, V Basile, and V Patti. 2019. Stance classification for rumour analysis in twitter: Exploiting affective information and conversation structure. In *2nd International Workshop on Rumours and Deception in Social Media (RDSM 2018)*, volume 2482, pages 1–7. CEUR-WS.

Andrey Sakhovskiy, Zulfat Miftahutdinov, and Elena Tutubalina. 2021. Kfu nlp team at smm4h 2021 tasks: Cross-lingual and cross-modal bert-based models for adverse drug effects. In *Proceedings of the Sixth Social Media Mining for Health (SMM4H) Workshop and Shared Task*, pages 39–43.

EV Tutubalina, Z Sh Miftahutdinov, RI Nugmanov, TI Madzhidov, SI Nikolenko, IS Alimova, and AE Tropsha. 2017. Using semantic analysis of texts for the identification of drugs with similar therapeutic effects. *Russian Chemical Bulletin*, 66(11):2180–2189.

Ashish Vaswani, Noam Shazeer, Niki Parmar, Jakob Uszkoreit, Llion Jones, Aidan N Gomez, Łukasz Kaiser, and Illia Polosukhin. 2017. Attention is all you need. *Advances in neural information processing systems*, 30.

Fraunhofer SIT@SMM4H'22: Learning to Predict Stances and Premises in Tweets related to COVID-19 Health Orders Using Generative Models

Raphael Antonius Frick and **Martin Steinebach**
Fraunhofer Institute for Secure Information Technology SIT |
ATHENE — National Research Center for Applied Cybersecurity
Rheinstrasse 75, Darmstadt, 64295, Germany
{raphael.frick, martin.steinebach}@sit.fraunhofer.de

Abstract

This paper describes the system used to predict stances towards health orders and to detect premises in Tweets as part of the Social Media Mining for Health 2022 (#SMM4H) shared task. It takes advantage of GPT-2 to generate new labeled data samples which are used together with pre-labeled and unlabeled data to fine-tune an ensemble of GAN-BERT models. First experiments on the validation set yielded good results, although it also revealed that the proposed architecture is more suited for sentiment analysis. The system achieved a score of 0.4258 for the stance and 0.3581 for the premise detection on the test set.

1 Introduction

During the pandemic many countries, such as the US, China, and Germany, introduced health mandates to cope against the fast spread of the coronavirus. In this time, social networks like Facebook and Twitter enabled users to share their opinion regarding the implemented health orders. These social media posts can be used as part of argument mining to find reasons for and against those. Further, it also allows analyzing whether any opposing opinions are based on false information and intentionally spread disinformation.

In Task 2. of the Social Media Mining for Health 2022 (#SMM4H) shared task (Weissenbacher et al., 2022), one objective was to develop a classification system that can predict the stance towards imposed health mandates during the pandemic, such as *face masks*, *school closures* and *stay at home orders*, in Tweets. Another was to detect whether a Tweet contained statements that could be used for reasoning.

In this paper, the proposed approach of our team, *Fraunhofer SIT*, is described. It takes advantage of two generative models to generate new labeled samples as well as a generative adversarial network for classification and to also fully utilize unlabeled samples during training. These are then incorporated into an ensemble classification scheme to improve the classification accuracy.

The paper is organized as follows: in Section 2 the proposed system developed for the shared task is presented. Section 3 showcases and discusses the experimental results achieved by the proposed components on the validation set. The paper then concludes in Section 4.

2 Proposed Approach

The main component of our classification system to estimate stances and to identify premises in Tweets was built around an ensemble of fine-tuned GAN-BERT (Croce et al., 2020) models. *GAN-BERT* is a generative adversarial network (Goodfellow et al., 2014) that allows to fine-tune models on labeled and unlabeled data. Its generator network produces artificially crafted embeddings, while the discriminator tries to distinguish them from real embeddings extracted from transformer models. Here, *BERT uncased*, *BERT cased* (Devlin et al., 2019) and *ALBERT v2* (Lan et al., 2019) served as transformer models due to their performance on the validation set. In the case of real samples, the discriminator also tries to predict their classes. The class-labels hereby consisted of L_{FAVOR} and $L_{AGAINST}$ for the stance estimation task and L_0 and L_1 for the premise detection task. The prediction probabilities of all the three fine-tuned GAN-BERT models were averaged to provide a unified classification decision.

To provide additional labeled training data and to prevent data scarcity having an impact on the classification accuracy, new samples were generated through *GPT-2* (Radford et al., 2019). To ensure that the generated data represent a particular class, one text-generation model was trained on data of a specific class. The data then underwent several preprocessing steps, where emojis were converted to their descriptive terms, URLs and user mentions

Proceedings of the 29th International Conference on Computational Linguistic, pages 111–113
October 12–17, 2022.

	Stance				Premise			
	$F1_{mask}$	$F1_{school}$	$F1_{home}$	$F1_{avg}$	$F1_{mask}$	$F1_{school}$	$F1_{home}$	$F1_{avg}$
bert-uncased	0.8845	0.8663	0.7397	0.8302	0.6994	**0.8235**	0.6415	0.7215
bert-uncased + PP	0.8348	0.8492	0.7500	0.8113	0.7322	0.7676	**0.7360**	**0.7453**
GAN + bert-uncased + PP	0.8974	0.8587	0.7879	0.8480	0.6993	0.8075	0.6604	0.7224
GAN + bert-uncased + PP +U	0.8945	0.8488	0.8235	0.8556	0.6107	0.7397	0.5934	0.6479
GAN + bert-cased + PP +U	0.8622	0.8353	0.8000	0.8325	0.6565	0.7200	0.5773	0.6513
GAN + albert-v2 + PP +U	0.8561	0.7836	0.6437	0.7615	0.7034	0.7590	0.6729	0.7118
GAN + bert-uncased + PP + U + GPT2	**0.9030**	**0.8667**	0.7887	0.8528	0.7226	0.7738	0.6606	0.7189
GAN + bert-cased + PP + U + GPT2	0.8496	0.8383	**0.8308**	0.8396	0.7143	0.7329	0.6909	0.7127
GAN + albert-v2 + PP + U + GPT2	0.8462	0.8457	0.7826	0.8248	0.7044	0.7297	0.6604	0.6982
GAN + ensemble + PP + U + GPT2	0.8945	0.8605	0.8182	**0.8577**	**0.7451**	0.7898	0.6916	0.7422

Table 1: F1-scores achieved by each model configuration on the validation set of each task. *PP* refers to preprocessed data and *U* to the usage of unlabeled data from Glandt et al. (2021) during training.

were exchanged with generic tokens.

3 Experimental Results

The classification system was implemented using Python. For data preprocessing, the *emoji*[1] package as well as libraries provided by Python were used.

The GPT-2 models were trained on the preprocessed train split of the dataset provided along the shared task (Davydova and Tutubalina, 2022) using the *aitextgen* framework[2]. To avoid overfitting, the models were only trained for 3500 steps with a learning rate of $1e^{-3}$. The words *the*, *face masks*, *school closures*, *stay at home orders*, *Corona* and *COVID19* served as prefixes for the text-generation. Further, various temperatures were considered, such as 0.8, 1.0 and 1.2. For each configuration consisting of a label, prefix and temperature, 50 samples were generated and occurring duplicates were removed. The following examples showcase Tweets generated by our trained model:

Stance — Favor: COVID19 @user What a frightening lot of people. No one is wearing a mask. and related ways of managing it. It's so disappointing.

Stance — Against: The death figures are inflated. A major percentage of deaths are those who died with Covid at home, and not from it.

For training the GAN-BERT models in a semi-supervised manner, data from the *Stance Detection in COVID-19 Tweets Dataset* (Glandt et al., 2021) served as unlabeled data. The sequence length for the embeddings was set to 280 and the models were trained using a batch size of 16 and learning rates of $5e^{-5}$ for both the generator and the discriminator. As optimizer *adam* (Kingma and Ba, 2015) was used to take advantage of adaptive learning rates and for regularization, a dropout rate of 0.3 was chosen. To further avoid overfitting, the training procedure took advantage of model checkpoints to keep only the best performing models while the models were trained for 20 epochs.

Table 1 showcases the F1-scores achieved by various model configurations. A *BERT uncased* model fine-tuned using *GAN-BERT* achieved overall higher F1-scores when trying to estimate stances than the same model trained traditionally with or without the application of preprocessing. When trying to classify premises using GAN-BERT, the performance on the validation set decreased. One reason for this could be, that the stance towards a particular topic can often be expressed using a single term (e.g., #WearAMask), while for premises, the structure, and wording of the entire sentence is of high importance. The generator of GAN-BERT may not be able to replicate sentence structures as accurately as wordings. Similar effects are also to be expected when using GPT-2 to craft new samples, as the synthesis is prone to grammatical errors. However, synthesizing new samples as well as averaging the classification probabilities of multiple transformer networks helped with increasing the classification accuracies. On the test set a score of 0.4258 for the stance and a score of 0.3581 for the premise detection was achieved.

4 Conclusion

In this paper, a classification system to estimate stances of health orders and to detect premises in Tweets is proposed. The system incorporates two types of generative models to allow training on unlabeled data and to generate new labeled data. The F1-scores achieved by the proposed model on the validation set yielded promising first results. However, it also showed, that the classification architecture is more suited for sentiment estimation rather than premise detection.

[1]https://github.com/carpedm20/emoji/

[2]https://github.com/minimaxir/aitextgen

Acknowledgements

This work was supported by the German Federal Ministry of Education and Research (BMBF) and the and the Hessian Ministry of Higher Education, Research, Science, and the Arts within their joint support of "ATHENE – Disinformation on Corona" and "Lernlabor Cybersicherheit".

References

Danilo Croce, Giuseppe Castellucci, and Roberto Basili. 2020. GAN-BERT: Generative adversarial learning for robust text classification with a bunch of labeled examples. In *Proceedings of the 58th Annual Meeting of the Association for Computational Linguistics*, pages 2114–2119, Online. Association for Computational Linguistics.

Vera Davydova and Elena Tutubalina. 2022. Smm4h 2022 task 2: Dataset for stance and premise detection in tweets about health mandates related to covid-19. In *Proceedings of the Seventh Social Media Mining for Health Applications (#SMM4H) Workshop & Shared Task*, pages –.

Jacob Devlin, Ming-Wei Chang, Kenton Lee, and Kristina Toutanova. 2019. Bert: Pre-training of deep bidirectional transformers for language understanding. *ArXiv*, abs/1810.04805.

Kyle Glandt, Sarthak Khanal, Yingjie Li, Doina Caragea, and Cornelia Caragea. 2021. Stance detection in COVID-19 tweets. In *Proceedings of the 59th Annual Meeting of the Association for Computational Linguistics and the 11th International Joint Conference on Natural Language Processing (Volume 1: Long Papers)*, pages 1596–1611, Online. Association for Computational Linguistics.

Ian Goodfellow, Jean Pouget-Abadie, Mehdi Mirza, Bing Xu, David Warde-Farley, Sherjil Ozair, Aaron Courville, and Yoshua Bengio. 2014. Generative adversarial nets. In *Advances in Neural Information Processing Systems*, volume 27. Curran Associates, Inc.

Diederik P. Kingma and Jimmy Ba. 2015. Adam: A method for stochastic optimization. *CoRR*, abs/1412.6980.

Zhenzhong Lan, Mingda Chen, Sebastian Goodman, Kevin Gimpel, Piyush Sharma, and Radu Soricut. 2019. ALBERT: A lite BERT for self-supervised learning of language representations. *CoRR*, abs/1909.11942.

Alec Radford, Jeff Wu, Rewon Child, David Luan, Dario Amodei, and Ilya Sutskever. 2019. Language models are unsupervised multitask learners.

Davy Weissenbacher, Ari Z. Klein, Luis Gascó, Darryl Estrada-Zavala, Martin Krallinger, Yuting Guo, Yao Ge, Abeed Sarker, Ana Lucia Schmidt, Raul Rodriguez-Esteban, Mathias Leddin, Arjun Magge, Juan M. Banda, Vera Davydova, Elena Tutubalina, and Graciela Gonzalez-Hernandez. 2022. Overview of the seventh social media mining for health applications #smm4h shared tasks at coling 2022. In *Proceedings of the Seventh Social Media Mining for Health Applications (#SMM4H) Workshop & Shared Task*, pages –.

UB Health Miners@SMM4H'22: Exploring Pre-processing Techniques To Classify Tweets Using Transformer Based Pipelines.

Roshan Vivek Khatri, Sougata Saha, Souvik Das, Rohini K. Srihari
Department of Computer Science and Engineering
University at Buffalo, Amherst, NY 14260
{roshanvi,sougatas,souvikda,rohini}@buffalo.edu

Abstract

Here we discuss our implementation of two tasks in the Social Media Mining for Health Applications (SMM4H) 2022 shared tasks – classification, detection, and normalization of Adverse Events (AE) mentioned in English tweets (Task 1) and classification of English tweets self-reporting exact age (Task 4). We have explored different methods and models for binary classification, multi-class classification, and named entity recognition (NER) for these tasks. We have also processed the provided dataset for noise, imbalance, and creative language expression from data. Using diverse NLP methods, we classified tweets for mentions of adverse drug effects (ADEs) and self-reporting the exact age in the tweets. Further, extracted reactions from the tweets and normalized these adverse effects to a standard concept ID in the MedDRA vocabulary.

1 Introduction

Since 2005 in the US, the percentage of American adults using social media has risen from 53% in 2012 to 72% in 2021 according to the research and tracking by Pew Research Center. Tremendous amounts of data are being generated on social media, where people express different aspects of their lives to their social circle. Many people also express their thoughts on various health-related topics. All this data presents significant opportunities for studying it to monitor people's health and the factors affecting their health. Our work is inspired by the current research using transformer architectures, and the results that Autobots Ensemble Team (Saha et al., 2020) had achieved at SMM4H 2020 and the KFU NLP Team (Miftahutdinov et al., 2019) had achieved at SMM4H 2019 using BERT (Kenton and Toutanova, 2019) as well as by the benchmark work and system configuration in ReportAGE (Klein et al., 2022).

Task 1 consists of 3 subtasks, Classification, Extraction, and Normalization. For a binary classification task, distinguishing tweets that report an adverse effect (AE) of medication are annotated as "1", from those that do not are annotated as "0", subtle causal language variation in the tweets needs to be addressed between AEs and indications. For extractions, the beginning and end offsets of these spans and the actually extracted spans are provided in the dataset for training the model and removing the spans for the unseen data. For the Normalization subtask, the extracted spans, and the normalized standard concept IDs of MedDRA[1] vocabulary are provided.

Task 4 is a binary classification task that automatically distinguishes tweets that self-report the user's exact age, annotated as "1", from those that do not, annotated as "0". Automatically identifying the actual age, rather than their age groups, would enable the large-scale use of social media data for applications that do not align with the predefined age groupings of extant models, including health applications such as identifying specific age-related risk factors for observational studies, or selecting age-based study populations.

2 Models

2.1 Task 1a: Classification of tweets that report adverse effects

The dataset (Magge et al., 2021) provided for this task has 17,120 annotated tweets for classifying tweets containing Adverse Drug Events (ADEs) related to drugs from the ones that do not contain ADEs. For classification, we used a deep neural network classifier based RoBERTa-based (Liu et al., 2019), pre-trained transformer model from Hugging Face[2] (Wolf et al., 2019).

For the RoBERTa-based classifier, we preprocessed the tweets by normalizing the hashtags to words, emoticons to expressions, and removing

[1] https://www.meddra.org/
[2] https://huggingface.co/

Proceedings of the 29th International Conference on Computational Linguistic, pages 114–117
October 12–17, 2022.

Model Type	Precision	Recall	F1-score
RoBERTa	0.74	0.43	0.54
RoBERTa + Downsampling	0.38	0.88	0.53
RoBERTa + Downsampling + Preprocessing	0.4	0.83	0.54
RoBERTa + Downsampling + Preprocessing + Data Augmentation	0.54	0.71	0.62
RoBERTa + Downsampling + Preprocessing + Data Augmentation + Paraphrasing	0.38	0.86	0.52
RoBERTa + Preprocessing + Data Augmentation	0.53	0.73	0.61

Table 1: Task 1a results for different system configurations on the validation set

the duplicates, and URLs from the dataset. After getting the embeddings of the pre-processed tweet, we took the mean of the sequence output to be the pooled output and passed it to the hidden layer, dropout layer (drop rate of 0.5), and a linear layer that predicts the class of each tweet. For training, we used Adam Optimizer (Kingma and Ba, 2014), 10 epochs, warmup of 0.2, the learning rate of 1e-5, and weight decay of 0.001.

As the data set was quite imbalanced, with approximately 10% data of "ADEs" class, and the remaining 90% data for the "non-ADEs" class, we have tried different techniques to balance the dataset by downsampling the major class, data augmentation of the positive class by back-translation (Aji et al., 2021) method to german as well as Russian, paraphrasing using the parrot paraphraser, and assigning weights to the positive class. Table 1 contains our result for task 1a and the best F1 score of 0.62 was obtained by downsampling, pre-processing, and data augmentation by back-translation methods from both German and Russian.

2.2 Task 1b: Extraction of spans in the tweets

For this task, only 1,708 tweets in the dataset contain an extraction of spans in the tweets containing ADEs. We trained a Named Entity Recognition model using this dataset. For this, we converted the dataset in a data frame to the Spacy required format containing the text and the entities containing the

Model Type	Precision	Recall	F1-score
Spacy NER	0.19	0.48	0.28

Table 2: Task 1b results on validation set.

Training Data	Precision	Recall	F1-score
2022 Dataset	0.03	0.09	0.03
2022 + 2020 Dataset	0.22	0.27	0.23
2022 + 2020 + CADEC Dataset	0.45	0.47	0.45

Table 3: Task 1c results for different datasets on the validation set

offset of the entities. We trained our model for the new entity over the "en_core_web_sm" model of Spacy and specifically the "ner", "trf_wordpiecer", "trf_tok2vec" pipelines of the model. Table 2 shows the achieved Relaxed F1 score of 0.28 for the model we trained.

2.3 Task 1c: Mapping the spans of adverse effects to standard concept IDs in the MedDRA vocabulary

The dataset contains 1,711 tweet spans and mapped MedDRA ids for training. For this mapping, we developed a multi-class classifier where the number of classes is equal to the number of unique MedDRA terms available in the dataset. This multi-class classifier is also a deep neural network classifier based on the RoBERTa-based pre-trained transformer model from Hugging Face. We have padded/truncated (Gattepaille, 2020) the embeddings to the length of 16 to get the best results. After getting these padded/truncated embeddings of the preprocessed spans, we took the mean of the sequence output to be the pooled output and passed it to the hidden layer, dropout layer (drop rate of 0.5), and a linear layer that predicts the class of each tweet. For training, we used Adam Optimizer, 16 epochs, warmup of 0.2, the learning rate of 2e-5, and weight decay of 0.001.

As the dataset was quite small, we also used the 2020 dataset along with the CSIRO Adverse Drug Event Corpus (CADEC) (Karimi et al., 2015) MedDRA dataset, which in total 8,815 for training. Table 3 shows the results for this task and the best F1 score achieved was 0.45.

2.4 Task 4: Classification of tweets self-reporting exact age

The provided dataset comprises 11,000 annotated tweets for the experiment. Amongst them are 8,800

Model Type	Precision	Recall	F1-score
RoBERTa	0.85	0.92	0.88
RoBERTa + Preprocessing	0.86	0.90	0.88
RoBERTa + Downsampling	0.75	0.97	0.85
RoBERTa + Downsampling + Preprocessing	0.77	0.94	0.85
RoBERTa + Preprocessing + Data Augmentation(german)	0.82	0.92	0.87
RoBERTa + Downsampling + Preprocessing + Data Augmentation + Paraphrasing	0.79	0.92	0.85

Table 4: Task 4 results for different system configurations on the validation set

tweets for training and 2,200 as validation dataset for the classification of "age" and "no age" tweets. For classification, we used a deep neural network classifier based on the RoBERTa pre-trained transformer model from Hugging Face.

For the RoBERTa-based classifier, we preprocessed the tweets by normalizing the hashtags to words, emoticons to expressions, and usernames to the "@USER_" special token, along with normalizing the URLs and removing the duplicates from the dataset. After getting the embeddings of the preprocessed tweet, the mean of the sequence output is considered as the pooled output and passed to one hidden layer, dropout layer (drop rate of 0.5), and a linear layer that predicts the class of each tweet. For training, we used Adam Optimizer, 10 epochs, warmup of 0.2, the learning rate of 1e-5, and weight decay of 0.001.

As the dataset was quite imbalanced amongst the classes, with 32.20% data for the "age" class and the remaining 67.80% data for the "no age" class, we have tried different techniques to balance the data set by downsampling, and data augmentation by back-translation method to german as well as Russian and paraphrasing using the parrot paraphraser. All these techniques did not help in increasing the F1 score of the model as seen in Table 4 and therefore, only by preprocessing the data did the best F1 score of 0.88 with the best precision of 0.86 obtained.

3 Error Analysis

We observe that the model performs poorly to classify the tweets containing drug names and other medical terms. For the Named Entity Recognition, the model using Spacy fails in the cases where multiple spans must be extracted from the tweet. For the Normalization, for improving the F1 score, more data is required as we have only 8,815 data points including the dataset of SMM4H 2022, SMM4H 2020, and CADEC corpus trained to classify a total of 1,023 unique MedDRA IDs.

4 Result and Analysis

The performance evaluation metrics for the tasks are precision (P), recall (R), and F1-score (F1) computed on the positive class which is the minority class. The above result tables report the various techniques and the scores of those respective models. Different hyperparameters were tested to get the optimum F1 score of each of the models and different padding/truncating length lines 40, 32, and 16 of the embedding for the Normalization task were explored to see if that affected the results for the Tasks and padding/truncating to embedding length of 16 gave the best results.

5 Conclusion

All the datasets that were provided were highly imbalanced amongst the classes and many techniques were explored to overcome these imbalances, namely Random downsampling of the majority class, increasing the number of minority classes by paraphrasing the data of those classes, data augmentation by a round trip translation from English to German and back to English and this was also tried with Russian to create more data points. From the whole project, it was learned that the balance of the data between classes does not guarantee the performance of the model. It can also be seen that the preprocessing step is also very important to preserve the context of the tokens with respect to each other. For future scope, a medical domain-based pre-trained language model, with the present ones in an ensemble architecture might work to improve the score of the system.

References

Alham Fikri Aji, Made Nindyatama Nityasya, Haryo Akbarianto Wibowo, Radityo Eko Prasojo, and Tirana Fatyanosa. 2021. Bert goes brrr: A venture towards the lesser error in classifying medical self-reporters on twitter. In *Proceedings of the Sixth Social Media Mining for Health (# SMM4H) Workshop and Shared Task*, pages 58–64.

Lucie Gattepaille. 2020. How far can we go with just out-of-the-box bert models? In *Proceedings of the Fifth Social Media Mining for Health Applications Workshop & Shared Task*, pages 95–100.

Sarvnaz Karimi, Alejandro Metke-Jimenez, Madonna Kemp, and Chen Wang. 2015. Cadec: A corpus of adverse drug event annotations. *Journal of biomedical informatics*, 55:73–81.

Jacob Devlin Ming-Wei Chang Kenton and Lee Kristina Toutanova. 2019. Bert: Pre-training of deep bidirectional transformers for language understanding. In *Proceedings of naacL-HLT*, pages 4171–4186.

Diederik P Kingma and Jimmy Ba. 2014. Adam: A method for stochastic optimization. *arXiv preprint arXiv:1412.6980*.

Ari Z Klein, Arjun Magge, and Graciela Gonzalez-Hernandez. 2022. Reportage: Automatically extracting the exact age of twitter users based on self-reports in tweets. *PloS one*, 17(1):e0262087.

Yinhan Liu, Myle Ott, Naman Goyal, Jingfei Du, Mandar Joshi, Danqi Chen, Omer Levy, Mike Lewis, Luke Zettlemoyer, and Veselin Stoyanov. 2019. Roberta: A robustly optimized bert pretraining approach. *arXiv preprint arXiv:1907.11692*.

Arjun Magge, Elena Tutubalina, Zulfat Miftahutdinov, Ilseyar Alimova, Anne Dirkson, Suzan Verberne, Davy Weissenbacher, and Graciela Gonzalez-Hernandez. 2021. Deepademiner: a deep learning pharmacovigilance pipeline for extraction and normalization of adverse drug event mentions on twitter. *Journal of the American Medical Informatics Association*, 28(10):2184–2192.

Zulfat Miftahutdinov, Ilseyar Alimova, and Elena Tutubalina. 2019. Kfu nlp team at smm4h 2019 tasks: Want to extract adverse drugs reactions from tweets? bert to the rescue. In *Proceedings of the fourth social media mining for health applications (# SMM4H) workshop & shared task*, pages 52–57.

Sougata Saha, Souvik Das, Prashi Khurana, and Rohini K Srihari. 2020. Autobots ensemble: Identifying and extracting adverse drug reaction from tweets using transformer based pipelines. In *Proceedings of the Fifth Social Media Mining for Health Applications Workshop & Shared Task*, pages 104–109.

Ashish Vaswani, Noam Shazeer, Niki Parmar, Jakob Uszkoreit, Llion Jones, Aidan N Gomez, Łukasz Kaiser, and Illia Polosukhin. 2017. Attention is all you need. *Advances in neural information processing systems*, 30.

Thomas Wolf, Lysandre Debut, Victor Sanh, Julien Chaumond, Clement Delangue, Anthony Moi, Pierric Cistac, Tim Rault, Rémi Louf, Morgan Funtowicz, et al. 2019. Huggingface's transformers: State-of-the-art natural language processing. *arXiv preprint arXiv:1910.03771*.

CSECU-DSG@SMM4H'22: Transformer based Unified Approach for Classification of Changes in Medication Treatments in Tweets and WebMD Reviews

Afrin Sultana*, Nihad Karim Chowdhury, and Abu Nowshed Chy
Department of Computer Science and Engineering
University of Chittagong, Chattogram-4331, Bangladesh
afrin.sultana.cu@gmail.com, nihad@cu.ac.bd, and nowshed@cu.ac.bd

Abstract

Medications play a vital role in medical treatment as medication non-adherence reduces clinical benefit, results in morbidity, and medication wastage. Self-declared changes in drug treatment and their reasons are automatically extracted from tweets and user reviews, helping to determine the effectiveness of drugs and improve treatment care. SMM4H 2022 Task 3 introduced a shared task focusing on the identification of non-persistent patients from tweets and WebMD reviews. In this paper, we present our participation in this task. We propose a neural approach that integrates the strengths of the transformer model, the Long Short-Term Memory (LSTM) model, and the fully connected layer into a unified architecture. Experimental results demonstrate the competitive performance of our system on test data with 61% F1-score on task 3a and 86% F1-score on task 3b. Our proposed neural approach ranked first in task 3b.

1 Introduction

User-generated contents on social media and online health forums represent a wide variety of facts, experiences, and opinions on various health topics including personal health issues, reviews on medication, side effects, and informal questions on health concerns. Shared information on social media or online health forums allows to detect users' self-declared changes in medication. Self-declared changes include stopping a treatment, changing a dose, or forgetting to take the drugs, etc. In this work, we focus on classifying patients making changes to their medication treatments (i.e non-persistent patients) as a part of our participation in the Social Media Mining for Health (SMM4H) 2022 shared task 3 (Davy et al., 2022). This task includes two subtasks. We have to detect changes in medication treatments from tweets and WebMD reviews in subtasks 3a and 3b, respectively. WebMD is one of the top healthcare websites and an online publisher of news and information pertaining to drugs, human health and well-being. Table 1 represents some examples of persistent and non-persistent tweets and reviews from the SMM4H 2022 task 3 dataset, respectively.

Tweet/Review	Label
Task 3a: tweet classification	
E#1: I broke out with a rash after starting this medication, is this normal?	Persistent
E#2: This medication caused me to cough a lot . Hated it.	Non-persistent
Task 3b: WebMD review classification	
E#3: It's totally normal to take a ambien on a 2 hour flight, right?	Persistent
E#4: I quit Lipitor all together	Non-persistent

Table 1: Examples of persistent and non-persistent tweets/reviews.

The major contribution of this paper is that we propose an approach to explore the robustness of the RoBERTa (Liu et al., 2019) model with the unification of features from LSTM (Hochreiter and Schmidhuber, 1997) and a fully connected layer where RoBERTa maps the input text into meaningful embeddings effectively, LSTM captures long-term dependencies, and fully connected linear layer selects effective features to encapsulate context.

The rest of the paper describes our proposed framework, experimental details, and evaluation.

2 Proposed Framework

An overview of our proposed framework is shown in Figure 1. For a given text, we use the FLAIR (Akbik et al., 2019) framework to extract document embedding features from the transformer model RoBERTa. These features are carried out

*Corresponding Author

Proceedings of the 29th International Conference on Computational Linguistic, pages 118–122
October 12–17, 2022.

through two different capsules. One is a stacked LSTM and the other is a simple linear layer. Then, we concatenate the features of the LSTM and linear architecture, which are then applied to a feed-forward fully-connected output layer to procure predicted labels.

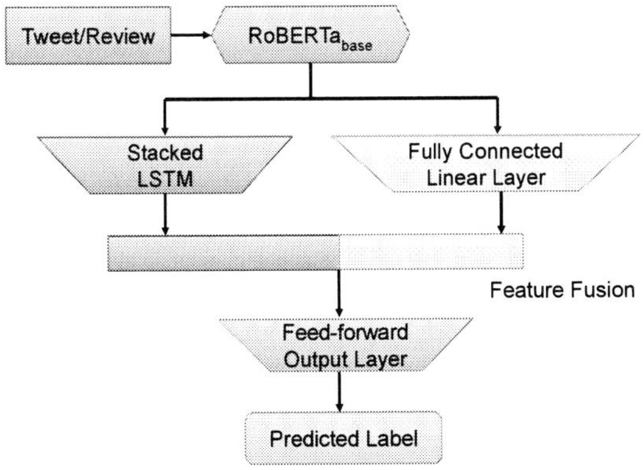

Figure 1: Proposed framework.

2.1 Transformer Document Embedding

Document Embedding provides an embed for the entire text. We leverage the FLAIR framework to generate document embeddings for each given text using a RoBERTa model from the Transformers (Vaswani et al., 2017) family.

RoBERTa: RoBERTa, a robustly optimized BERT pre-training method, is an extension to the original BERT (Devlin et al., 2018) model that trains the encoding vocabulary using larger byte-level bytes containing 50K subword units. It improves BERT by removing next-sentence prediction targets, dynamically changing mask patterns, and training the model with larger batches on more data (Liu et al., 2019).

2.2 Stacked LSTM

Document embedding vectors are fed into stacked long short-term memory (LSTM) capsules, as LSTMs capture long-term dependencies by storing previous information. Furthermore, stacked LSTM is an extension of the LSTM model that has multiple hidden LSTM layers, where each layer contains multiple memory cells. A stacked LSTM is utilized in order to increase capacity (Staudemeyer and Morris, 2019) and depth of the model (Graves et al., 2013). The stacked LSTM capsule generates an m-dimensional feature vector where 'm' is the hidden size of LSTM capsule.

2.3 Fully Connected Linear Layer

The document embedding vector obtained from RoBERTa is passed through a fully connected linear layer. Fully connected layers learn from high-level features and provide efficient feature representations that encapsulate the essence of a given high-level feature. An n-dimensional feature vector is produced from the document embedding vector where 'n' is the number of neurons of fully connected linear layer.

2.4 Output Layer

We concatenated the m-dimensional feature vector of the stacked LSTM capsule and the n-dimensional feature vector of the fully connected linear layer. Later, the fusion vector passes through a feed-forward fully connected linear layer operating as an output layer to obtain predicted labels. We utilize the SoftMax activation function in the output layer to normalize the features to probabilities, considering the highest probability class as the predicted label.

3 Experiments and Evaluations

3.1 Dataset Description and Evaluation Measures

Category	Train	Dev	Test
Task 3a: tweet classification			
Persistent	5380	1434	-
Non-persistent	518	138	-
Total	5898	1572	2360
Task 3b: WebMD review classification			
Persistent	4632	556	-
Non-persistent	5746	741	-
Total	10378	1297	1297

Table 2: The statistics of SMM4H 2022 task 3 dataset.

The organizers of the SMM4H 2022 task 3 (Davy et al., 2022) provided a benchmark dataset that consists of two corpora: a set of tweets with imbalanced positive and negative tweets, and a set of drug reviews from WebMD.com with balanced positive and negative reviews. Persistent data is labeled with '0' whereas non-persistent data is labeled with '1'. Table 2 shows the statistics of the used dataset.

To evaluate the performance of participants' systems, SMM4H 2022 task 3 organizers employed standard strategies for sub-tasks 3a and 3b. Standard evaluation metrics, including precision, recall, and F1-score for non-persistent classes, were used to evaluate the performance of the system.

3.2 Experimental Settings

In our CSECU-DSG system, we use the hashtag, URL, and username stripping techniques for tweets. We utilize the Huggingface (Wolf et al., 2019) transformer model RoBERTa (Liu et al., 2019) and fine-tune it with PyTorch (Paszke et al., 2019). Using the average of the top four layers, a 768-dimensional document embedding is generated from RoBERTa, which is forwarded to two stacked layers of the LSTM module and a fully connected linear architecture, yielding 1024- and 512-dimensional feature vectors, respectively. A 1536-dimensional fusion vector is generated by concatenating the output vector of the stacked LSTM and the linear architecture. Another feed-forward fully connected linear layer produces output from the 1536-dimensional fusion vector.

Parameter & Embeddings	Settings	
	Task 3a	Task 3b
learning_rate	1e-5	2e-5
max_epoch	15	4
batch_size	4	4
anneal_factor	0.5	0.5
patience	3	2
Transformer document embeddings	"roberta-base", layers = "-1,-2,-3,-4", layer_mean = True	
LSTM	input_size = 768, hidden_size = 1024, num_layers = 2, bidirectional = False	
Linear architecture	input_size = 768, output_size = 512	

Table 3: Model configuration and settings.

We use the FLAIR (Akbik et al., 2019) framework to implement our system. We train our system with the provided training data. We use Google Colab's GPU and set the set_seed = 42 to generate the reproducible results. We experiment with epochs in range [4,8,12,15] with different values of patience, batch size in range [4,8], learning rate in range [1e-5, 2e-5, 2e-4, 3e-4]. For the embedding settings, we consider the last layer, the average of the last four layers in RoBERTa, LSTM hidden_size 256,512,1024, LSTM num_layers 1,2,3, and linear layer feature value 256,512. Table 3 describes the set of optimal parameters used to design our proposed model. Default settings are used for the other parameters.

3.3 Results and Analysis

Category	F1-Score	Precision	Recall
Task 3a: tweet classification			
Dev set	0.6320	0.6489	0.6159
Test set	0.6082	0.6555	0.5673
Test (Median)	0.5859	0.6170	0.5577
Task 3b: WebMD review classification			
Dev set	0.9031	0.8767	0.9312
Test set	0.8608	0.8472	0.8748
Test (Median)	0.8432	0.8436	0.8646

Table 4: Results on the development and test sets.

Table 4 shows the performance of our best-performing system based on the development set and the official results on test data in the individual sub-task. Compared to the official median scores on the test set computed using the participants' submissions, our system scored 61% and 86% F1-scores in subtasks 3a and 3b, respectively. With an 86% F1-score on WebMD review classification, we ranked first at task 3b.

3.4 Discussion

To qualitatively analyze the effectiveness of components utilized in our system, we perform an ablation study on the development set of task 3a and 3b. The experimental results are summarized in Table 5. The experimental result shows that the RoBERTa model performs pretty well, but combining RoBERTa with the LSTM layer increases the F1-score whereas incorporating stacked LSTM with RoBERTa brings a shrink in the F1-score. RoBERTa with LSTM and fully connected layer provides 3.37% and 0.22% growth in F1-score, in contrast, our proposed unified approach of RoBERTa, stacked LSTM, and fully connected linear architecture lead to a rise of 5.53% and 1.45% in F1-score on task 3a and 3b, respectively. Though RoBERTa with stacked LSTM lessens the F1-score, competitive performance appears for RoBERTa with stacked LSTM and fully connected layer inferring the efficacy of fusion of stacked LSTM and fully connected layer.

Method	Task 3a: tweet data			Task 3b: WebMD review		
	F1-score	Precision	Recall	F1-score	Precision	Recall
RoBERTa	0.5785	0.6731	0.5072	0.8886	0.8735	0.9042
RoBERTa + LSTM	0.5847	**0.7041**	0.5000	0.8903	0.8670	0.9150
RoBERTa + Stacked LSTM	0.5283	0.5512	0.5072	0.8830	0.8702	0.8961
RoBERTa + LSTM + Fully Connected Layer	0.6122	0.7009	0.5435	0.8908	**0.8896**	0.8920
Proposed Method: RoBERTa + Stacked LSTM + Fully Connected Layer	**0.6320**	0.6489	**0.6159**	**0.9031**	0.8767	**0.9312**

Table 5: Ablation study on the development set of task 3a and 3b. The best results are highlighted in boldface.

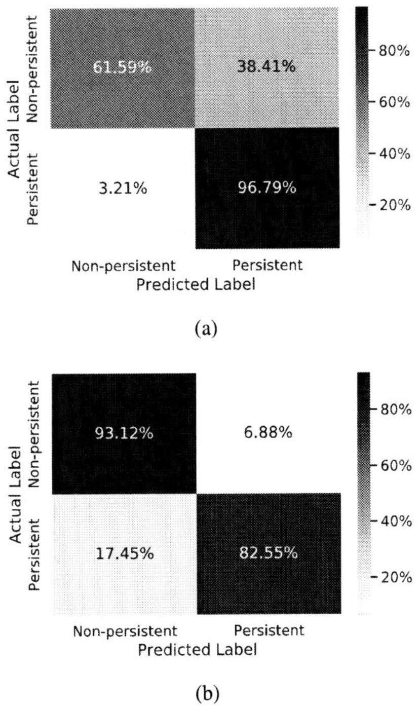

Figure 2: Confusion matrix of task 3a and 3b.

Tweet/Review	Actual	Predicted
Task 3a: tweet classification		
E#1: So for all your information, no I did not take zofran when I was pregnant.	0	1
E#2: I'll be back on Lipitor tomorrow, I'm sure.	1	0
Task 3b: WebMD review classification		
E#3: I had several debillatating side effects .	0	1
E#4: This med is useless for pain. The only pain meds that works is hydrocodone.	1	0

Table 6: Erroneous prediction of proposed method.

4 Error Analysis

Figure 2 shows the confusion matrix for subtasks 3a and 3b on the development set. We observe a relatively high proportion of misclassified non-persistent tweets, while the system performs fairly well in detecting non-persistent reviews.

Further, we articulate some erroneous predictions of our system in Table 6 to look into the reasons for misclassification. The first example (E#1) expresses a person's loftiness for not taking medicine, but our system misinterprets it as non-adherence to medication. The second example (E#2) implicitly conveys a patient's intention to skip medication dose today and stick to it from tomorrow. Due to the implicit nature of meaning, our system fails to detect the non-adherence characteristics of the tweet. On the contrary, for WebMD review classification, sample E#3 indicates aftereffects of consuming a drug that is misapprehended by our system. The first part of the sample E#4 shows the inefficacy of a drug whereas the second part reveals the effectiveness of another drug. Owing to the contradictory meaning, our system fails to classify the review correctly. Besides, the dataset is quite imbalanced as depicted in Table 2. Therefore, an appropriate strategy to handle implicitness, brevity, ambiguity, and unusual structure of text will improve the performance of our system.

5 Conclusion

In this paper, we introduce an approach to identify the self-declared drug-changing information by integrating RoBERTa, LSTM, and fully connected linear layers. In the future, we intend to explore biomedical-based transformers and model ensembling to distill better feature representation.

References

Alan Akbik, Tanja Bergmann, Duncan Blythe, Kashif Rasul, Stefan Schweter, and Roland Vollgraf. 2019. Flair: An easy-to-use framework for state-of-the-art nlp. In *Proceedings of the 2019 conference of the North American chapter of the association for computational linguistics (demonstrations)*, pages 54–59.

Weissenbacher Davy, Z. Klein Ari, Gascó Luis, Estrada-Zavala Darryl, Krallinger Martin, Guo Yuting, Ge Yao, Sarker Abeed, Lucia Schmidt Ana, Rodriguez-Esteban Raul, Leddin Mathias, Magge Arjun, M. Banda Juan, Davydova Vera, Tutubalina Elena, and Gonzalez-Hernandez Graciela. 2022. Overview of the seventh social media mining for health applications #smm4h shared tasks at coling 2022. In *Proceedings of the Seventh Social Media Mining for Health (#SMM4H) Workshop and Shared Task*.

Jacob Devlin, Ming-Wei Chang, Kenton Lee, and Kristina Toutanova. 2018. Bert: Pre-training of deep bidirectional transformers for language understanding. *arXiv preprint arXiv:1810.04805*.

Alex Graves, Abdel-rahman Mohamed, and Geoffrey Hinton. 2013. Speech recognition with deep recurrent neural networks. In *2013 IEEE international conference on acoustics, speech and signal processing*, pages 6645–6649. Ieee.

Sepp Hochreiter and Jürgen Schmidhuber. 1997. Long short-term memory. *Neural computation*, 9(8):1735–1780.

Yinhan Liu, Myle Ott, Naman Goyal, Jingfei Du, Mandar Joshi, Danqi Chen, Omer Levy, Mike Lewis, Luke Zettlemoyer, and Veselin Stoyanov. 2019. Roberta: A robustly optimized bert pretraining approach. *arXiv preprint arXiv:1907.11692*.

Adam Paszke, Sam Gross, Francisco Massa, Adam Lerer, James Bradbury, Gregory Chanan, Trevor Killeen, Zeming Lin, Natalia Gimelshein, Luca Antiga, Alban Desmaison, Andreas Kopf, Edward Yang, Zachary DeVito, Martin Raison, Alykhan Tejani, Sasank Chilamkurthy, Benoit Steiner, Lu Fang, Junjie Bai, and Soumith Chintala. 2019. Pytorch: An imperative style, high-performance deep learning library. In *Advances in Neural Information Processing Systems 32*, pages 8024–8035. Curran Associates, Inc.

Ralf C Staudemeyer and Eric Rothstein Morris. 2019. Understanding lstm–a tutorial into long short-term memory recurrent neural networks. *arXiv preprint arXiv:1909.09586*.

Ashish Vaswani, Noam Shazeer, Niki Parmar, Jakob Uszkoreit, Llion Jones, Aidan N Gomez, Łukasz Kaiser, and Illia Polosukhin. 2017. Attention is all you need. *Advances in neural information processing systems*, 30.

Thomas Wolf, Lysandre Debut, Victor Sanh, Julien Chaumond, Clement Delangue, Anthony Moi, Pierric Cistac, Tim Rault, Rémi Louf, Morgan Funtowicz, et al. 2019. Huggingface's transformers: State-of-the-art natural language processing. *arXiv preprint arXiv:1910.03771*.

Innovators@SMM4H'22: An Ensembles Approach for self-reporting of COVID-19 Vaccination Status Tweets

Mohammad Zohair[ϕ], **Nidhir Bhavsar**[&], **Aakash Bhatnagar**[$], **Muskaan Singh**[#], **Petr Motlicek**[#]
[ϕ]Jamia Millia Islamia, New Delhi, India
[$]Boston University, Boston, Massachusetts
[&]University of Potsdam, Germany
[#] Speech and Audio Processing Group, IDIAP Research Institute, Martigny, Switzerland
mohammadzohair2002@gmail.com,aakash07@bu.edu, nidbhavsar989@gmail.com, (msingh,petr.motlicek)@idiap.ch

Abstract

With the Surge in COVID-19, the number of social media postings related to the vaccine has grown, specifically tracing the confirmed reports by the users regarding the COVID-19 vaccine dose termed as *Vaccine Surveillance*. To mitigate this research problem, we present our novel ensembled approach for self-reporting COVID-19 vaccination status tweets into two labels, namely *Vaccine Chatter* and *Self Report*. We utilize state-of-the-art models, namely BERT, RoBERTa, and XLNet. Our model provides promising results with 0.77, 0.93, and 0.66 as precision, recall, and F1-score (respectively), comparable to the corresponding median scores of 0.77, 0.9, and 0.68 (respectively). The model gave an overall accuracy of 93.43. We also present an empirical analysis of the results to present how well the tweet was able to classify and report. We release our code base here `https://github.com/Zohair0209/SMM4H-2022-Task6.git`

1 Introduction and Motivation

COVID-19 pandemic has unprecedented challenges and also advances to humanity. It provides the scientific community to advance science with wider access to openly available data(Banda et al., 2021). Vaccine surveillance became a pressing research issue with the widespread roll-out of COVID-19 vaccines. Social media platforms such as Twitter and Facebook contain abundant text data that can be utilized for research purposes(Banda et al., 2021).

In this paper, we describe our submission for the shared task [1], for vaccine surveillance which is a very pressing issue. We aim to target people who report adverse events via their healthcare providers to systems like Vaccine Adverse Event Reporting System (VAERS)(Chen et al., 1994), or are found documented in their electronic health record (EHR), a more robust and convenient method could be devised using self-reports from social media. In this task, we experiment with a dataset of Twitter users personally reporting vaccination status and users discussing vaccination status but not revealing their own. This task is tricky because users discuss the vaccination status of others or from news reports in similar ways than they discuss their own at a higher rate (1 to 8 on average). The dataset presents as the positive class, unambiguous tweets of users clearly stating that they have been vaccinated. All other tweets are of users discussing vaccination status. This task involves the identification of self-reported COVID-19 vaccination status in English tweets. As a two-way classification task, the two classes in the training dataset correspond to vaccination confirmation and vaccine-related chatter.

In this task, there is a class imbalance of roughly 1 to 8, implying that 1,496 tweets correspond to vaccination confirmation, and 12,197 tweets correspond to vaccine chatter.

1.1 Proposed System Architecture

This section describes the proposed methodology for classifying tweets as Self-reports or Vaccine Chatter in the context of COVID-19 vaccines. As presented in the 1, we proposed an ensembles approach for this multi-class classification problem. For this purpose, we used multiple pre-trained transformer models, which provide good results based on the dataset. Since transformer models are based on the mechanism of self-attention and differentially weight the significance of each part of the input data, they serve as an ideal option for this task. Initially, split the training dataset in the ratio of 80:20 to execute the classification task

[1]`https://healthlanguageprocessing.org/smm4h-2022/`

Proceedings of the 29th International Conference on Computational Linguistic, pages 123–125
October 12–17, 2022.

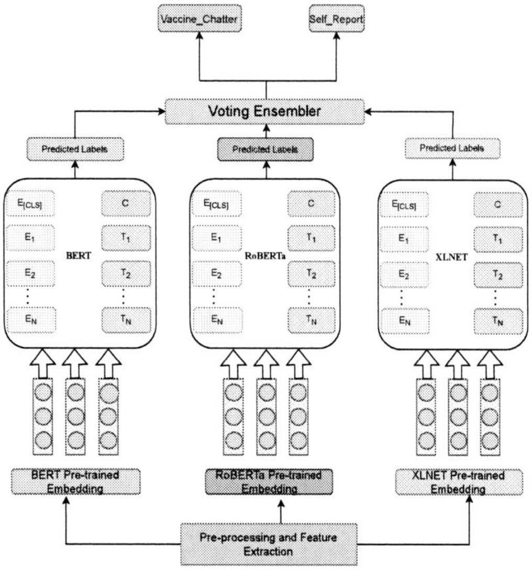

Figure 1: Proposed Architecture

and check the performance of the respective models. We had fine-tuned BERT (Sun et al., 2019), Roberta (Liu et al., 2019), and XLNet (Topal et al., 2021) models, respectively, for this dataset. As a part of the fine-tuning procedure, we tried different combinations of the optimizers, like AdamW and AdaFactor, along with other schedulers, including LinearScheduler, and CosineScheduler, PolyScheduler, etc. The loss function used in the models is Cross-Entropy loss (Ho and Wookey, 2019). Having tried various combinations, we selected the most optimum combination giving the most accurate results. Next, we ensembled these models, incorporating the voting method, which gave the combined label predictions for the respective tweets. Compared to the performance of the ensembles model with the performance of the separate individual models.

1.2 Hyperparameter

Finally, the ensemble model was considered and used for predicting the label class on the test Set as part of this task. We used the Tesla T4 GPU with 16GB RAM alongside 2560 CUDA cores for the experimental setup. We divide the entire dataset into an 80:20 training and validation split of 8 batches, with a learning rate (1e-5) and Adam optimizer(Bock and Weiß, 2019) with epsilon (1e-8). We feed a seed_val of 42. To calculate the training loss over all the batches, we use gradient descents (Ruder, 2016) clipping the norm to 1.0 to avoid exploding gradient problems.

Model	Precision	Recall	F1
Baseline	0.83	0.9	0.77
Median	0.77	0.9	0.68
Ours	0.77	0.93	0.66

Table 1: Results for our proposed ensembled approach with the median and baseline scores

Tweets	Actual	Predict
A few months ago he said masks don't work go figure. He's an idiot. I wonder how many people know we had the coronavirus last year. This one has been (I think) genetically engineered by China to be more lethal. A vaccine this year will not work next yr. Viruses mutate.	Vaccine_chatter	Vaccine_chatter
@DrAnthonyF The new coronavirus has been in081 the United States for seven months since March. I heard that the vaccine in the United States may not be available until March next year. I believe you and hope you can lead the Americans through the epidemic sa	Vaccine_chatter	Vaccine_chatter
I fi- nally got my Covid vaccine today! And so far, I'm feeling fine. I'm going back on April 16th for my second dose. Covid-Vaccine COVID19Vaccination COVID19Vaccine COVID19	Self_report	Self_report
@GovMikeDeWine Got my sec- ond covid-19 vaccine yesterday! Feeling general malaise and sore arm other than that I'm doing fine. The vaccine is safe and effective. The more that are vccinated the faster we can get back to normal.	Self_report	Self_report
Every- one in my household apart from me has had the vaccine and none of grown extra heads or got covid. Grow up and take it so we can get out of this pan- demic and donâC™t head into a third wave	Self_report	Vaccine_chatter
Because they havenâC™t got their vaccine yet. Why? Because itâC™s not their turn yet. Until when? Until when? Harini je dah ada a few under thirties without any comorbidity issues died. And yet there are elderlies yang tak dapat vaccination date lagi.	Vaccine_chatter	Self_report

Table 2: Result analysis for empirical evaluation of predicted results from our proposed system

2 Results

We describe the results in Table 2. The baseline model achieved 0.83, 0.9, and 0.77 precision, recall, and F1-score (respectively). Our model gave promising results with the precision, recall, and F1-score, standing at 0.77, 0.93, and 0.66 (respectively), which was comparable to the corresponding median scores of 0.77, 0.9, and 0.68 (respectively). The proposed model achieved an overall accuracy of 93.43%.

2.1 Analysis

In this section, we present the empirical analysis of our results in Table: 2. The first and second instances are correctly labeled as Vaccine Chatter, as they do not contain any relevant phrase indicating the user's undertaking of any vaccine dose. These results suggest the perfect predictions of the model. Similarly, the third and fourth instances show that the user has taken a dose of the COVID-19 vaccine. Hence, they have been correctly classified under the Self Report label, re-emphasizing the correct working of the model.

Contrary to these, the fifth instance was initially labeled as Self Report. However, due to the ambiguous declaration of the undertaking of COVID-19 vaccine dose, the model fails to classify it correctly and instead predicts it under the label of Vaccine Chatter. The sixth instance, on the other hand, was initially labeled as Vaccine Chatter. However, the model mislabels it due to the indecisive phrases featured in the separate tweet.

3 Conclusion

In this paper, we present our system paper submission for Innovators @ SMM4H'22. We aim to classify user tweets, indicating whether they represent Self Reports for the COVID-19 vaccines or are a part of general Vaccine Chatter. The proposed system is an ensembled voting model with fine-tuned BERT, RoBERTa, and XLNet. Given tweets in English, the submitted model classifies each vaccine-related tweet into one of the two labels: "Vaccine Chatter" and "Self Report." The system performs quite well to accomplish the desired task with an accuracy of 93.43%. In the future, we intend to work on a multitask learning framework to handle social media postings related to other medical advancements, apart from the COVID-19 vaccines. We also aim to develop models for multi-lingual postings featuring similar scenarios.

4 Acknowledgements

This work was supported by the European Union's Horizon 2020 research and innovation program under grant agreement No. 833635 (project ROXANNE: Real-time network, text, and speaker analytics for combating organized crime, 2019-2022).

References

Juan M. Banda, Ramya Tekumalla, Guanyu Wang, Jingyuan Yu, Tuo Liu, Yuning Ding, Ekaterina Artemova, Elena Tutubalina, and Gerardo Chowell. 2021. A large-scale covid-19 twitter chatter dataset for open scientific research—an international collaboration. *Epidemiologia*, 2(3):315–324.

Sebastian Bock and Martin Weiß. 2019. A proof of local convergence for the adam optimizer. In *2019 International Joint Conference on Neural Networks (IJCNN)*, pages 1–8. IEEE.

Robert T Chen, Suresh C Rastogi, John R Mullen, Scott W Hayes, Stephen L Cochi, Jerome A Donlon, and Steven G Wassilak. 1994. The vaccine adverse event reporting system (vaers). *Vaccine*, 12(6):542–550.

Yaoshiang Ho and Samuel Wookey. 2019. The real-world-weight cross-entropy loss function: Modeling the costs of mislabeling. *IEEE Access*, 8:4806–4813.

Yinhan Liu, Myle Ott, Naman Goyal, Jingfei Du, Mandar Joshi, Danqi Chen, Omer Levy, Mike Lewis, Luke Zettlemoyer, and Veselin Stoyanov. 2019. Roberta: A robustly optimized bert pretraining approach. *arXiv preprint arXiv:1907.11692*.

Sebastian Ruder. 2016. An overview of gradient descent optimization algorithms. *arXiv preprint arXiv:1609.04747*.

Chi Sun, Xipeng Qiu, Yige Xu, and Xuanjing Huang. 2019. How to fine-tune bert for text classification? In *China national conference on Chinese computational linguistics*, pages 194–206. Springer.

M Onat Topal, Anil Bas, and Imke van Heerden. 2021. Exploring transformers in natural language generation: Gpt, bert, and xlnet. *arXiv preprint arXiv:2102.08036*.

Innovators@SMM4H'22: An Ensembles Approach for Stance and Premise Classification of COVID-19 Health Mandates Tweets

Vatsal Savaliya[ϕ], Aakash Bhatnagar[$], Nidhir Bhavsar[&], Muskaan Singh[#], Petr Motlicek[#]
[ϕ]Indian Institute of Information Technology Lucknow, Indian
[$]Boston University, Boston, Massachusetts
[&]University of Potsdam, Germany
[#] Speech and Audio Processing Group, IDIAP Research Institute, Martigny, Switzerland
lci2020013@iiitl.ac.in, aakash07@bu.edu, nidbhavsar989@gmail.com,
(msingh,petr.motlicek)@idiap.ch

Abstract

This paper presents our submission for the Shared Task-2 of classification of stance and premise in tweets about health mandates related to COVID-19 at the Social Media Mining for Health 2022. There have been a plethora of tweets about people expressing their opinions on the COVID-19 epidemic since it first emerged. The shared task emphasizes finding the level of cooperation within the mandates for their stance towards the health orders of the pandemic. Overall the shared subjects the participants to propose system's that can efficiently perform 1) Stance Detection, which focuses on determining the author's point of view in the text. 2) Premise Classification, which indicates whether or not the text has arguments. Through this paper we propose an orchestration of multiple transformer based encoders to derive the output for stance and premise classification. Our best model achieves a F1 score of 0.771 for Premise Classification and an aggregate macro-F1 score of 0.661 for Stance Detection. We have made our code public here

1 Introduction

Users actively share their opinions on numerous problems on social media. These concerns are now frequently linked to the COVID-19 pandemic. Users, for example, share their feelings about quarantines and wearing masks in public places. Some comments are supported by arguments, while others are simply emotional claims.

Using Twitter tweets, automated algorithms for assessing people's attitudes regarding COVID-19 health orders can assist in determining the extent of cooperation with the mandates.

In this paper, we try to provide a feasible solution towards the task of Stance-Detection(Glandt et al., 2021). We are extracting arguments from COVID-related tweets. According to argumentation theory, an argument must include a claim containing a stance towards some topic or object, and at least one premise supporting this stance. So basically, we are finding whether the given tweet contains a stance (favour, against, or none) and also if it contains a premise (reason to support the stance).

Model	Precision	Recall	F1
BERT	0.672	0.667	0.659
RoBERTa	0.671	0.616	0.599
PubMedBERT	0.643	0.637	0.639
SciBERT	0.663	0.66	0.661
SPECTER	0.596	0.566	0.572
Ensemble	0.717	0.724	0.716

Table 1: Shows the evaluation scores obtained on the test-set of the Stance Detection dataset measured across the metrics precision, recall, and f1. We represent each individual score obtained over fine-tuning each of the previously mentioned pre-trained language models.

The first subtask of the shared task is to design a system that can determine the point of view (stance) of the text's author in relation to the given claim (e.g., wearing a face mask).Given a tweet, there can be 3 different values for stance *FAVOR* – positive stance *AGAINST* – negative stance *NEITHER* – neutral/unclear/irrelevant stance

The second subtask is to predict whether at least one premise or argument is mentioned in the text. A given tweet is considered to have a premise if it contains a statement that can be used as an argument in a discussion. There are 2 values for premise. **1** – tweet contains a premise (argument) **0** – tweet doesn't contain a premise (argument).

1.1 Motivation

Millions of users share and read tweets actively. During the COVID-19 pandemic, many users shared information about health mandates such as face masks, stay-at-home orders, and school closures. So basically, every tweet claims some particular topic.

The first subtask is to determine the point of view (stance) of the text's author in relation to the given claim. This means that stance detection is very important for a tweet, especially when the

Proceedings of the 29th International Conference on Computational Linguistic, pages 126–129
October 12–17, 2022.

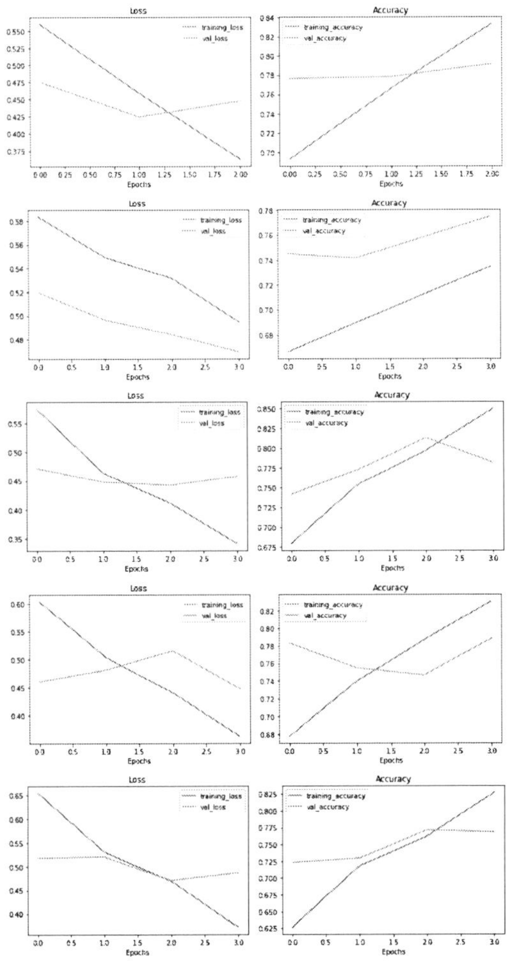

Figure 1: Graphical presentation for the loss and accuracy plot for (a)BERT (b)RoBERTa (c) PubMedBERT (d) SciBERT (e)SPECTER

tweet is done by a renowned personality. The second subtask is to predict whether at least one premise/argument is mentioned in the tweet. It's also important to find out whether the tweet is supported by an argument or not. This way, we can show tweets to users which are actually supported by an argument.

2 System Architecture

The main task was divided into two subtasks. Subtasks (a)STANCE DETECTION and (b)PREMISE CLASSIFICATION, respectively. Both subtasks have classification problems, with subtask (a) having 3 classes (favour, against, or none) and subtask (b) having 2 classes (1-tweet contains a premise , 2-tweet does not contain a premise).

We try out five different transformer models: BERT, PubMedBERT, ROBERTS, SciBERT, and SPECTER. Finally, we combine all five models' predictions with a voting classifier.

n ensemble modeling, multiple diverse models are created to predict an outcome. In our case, we use a voting classifier to ensemble the predictions of all five models (BERT, RoBERTa, PubMedBERT, SciBERT and SPECTER). As a result, the overall accuracy and f1 scores increased for stance detection and premise classification.

- BERT(Devlin et al., 2018)is pretrained language model trained specifically for the task of masked language modelling (MLM) and next sentence prediction (NSP).

- RoBERTa(Liu et al., 2019) is a transformers based pretrained language model trained on a large corpus of English data in a self-supervised fashion. The model follows the pre-training architecture proposed by Devlin et al. (2018) tries to overcome the limitation relative to Next Sentence Prediction (NSP). It also states the impact of hyperparameter tuning and training data size.

- PubMedBERT(Gu et al., 2020) is a pretrained biomedical domain-specific model which follows the BERT architecture. It is trained on the publicly available PubMed and full-text articles from PubMedCentral.

- SciBERT(Beltagy et al., 2019) follows the BERT architecture and is pretrained on scientific training corpus of papers from the Semantic Scholar and constitutes of both abstracts and full-texts. The language model follows a different wordpiece vocabulary "scivocab" built specifically to match the training data.

- SPECTER(Cohan et al., 2020) is a pre-trained model to generate document-level embeddings of scientific texts. It is pretrained on a powerful signal of document-level relatedness also refereed as citation graphs. Furthermore, the language model incorporates SciBERT, and can be applied easily to downstream scientific domain specific tasks.

3 Results

We present our results in Table: 1 and 2 across five different models. BERT, PubMedBERT, RoBERTa, SciBERT, SPECTER which were trained and evaluated on validation and testing data for both the

Model	Precision	Recall	F1	Accuracy
BERT	0.7191	0.7681	0.7428	0.805
RoBERTa	0.6313	0.8409	0.7212	0.7616
PubMedBERT	0.75	0.709	0.7289	0.8066
SciBERT	0.7759	0.6454	0.7047	0.8016
SPECTER	0.6762	0.75	0.7112	0.7766
ensemble	0.761	0.7818	0.7713	0.83

Table 2: Shows the evaluation scores obtained on the validation-set of the Premise Classification dataset measured across the metrics precision, recall, f1, and Accuracy. We represent each individual score obtained over fine-tuning each of the previously mentioned pre-trained language models.

	F1 Stance Detection	F1 Premise Classification
Ours	0.4816236592	0.64073476
Mean	0.4913350946	0.5739699138
Median	0.5501048571	0.6472029954

Table 3: Compares the F1 scores obtained by the best performing system at the SMMH'22 Task 2. It also provides and analysis over the Mean and Median of submission under the same task

subtasks. Our proposed approach of ensembled voting classifier, performs best (in terms of F1-scores of **0.716** for stance detection and **0.83** for premise classification.

4 Analysis

The overall dataset contains 6269 tweets, 3669 in training, 600 in validation and 2000 in testing. The tweets were further divided into three categories based on their claim, (i) face masks (ii) stay at home orders (iii) school closures.

As previously indicated, our proposed system employs an ensemble voting classifier, which assigns distinct weights to each of the predictions from the 5 included models during inference and determines the output label appropriately. Table 1 provides separate results obtained by each of our fine-tuned models on the stance detection dataset. It is clearly justifiable that the ensemble approach easily outmatch the individual models. Next, the individual results from each of our trained models on the given premise classification dataset are shown in Table 2. Although the SciBERT model produces a higher precision-score of 0.7759, the total f1-score obtained by ensemble surpasses the same by a factor of 0.01 compared to the prior method, yielding better results overall.

Table 4 compares our models prediction against the ground truth labels. As it is clearly seen, our system can easily classify both stance and premise appropriately. For better clarification, we also provide the associated claims alongside the given tweets. The proposed model may be partially treating all of the tweets that express negativity as the stance *AGAINST*, which is indirectly related to premise being mis-classified, and as a result, the models may occasionally discover the presence of premise for assertions like the `stay at home orders`.

Table 3 shows the complete task statistics for the test dataset for both stance detection and premise classification. Our premise classification proposal received a f1 score of 0.64, which is significantly higher than the typical submission for this subtask. However, for stance detection, our model underperformed the average submission, while the median shows that the majority of recursive submissions earned a f1 score of 0.51, showing that our suggested approach has to be improved.

5 Conclusions and Future work

We explored the use of a simple transformer based architecture for task 2 proposed by SMM4H'22. The five different transformer models (BERT, RoBERTa, PubMed-BERT, SciBERT, and SPECTER) were trained separately on both subtasks, and their results were then combined to create a voting classifier. As expected, the overall accuracy and f1 scores of the voting ensemble classifier were better as compared to individual model prediction for both the aforementioned subtasks. However, there are still greater possibilities towards improving the overall accuracy. We intent to mine crucial medical information from the text and train a separate classifier on them in this manner we'll have a more precise and robust data to infer from. Also, due to time constraint we were restricted towards training a monolingual model, however the given data also entail written text in languages other than English which are ignored by our proposed system. Recently proposed Language-Models like Bloom[1] can help the model in dealing with the multilingual data.

6 Acknowledgements

This work was supported by the European Union's Horizon 2020 research and innovation program under grant agreement No. 833635 (project ROXANNE: Real-time network, text, and speaker analytics for combating organized crime, 2019-2022).

[1] `https://huggingface.co/bigscience/bloom`

References

Iz Beltagy, Arman Cohan, and Kyle Lo. 2019. Scibert: Pretrained contextualized embeddings for scientific text. *CoRR*, abs/1903.10676.

Arman Cohan, Sergey Feldman, Iz Beltagy, Doug Downey, and Daniel S. Weld. 2020. SPECTER: document-level representation learning using citation-informed transformers. *CoRR*, abs/2004.07180.

Jacob Devlin, Ming-Wei Chang, Kenton Lee, and Kristina Toutanova. 2018. BERT: pre-training of deep bidirectional transformers for language understanding. *CoRR*, abs/1810.04805.

Kyle Glandt, Sarthak Khanal, Yingjie Li, Doina Caragea, and Cornelia Caragea. 2021. Stance detection in COVID-19 tweets. In *Proceedings of the 59th Annual Meeting of the Association for Computational Linguistics and the 11th International Joint Conference on Natural Language Processing (Volume 1: Long Papers)*, pages 1596–1611, Online. Association for Computational Linguistics.

Yu Gu, Robert Tinn, Hao Cheng, Michael Lucas, Naoto Usuyama, Xiaodong Liu, Tristan Naumann, Jianfeng Gao, and Hoifung Poon. 2020. Domain-specific language model pretraining for biomedical natural language processing. *CoRR*, abs/2007.15779.

Yinhan Liu, Myle Ott, Naman Goyal, Jingfei Du, Mandar Joshi, Danqi Chen, Omer Levy, Mike Lewis, Luke Zettlemoyer, and Veselin Stoyanov. 2019. Roberta: A robustly optimized BERT pretraining approach. *CoRR*, abs/1907.11692.

A Appendix

Tweet	Claim	Actual		Predicted	
		Stance	Premise	Stance	Premise
Ordered a mask that had a cute chain is attached so.you don't lose it during the day: raising_hands_medium-light_skin_tone	face masks	Favour	0	Favour	0
How many of #US would've.agreed to #shutdown had we known it would be for 4+ months !?#California rules regarding #COVID19 #lockdown are completely ridiculous. Many family businesses & corporate #Restaurants will file #bankruptcy due to prolonged compliance burden.	stay at home orders	Against	1	Against	1
@ananavarro That representative from Texas said almost the same crap in his mind, refused to wear a mask when asked...Guess what, God doesn't appreciate that crap.. he's now dealing with Covid-19 wherever he is.	face masks	Favour	1	Favour	0
we gotta go back to school, i bet once they start pushing kids back to school, there will be a massive increase in cases. Kinda sad how they still don't realize we are in the middle of a pandemic where cases are still rising erryday. #SchoolReopening	school closures	Favour	0	Favour	1
Ok, so we can't trust educators to safe distance or wear masks? Why in the hell are we letting them teach our children?	school closures	Against	1	Favour	1
@realDonaldTrump And now I bet they're wishing they had stayed in space #TrumpVirus #TrumpFailsAmerica	face masks	None	0	None	0
@garethicke In last week, 2 of my family have been made redundant and 1 has attempted suicide. #endthelockdownuk #EndTheNightmare	stay at home orders	Against	1	None	0
@BetsyDeVosED you are crazy if you think the NBA and all it's resources in a bubble aren't able to keep Covid out, but the extremely underfunded schools are going to be able to.	school closures	Favour	1	Favour	1
@ottawacity The TRUTH @ottawahealth is hiding from you: Once the temperature reaches above 74F degrees 23.3C the virus CANNOT survive!. Why the masks? (mind control) Why the increase in numbers? (more mind control) Why distancing? (mind control) Awake now? #OttawaMaskContest	face masks	Against	1	Against	1
What's more important? A students education? Or the wellness of the student and staff? Closing schools for a day may work for bad weather, but not for the coronavirus. Districts all around need to open their eyes and realize one day closures do nothing to help!	school closures	Favour	1	Favour	1

Table 4: Result Analysis for empirical

AILAB-Udine@SMM4H'22: Limits of Transformers and BERT Ensembles

Beatrice Portelli
portelli.beatrice@spes.uniud.it
University of Udine, Italy
University of Naples Federico II , Italy

Simone Scaboro
scaboro.simone@spes.uniud.it
University of Udine, Italy

Emmanuele Chersoni
emmanuele.chersoni@polyu.edu.hk
The Hong Kong Polytechnic University, Hong Kong

Enrico Santus
esantus@gmail.com
DSIG - Bayer Pharmaceuticals, New Jersey, USA

Giuseppe Serra
giuseppe.serra@uniud.it
University of Udine, Italy

Abstract

This paper describes the models developed by the AILAB-Udine team for the SMM4H'22 Shared Task. We explored the limits of Transformer based models on text classification, entity extraction and entity normalization, tackling Tasks 1, 2, 5, 6 and 10. The main takeaways we got from participating in different tasks are: the overwhelming positive effects of combining different architectures when using ensemble learning, and the great potential of generative models for term normalization.

1 Introduction

Transformer-based models are the backbone of state-of-the-art solutions for a lot of NLP tasks. The real strength of these models (like BERT Devlin et al., 2018, GPT Radford et al., 2019, and their variants Joshi et al., 2019; Gu et al., 2020) stands in the pre-training phase which permits them to have extensive language knowledge. This is particularly helpful in tasks where the amount of training data is restricted, like the ones addressed in this workshop (Weissenbacher et al., 2022).

In this work we used a variety of pretrained Transformers models for the tasks. We refer to Table 1 for a summary of their names in the Huggingface library and the shorthand version of their name used in this report.

Short name	Model name in the Huggingface library	Reference
GPT-2	gpt2	(Radford et al., 2019)
$BERT_{Eng}$	bert-base-uncased	(Devlin et al., 2018)
$BERT_{Mul}$	bert-base-multilingual-uncased	(Devlin et al., 2018)
$BERT_{Med}$	microsoft/BiomedNLP-PubMedBERT-base-uncased-abstract	(Gu et al., 2020)
$BERT_{Span}$	SpanBERT/spanbert-base-cased	(Joshi et al., 2019)
$RoBERTa_{Twi}$	cardiffnlp/twitter-roberta-base-sentiment	(Barbieri et al., 2020)
$RoBERTa_{XML}$	xlm-roberta-base	(Conneau et al., 2019)

Table 1: List of pretrained models used for the tasks and the abbreviations used in this report.

2 Simple Classification (Task 5 / 6)

Task 5 and 6 both entail the simple classification of tweets (binary or ternary) in a class-unbalanced setting. Task 5 consists in the ternary classification of Spanish tweets about COVID-19 symptoms as: containing literature/news reports (News, 60%), containing personal reports (Pers, 16%) or reporting about someone else's symptoms (Non-Pers, 24%). Task 6 consists in the binary classification of English tweets regarding COVID-19 vaccinations as: general vaccine chatter (Chatter, 89%) or personal reports confirming the vaccination status of the user (Pers, 11%). For both tasks the text preprocessing consisted in replacing all usernames with "user" and all URLs with "(see url)" or "(ver url)" ("(see url)" in Spanish).

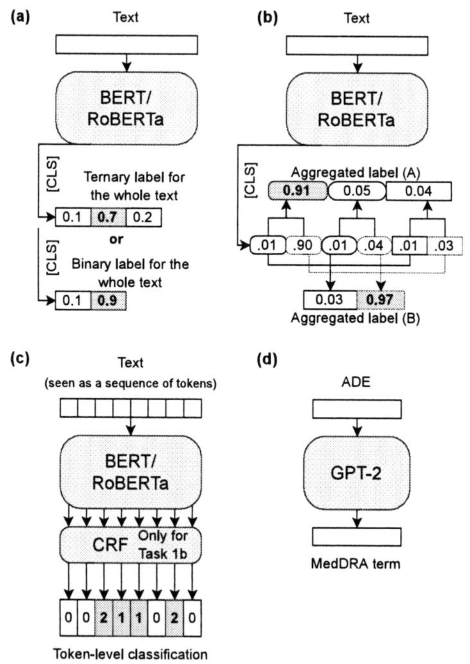

Figure 1: Summary of the model architectures used in the different tasks.

Proceedings of the 29th International Conference on Computational Linguistic, pages 130–134
October 12–17, 2022.

2.1 Models

We tackled the tasks using a simple Transformer-based model with a classification head, that is a linear layer applied to the output embedding of the [CLS] token (see Figure 1a). This layer maps the embedding to either two or three classes depending on the task. The kind of pretrained model was chosen based on the characteristics of the input text, such as language. For each task we selected two kinds of pretrained models with different characteristics to try and combine them, and ensembled their predictions. The selected models were: two BERT_{Mul} and one $\text{RoBERTa}_{\text{XML}}$ for Task 5, two BERT_{Mul} and one BERT_{Eng} for Task 6. They were chosen by training and evaluating 5 models of each kind locally on a 70-30 random split of the training data, and selecting the ones with the higher performance on their respective test fold. In Task 5, BERT_{Mul} models were trained for 10 epochs while $\text{RoBERTa}_{\text{XML}}$ models for 15 epochs. In Task 6, BERT_{Mul} models were trained for 5 epochs while BERT_{Eng} models for 4 epochs. The final label for the task was chosen via majority vote.

2.2 Results

Table 2 shows the results of the base models and their ensemble for both tasks on the validation set.

Task	Model	P	R	F1
5	BERT_{Mul} (1)	0.826	0.826	0.826
5	BERT_{Mul} (2)	0.833	0.833	0.833
5	$\text{RoBERTa}_{\text{XML}}$	0.823	0.823	0.823
5	**Ensemble**	**0.838**	**0.838**	**0.838**
6	BERT_{Mul} (1)	0.875	0.734	0.799
6	BERT_{Mul} (2)	0.946	**0.748**	**0.835**
6	BERT_{Eng}	0.928	0.715	0.807
6	**Ensemble**	**0.954**	0.741	0.834

Table 2: Task 5 and Task 6 results on the validation set.

Ensembling different model typologies had a positive effect on Task 5, as the overall performance is higher than any model on its own. Looking at the confusion matrices in Figure 2, we see that most of the improvements come from a better accuracy in classifying Pers samples and distinguishing them from Non-Pers ones. The precision on Pers class goes from 0.68 (single models) to 0.72 (Ensemble) and the percentage of Pers samples classified as Non-Pers lowers from 0.27 to 0.23.

As regards Task 6, the Ensemble has a higher precision than each individual model, but a lower recall. BERT_{Mul} (2) had significantly higher metrics compared to the other two models, and the majority

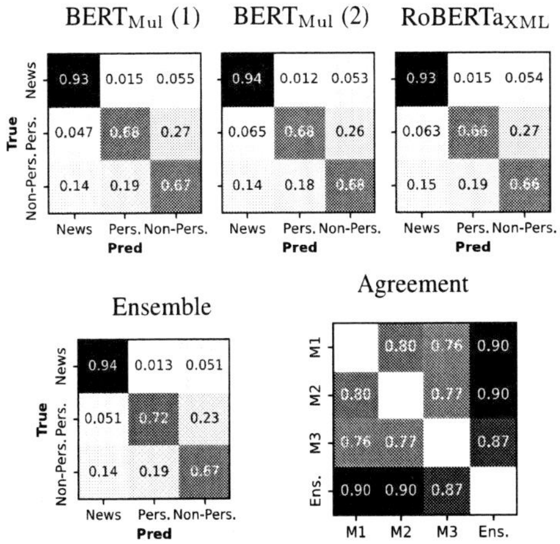

Figure 2: Task 5. Confusion matrices for the three separate models and their ensemble, and agreement matrix.

vote might have favored the most frequent (incorrect) prediction of the other two models. Looking at the confusion matrices in Figure 3, we can see that the Ensemble model has a higher precision on the most frequent class (Chatter) compared to the single models, but the two weaker models severely hampered the performance of BERT_{Mul} (2).

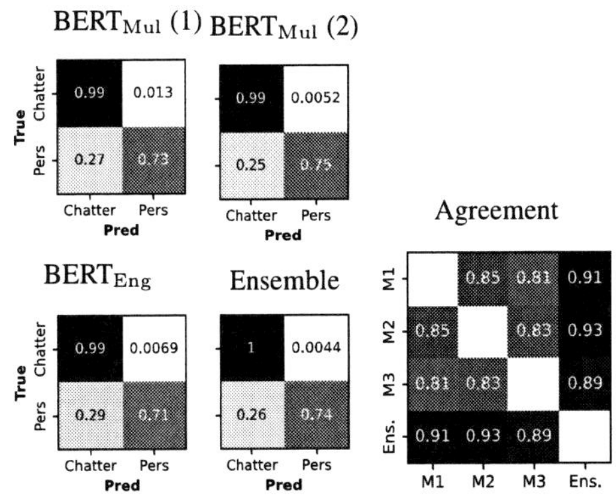

Figure 3: Task 6. Confusion matrices for the three separate models and their ensemble, and agreement matrix.

The main differences between the models ensembled for Task 5 and Task 6 is that the models used for Task 5 had different base architectures (BERT vs RoBERTa), while the ones for Task 6 were all based on BERT. If we calculate the agreements between the models using Cohen's Kappa Cohen, 1960, we see that the models used for Task 6 had higher agreement than the ones in Task 5 (compare the agreement matrices in Figures 2 and 3). The lower agreement for Task 5 is likely caused by the use of different model architectures, and it might lead to higher performance when combining the predictions.

3 Multitask Classification (Task 2a+2b)

Task 2a and 2b (Davydova and Tutubalina, 2022) consist in the classification of English tweets containing opinions about mandates during the COVID-19 pandemic. The tweets can deal with three topics: Face Masks (M), Stay At Home Orders (H) and School Closures (S). Task 2a is a ternary stance classification (Against, None, Favor), while Task 2b is a binary premise classification (1, 0) to determine whether the tweet is argumentative or not.

3.1 Models

We use the same architecture to solve both tasks, reformulating them as a single 6-way classification with labels: Against-1, Against-0, None-1, None-0, Favor-1 and Favor-0. We use a simple Transformer-based model with a classification head on top of the [CLS] embedding. The output of the classification head is a probability distribution over the six labels. At inference time, the probabilities are aggregated in three or two classes according to the task (e.g., Against=Against-1 + Against-0 for Task 2a or 1 = Against-1 + None-1 + Favor-1 for Task 2b). This process is illustrated in Figure 1b.

The text preprocessing is the same as Task 6. The input for the models was formatted as "About CLAIM. [SEP] TWEET_TEXT", where CLAIM is one of the three topics. We finetuned a BERT_{Eng} model for 4 epochs on all training data. To increase the robustness of the model for Task 2a, we also finetuned two $\text{RoBERTa}_{\text{Twi}}$ models for 5 epochs on the three-way classification Against/None/Favor, leveraging the model's pretrained weights for sentiment classification on Twitter. We then ensemble the predictions of the two $\text{RoBERTa}_{\text{Twi}}$ models and the BERT_{Eng} model for Task 2a (similarly to Task 5/6).

3.2 Results

Table 3 reports the metrics for Task 2a and 2b on the validation set. The Ensemble for Task 2a achieves higher metrics compared to the single models in two out of the three topics (M and H), as well as on the overall F1 score. The F1 score of the single models differ up to 7-9 points between each other, yet their interaction leads to a higher score overall. Figure 4 shows the agreement between the three models, which is even lower than the one recorded for Task 5. This further strengthens the hypothesis that using different architectures and models with high disagreements leads to an ensemble with higher performance.

Task	Model	$\mathbf{F1}_M$	$\mathbf{F1}_S$	$\mathbf{F1}_H$	**F1**
2a	$\text{RoBERTa}_{\text{Twi}}$ (1)	0.840	0.677	0.819	0.779
2a	$\text{RoBERTa}_{\text{Twi}}$ (2)	0.816	0.700	**0.821**	0.779
2a	BERT_{Eng}	0.749	0.610	0.742	0.700
2a	**Ensemble**	**0.855**	**0.722**	0.807	**0.795**
2b	$\mathbf{BERT}_{\mathbf{Eng}}$	**0.786**	**0.818**	**0.814**	**0.806**

Table 3: Task 2a and 2b results on the validation set.

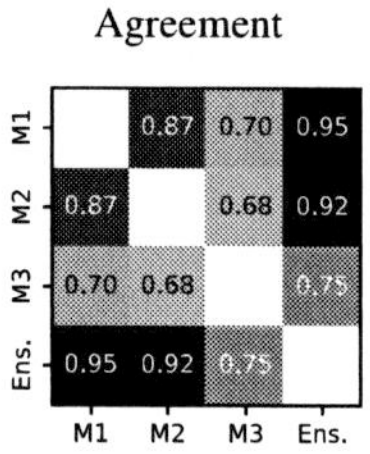

Figure 4: Task 2a. Agreement matrix for the three models and their ensemble.

4 Disease Extraction (Task 10)

Task 10 (Gasco et al., 2022) consists in extracting disease mentions from Spanish tweets.

4.1 Models

We solved this task using a simple Transformer-based model with a token-classification head, that is a linear layer applied to the output embedding of each token (see Figure 1c) with three output classes. These represent the BIO tagging scheme (Begin-Inside-Outside), commonly used to mark the presence of Named Entities in NER tasks. This straightforward method has been previously used in SMM4H Tasks for ADE extraction (Portelli et al., 2022), and we were interested in testing if it was possible to adapt it to Disease Extraction with minimal changes. The text preprocessing is the same as Task 5. We used a BERT_{Mul} model (with multilingual pretraining) and trained it for 5 epochs on the training data of SocialDisNER without using additional resources.

4.2 Results

Table 4 reports the strict and relaxed metrics on the validation set. There is a gap of 40 points between the two, which is almost double what is usually reported in ADE extraction tasks. This means that the model was able to identify the broad area of text containing the disease, but not to pinpoint it. This goes to show that a Disease extraction system needs more mechanisms in place to precisely extract the relevant text.

Task	Model	P	R	F1
10	BERT_{Mul} (Strict)	0.543	0.498	0.520
10	BERT_{Mul} (Relaxed)	0.946	0.865	0.904

Table 4: Task 10 results on the validation set.

5 ADE Normalization (Task 1c)

Task 1c consists in mapping ADE mentions from English tweets to their corresponding MedDRA terms (formal medical terms). The dataset (Magge et al., 2021) was developed with three sub-tasks in mind, so to perform Task 1c it is necessary to complete the two preliminary Tasks 1a (binary classification ADE/noADE) and 1b (ADE extraction).

5.1 Models

Task 1a and Task 1b were not the main focus of our work, so we tackled them with simple and effective strategies seen in other tasks. Task 1a was solved with a BERT_{Med} model with a binary classification head, trained as seen in Task 6, without model ensembling. For Task 1b we used a model previously developed for the same task the SMM4H'19 Shared Task Portelli et al. (2021, 2022). It consists of a $\text{BERT}_{\text{Span}}$ for token classification (see model for Task 10) combined with a Conditional Random Field (CRF) module (see Figure 1c).

For Task 1c, we used a GPT-2 model, trained to take as input a ADE and generate the string corresponding to the correct MedDRA term (e.g., "feel like crap" → "malaise"). GPT-2 was trained on the whole training set for 15 epochs.

5.2 Results

Table 5 reports the results of the models on the validation set. The models for Tasks 1a and 1b achieve average performance. We report the metrics for Task 1c in two ways: calculating them on the output of $\text{BERT}_{\text{Span}}$ (same procedures used on the blind test set); and using an oracle for Task 1b (that is, giving as input to GPT-2 only the correct ADEs). Using the oracle, we see that the model developed for Task 1c has a very high accuracy (0.759) if given the correct ADEs. Applying GPT-2 to the predictions of $\text{BERT}_{\text{Span}}$ leads to lower metrics due to the low quality of the preceding steps. The proposed model for Task 1c also performed extremely well on the blind test set, where it achieved results well over the average despite the low performance reached on Task 1b (see Table 6).

Task	Model	P	R	F1
1a	BERT_{Med}	0.663	0.477	0.544
1b	$\text{BERT}_{\text{Span}}$	0.295	0.851	0.438
1c	GPT-2 (from $\text{BERT}_{\text{Span}}$)	0.219	0.632	0.325
1c	GPT-2 (from oracle)	0.759	0.759	0.759

Table 5: Task 1 results on the validation set.

6 Results on the Test Set

The following table reports the metrics of all presented models on the test set, together with the reference scores supplied by organizers. Models were *not* re-trained using validation data, with the exception of Tasks 1a and 1b.

		Strict			Relaxed		
Task	Model	P	R	F1	P	R	F1
1a	Our	.607	.386	.472			
1a	Average	**.646**	**.497**	**.562**			
1b	Our	**.360**	.254	.298	.489	.344	.404
1b	Average	.344	**.339**	**.341**	**.539**	**.517**	**.527**
1c	Our	**.243**	**.171**	**.201**	**.294**	**.207**	**.243**
1c	Average	.085	.082	.083	.120	.112	.116
2a	Our			.529			
2a	Average			.491			
2a	Median			**.550**			
2b	Our			**.649**			
2b	Average			.574			
2b	Median			.647			
5	Our	.840	.840	.840			
5	Median	.840	.840	.840			
5	Baseline	**.900**	**.900**	**.900**			
6	Our	**.930**	.750	**.830**			
6	Median	.900	.680	.770			
6	Baseline	.900	**.770**	**.830**			
10	Our	.504	.461	.481			
10	Average	.680	.677	.675			
10	Median	**.758**	**.780**	**.761**			

Table 6: Results for all tasks on the blind test set.

7 Conclusions

We explored the use of simple Transformer-based architectures for several tasks proposed by SMM4H'22. The most noticeable phenomena we encountered were: the collaborative effect of different architectures (e.g., BERT and RoBERTa) when used in ensemble learning (Task 2 and 5, as opposed to Task 6); the efficacy of generative models for term normalization (Task 1c); and the low transferability of methods developed for ADE extraction to Disease detection (Task 10).

References

Francesco Barbieri, Jose Camacho-Collados, Luis Espinosa Anke, and Leonardo Neves. 2020. TweetEval: Unified benchmark and comparative evaluation for tweet classification. In *Findings of the Association for Computational Linguistics: EMNLP 2020*, pages 1644–1650, Online. Association for Computational Linguistics.

Jacob Cohen. 1960. A Coefficient of Agreement for Nominal Scales. *Educational and Psychological Measurement*, 20(1):37–46.

Alexis Conneau, Kartikay Khandelwal, Naman Goyal, Vishrav Chaudhary, Guillaume Wenzek, Francisco Guzmán, Edouard Grave, Myle Ott, Luke Zettlemoyer, and Veselin Stoyanov. 2019. Unsupervised Cross-lingual Representation Learning at Scale. *CoRR*, abs/1911.02116.

Vera Davydova and Elena Tutubalina. 2022. Smm4h 2022 task 2: Dataset for stance and premise detection in tweets about health mandates related to covid-19. In *Proceedings of the Seventh Social Media Mining for Health Applications (#SMM4H) Workshop & Shared Task*, pages –.

Jacob Devlin, Ming-Wei Chang, Kenton Lee, and Kristina Toutanova. 2018. BERT: Pre-training of Deep Bidirectional Transformers for Language Understanding.

Luis Gasco, Darryl Estrada-Zavala, Eulàlia Farré-Maduell, Salvador Lima-López, Antonio Miranda-Escalada, and Martin Krallinger. 2022. Overview of the SocialDisNER shared task on detection of diseases mentions from healthcare related and patient generated social media content: methods, evaluation and corpora. In *Proceedings of the Seventh Social Media Mining for Health (#SMM4H) Workshop and Shared Task.*

Yu Gu, Robert Tinn, Hao Cheng, Michael Lucas, Naoto Usuyama, Xiaodong Liu, Tristan Naumann, Jianfeng Gao, and Hoifung Poon. 2020. Domain-Specific Language Model Pretraining for Biomedical Natural Language Processing.

Mandar Joshi, Danqi Chen, Yinhan Liu, Daniel S. Weld, Luke Zettlemoyer, and Omer Levy. 2019. SpanBERT: Improving Pre-training by Representing and Predicting Spans.

Arjun Magge, Elena Tutubalina, Zulfat Miftahutdinov, Ilseyar Alimova, Anne Dirkson, Suzan Verberne, Davy Weissenbacher, and Graciela Gonzalez-Hernandez. 2021. DeepADEMiner: a deep learning pharmacovigilance pipeline for extraction and normalization of adverse drug event mentions on Twitter. *Journal of the American Medical Informatics Association*, 28(10):2184–2192.

Beatrice Portelli, Edoardo Lenzi, Emmanuele Chersoni, Giuseppe Serra, and Enrico Santus. 2021. Bert prescriptions to avoid unwanted headaches: a comparison of transformer architectures for adverse drug event detection. In *Proceedings of the 16th Conference of the European Chapter of the Association for Computational Linguistics: Main Volume*, pages 1740–1747.

Beatrice Portelli, Daniele Passabì, Edoardo Lenzi, Giuseppe Serra, Enrico Santus, and Emmanuele Chersoni. 2022. *Improving Adverse Drug Event Extraction with SpanBERT on Different Text Typologies*, pages 87–99. Springer International Publishing, Cham.

Alec Radford, Jeffrey Wu, Rewon Child, David Luan, Dario Amodei, Ilya Sutskever, et al. 2019. Language models are unsupervised multitask learners. *OpenAI blog*, 1(8):9.

Davy Weissenbacher, Ari Z. Klein, Luis Gascó, Darryl Estrada-Zavala, Martin Krallinger, Yuting Guo, Yao Ge, Abeed Sarker, Ana Lucia Schmidt, Raul Rodriguez-Esteban, Mathias Leddin, Arjun Magge, Juan M. Banda, Vera Davydova, Elena Tutubalina, and Graciela Gonzalez-Hernandez. 2022. Overview of the seventh social media mining for health applications #smm4h shared tasks at coling 2022. In *Proceedings of the Seventh Social Media Mining for Health Applications (#SMM4H) Workshop & Shared Task*, pages –.

AIR-JPMC@SMM4H'22: Classifying Self-Reported Intimate Partner Violence in Tweets with Multiple BERT-based Models

Alec Candidato, Akshat Gupta, Xiaomo Liu, Sameena Shah
J.P. Morgan AI Research
aleccandidato@gmail.com, {akshat.x.gupta, xiaomo.liu, sameena.shah}@jpmchase.com

Abstract

This paper presents our submission for the SMM4H 2022-Shared Task on the classification of self-reported intimate partner violence on Twitter (in English). The goal of this task was to accurately determine if the contents of a given tweet demonstrated someone reporting their own experience with intimate partner violence. The submitted system is an ensemble of five RoBERTa models each weighted by their respective F1-scores on the validation data-set. This system performed 13% better than the baseline and was the best performing system overall for this shared task.

1 Introduction

Intimate Partner Violence (IPV) refers to any form of physical or emotional abuse or aggression within a romantic relationship. Many people use social media, particularly twitter, as a means of self-reporting their experiences with IPV. Evidently, being able to parse through social media for these instances of self-reported IPV domestic violence is vital for finding victims most in need of resources and support. While scraping tweets for IPV-related content is relatively simple by keyword search, properly labeling and characterizing whether these tweets refer to self-reported IPV proves to be more challenging.

It is already known that using BERT models or transformer-based models already produces state-of-the-art results for the field of Natural Language Understanding (Devlin et al., 2019). In fact, shared tasks from previous years have already demonstrated successful results incorporating this technology (Magge et al., 2021). This paper elaborates on our submission for the task of classifying self-reported intimate partner violence on Twitter (in English). The existing research surrounding the provided data-set also demonstrates the effectiveness of RoBERTa models (Liu et al., 2019). Our

Split	Non-self-reported IPV	Self-reported IPV	Total
Train	4042	481	4523
Validation	480	54	534
Test	—	—	1291

Table 1: Frequency distribution for each dataset. The test dataset was unlabeled so the distribution is unknown.

final submission for this task is composed of a weighted ensemble of five RoBERTa-large models.

2 Dataset

There were three different data-sets provided: training, validation, and test data-sets. The training and validation data-sets were labeled while the test data-set was not. All data-sets are composed entirely of tweets that are related to the topic of domestic violence. Among these tweets, 11% are identified as self-reported IPV (Al-Garadi et al., 2022). Table 1 shows the tweet distributions.

3 Method

This submission utilizes an ensembling of multiple trained RoBERTa models which each used their best epoch by means of F1-score to outperform the previous results by almost a tenth of a point. The majority ensemble takes the mode model prediction of all five RoBERTa guesses for a given tweet. The weighted ensemble takes a precise linear combination of all five RoBERTa guesses based on each model's initial performance with the validation data-set. We have two defined classes: non-self-reported (0) and self-reported (1). Let us define $P_e(y = 1 \mid x)$ as the probability that the ensemble predicts class $y = 1$ for a given tweet, x. We can represent this explicitly as:

$$P_e(y = 1 \mid x) = \frac{\sum_{i=1}^{n} a_i P_i(y = 1 \mid x)}{\sum_{i=1}^{n} a_i}$$

Proceedings of the 29th International Conference on Computational Linguistic, pages 135–137
October 12–17, 2022.

where P_i represents an individual model's probability for a given tweet, a_i represents the corresponding weight, and n is the number of models being included in the ensemble. In the case of a majority vote, $a_i = 1$. Our submission uses $a_i =$ F1-score. Consequently, we can represent the probability for class 0 as $P_e(y = 0 \mid x) = 1 - P_e(y = 1 \mid x)$. The final prediction chooses the class with the highest probability. The task submission uses this weighted ensembling method because it can easily be combined with other models. Other non-transformer models were initially developed to being included with the ensemble but not ultimately used for the final submission.

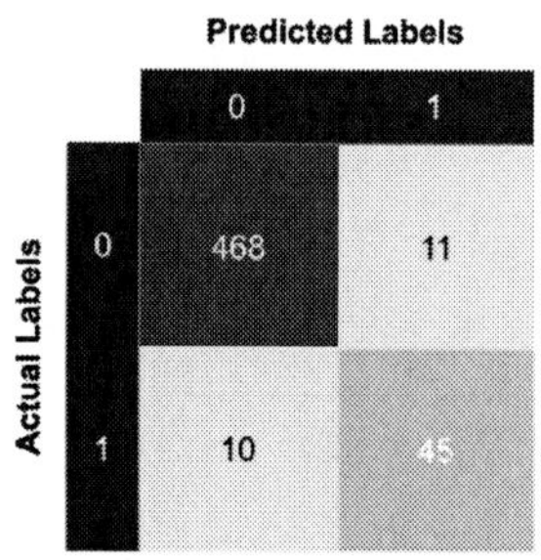

Figure 1: Confusion matrix for the RoBERTa-large ensembles on the validation data-set. This matrix represents the results for both the majority voting ensemble and weighted ensemble, since they performed identically to each other.

4 Preliminary Experiments

	BERT-base	RoBERTa-base	BERT-large	RoBERTa-large
Mean F1	0.718	0.749	0.710	0.779
Stdev	0.013	0.017	0.014	0.013

Table 2: Performance of BERT and RoBERTa classifiers on validation data. The F1 scores are averaged among all five models for each category. The standard deviation is also provided.

Five BERT-base (Devlin et al., 2019) and five RoBERTa-base models (Liu et al., 2019) were first trained to test the initial effectiveness of transformer classifiers. We saved the best performing epoch for each model, based on a general accuracy metric. Then, five BERT-large and five RoBERTa-large models were trained. These models saved their best performing epoch in terms of F1-score instead. The F1-score for each model was determined based on their performance with the validation dataset. The results are shown in Table 2.

The RoBERTa-large models performed significantly better than any other model we considered. Results from a majority voting ensemble and a weighted ensemble of five RoBERTa-large models were also calculated. They performed identically. Figure 1 shows the corresponding confusion matrix.

5 Results

The performances of the designed ensemble compared to the median Codalab results and the best performing model from the original research are shown on Table 3. The RoBERTa ensemble performs significantly better than all other models by all metrics. The system that the median performer

Classifier	F1-score	Precision	Recall
Baseline (Al-Garadi et al., 2022)	0.756	0.823	0.699
Median Task Performance	0.763	0.790	0.716
Our System	**0.851**	**0.860**	**0.841**

Table 3: Performance results for classifiers.

used is unknown at this time so no conclusions can be made comparing the results of the RoBERTa ensemble with that system. The best classifier from previous work was also built off of RoBERTa-large. Yet, this classifier achieved a 0.1 F1-score improvement. This may be because these RoBERTa models saved the epoch with the best F1-score rather than the best accuracy. Additionally, ensembling multiple transformer-based models has already proven to be effective (Jayanthi and Gupta, 2021).

6 Conclusions

This study implements ensembling with transformer-based language models to drastically improve precision and recall for this task. This overall system outperforms previous metrics by nearly 13%. This study also included initial preprocessing into part-of-speech frequency and the significance of narrative voice for self-reporting in tweets. Initial results suggests that tweets written in second or third person can help fine-tune against false positives. However, the final submission did not utilize any systems built off of this due to time limitations. Thus, these results are omitted from this paper. (Eisenberg and Finlayson, 2016) have already examined systems for narrative diegesis and point of view analysis. Ensembling transformer-based models with models trained off of narrative diegesis and point of view data may further improve results.

References

Mohammed Ali Al-Garadi, Sangmi Kim, Yuting Guo, Elise Warren, Yuan-Chi Yang, Sahithi Lakamana, and Abeed Sarker. 2022. Natural language model for automatic identification of intimate partner violence reports from twitter. *Array*, page 100217.

Jacob Devlin, Ming-Wei Chang, Kenton Lee, and Kristina Toutanova. 2019. BERT: Pre-training of deep bidirectional transformers for language understanding. In *Proceedings of the 2019 Conference of the North American Chapter of the Association for Computational Linguistics: Human Language Technologies, Volume 1 (Long and Short Papers)*, pages 4171–4186, Minneapolis, Minnesota. Association for Computational Linguistics.

Joshua Eisenberg and Mark Finlayson. 2016. Automatic identification of narrative diegesis and point of view. pages 36–46.

Sai Muralidhar Jayanthi and Akshat Gupta. 2021. SJ_AJ@DravidianLangTech-EACL2021: Task-adaptive pre-training of multilingual BERT models for offensive language identification. In *Proceedings of the First Workshop on Speech and Language Technologies for Dravidian Languages*, pages 307–312, Kyiv. Association for Computational Linguistics.

Yinhan Liu, Myle Ott, Naman Goyal, Jingfei Du, Mandar Joshi, Danqi Chen, Omer Levy, Mike Lewis, Luke Zettlemoyer, and Veselin Stoyanov. 2019. Roberta: A robustly optimized bert pretraining approach.

Arjun Magge, Ari Klein, Antonio Miranda-Escalada, Mohammed Ali Al-Garadi, Ilseyar Alimova, Zulfat Miftahutdinov, Eulalia Farre, Salvador Lima López, Ivan Flores, Karen O'Connor, Davy Weissenbacher, Elena Tutubalina, Abeed Sarker, Juan Banda, Martin Krallinger, and Graciela Gonzalez-Hernandez. 2021. Overview of the sixth social media mining for health applications (#SMM4H) shared tasks at NAACL 2021. In *Proceedings of the Sixth Social Media Mining for Health (#SMM4H) Workshop and Shared Task*, pages 21–32, Mexico City, Mexico. Association for Computational Linguistics.

ARGUABLY@SMM4H'22: Classification of Health Related Tweets Using Ensemble, Zero-Shot and Fine-Tuned Language Model

Prabsimran Kaur
Thapar University, Patiala, India
pkaur_be18@thapar.edu

Guneet Singh Kohli
Thapar University, Patiala, India
guneetsk99@gmail.com

Jatin Bedi
Thapar University, Patiala, India
jatin.bedi@thapar.edu

Abstract

With the increase in the use of social media, people have become more outspoken and are using platforms like Reddit, Facebook, and Twitter to express their views and share the medical challenges they are facing. This data is a valuable source of medical insight and is often used for healthcare research. This paper describes our participation in Task 1a, 2a, 2b, 3, 5, 6, 7, and 9 organized by SMM4H 2022. We have proposed two transformer-based approaches to handle the classification tasks. The first approach is fine-tuning single language models. The second approach is ensembling the results of BERT, RoBERTa, and ERNIE 2.0.

1 Introduction

A rapid increase in the use of social media has been seen in the past decade. Social media platforms such as Twitter, Reddit, and Facebook have become a place for people to articulate their views and emotions. Twitter especially has become a medium for people to share their medical lifestyle and the health-related problems that they are facing. Thus making Twitter an essential resource for extracting meaningful data that can help better understand and improve health services. The advancement of Natural Language Processing (NLP) in deep neural models and its ability to effectively process and understand data has attracted the attention of the healthcare research community. These healthcare researchers have developed a keen interest in processing this available data efficiently using deep learning.

The Social Media Mining for Health Applications (SMM4H) (Davy Weissenbacher, 2022) aims to bring researchers worldwide for the mining, representation, and analysis of data related to health, such as updates regarding COVID-19 and its vaccination status, drugs, and medical treatments that can help gain medical insights. This year SMM4H has proposed ten tasks that involve data classification, extraction, and Named Entity Recognition. Our team has participated in various classification tasks, namely, Task 1a, 2a, 2b, 3, 5, 6, 7, and 9.

Task 1 focuses on better understanding Adverse Drug Reactions (ADRs). The aim of Task 1a was to distinguish tweets mentioning adverse drug effects (ADE) from other tweets (NoADE). Task 2a focused on determining an author's stance toward various issues related to COVID-19. The training data were annotated for perspective according to three categories: favor (positive stance), against (negative stance), and neither (neutral stance). Task 2b focused on identifying whether a tweet contains a premise (a statement that can be used as an argument in a discussion), where "1" indicates that the tweet has a premise (argument), and "0" means that the tweet doesn't contain a premise. Task 3 focused on designing a binary classifier to detect Twitter users who self-declare that they are changing their treatment medications despite being advised by a health care professional to follow the prescription.

Task 5 deals with identifying personal mentions of COVID-19 symptoms that have been tweeted in Spanish. The dataset needed to be classified into non-personal_reports for non-personal reports, Lit-New/s_mentions for news and literature mentions, and Self_reports for self-reports. Task 6 focuses on distinguishing the self-reported COVID-19 vaccination status, which is labeled as "Self_reports," and users discussing vaccination status in general, labeled as "Vaccine_chatter". Task 7 (Al-Garadi et al., 2022) deals with identifying victims of Intimate partner violence (IPV) who seek help on social media like Twitter. The label "1" indicates a self-reported IPV, and "0" shows an average Tweet about domestic violence. Task 9 (Schmidt et al., 2022) focuses on the detection of demographic information on social media and distinguishing the Reddit posts that self-report the exact age of the social media user at the time of posting (annotated as "1") from those that do not (annotated as "0").

Proceedings of the 29th International Conference on Computational Linguistic, pages 138–142
October 12–17, 2022.

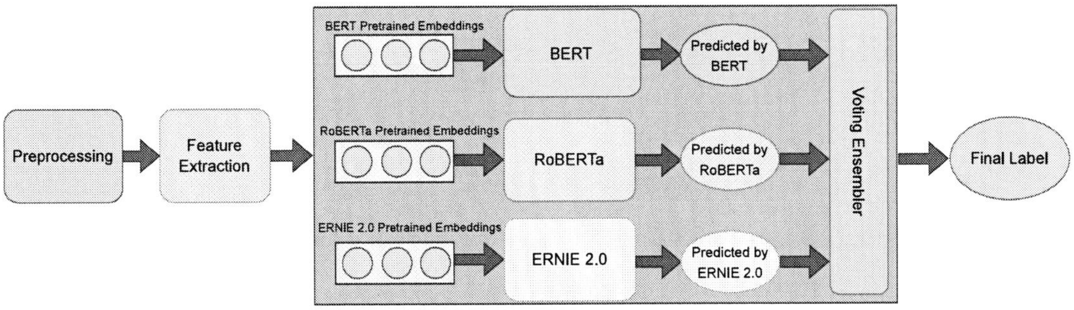

Figure 1: Architecture of Boosted Voting Ensembler

We propose two transformer-based (Vaswani et al., 2017) approaches for the classification of all the aforementioned tasks. The first approach is fine-tuning existing transformer models. The second approach uses a voting-ensemble model that comprises fine-tuned BERT (Devlin et al., 2018), RoBERTa (Liu et al., 2019), and ERNIE 2.0 (Sun et al., 2020). The XNLI (Conneau et al., 2018) model (zero-shot) was used to address the data imbalance in Task 1a. The multilingualism of the Spanish dataset in Task 5 was handled using XLM-RoBERTa model (Conneau et al., 2019).

2 Methodology

This section describes a detailed explanation of the approaches we have used for handling all the classification tasks. This paper proposes two architectures, all of which follow a common data preprocessing (Section 2.1).

2.1 Data Preprocessing

Social Media comments often consist of unstructured data containing special characters and emojis. Thus, a basic preprocessing involving the removal of stop words, punctuations, and emojis using the NLTK (Loper and Bird, 2002) library was performed for all the four methodologies mentioned. In case of twitter, the tweets tends to include lots of noise because of the usernames, keywords like *RT,FAV*, mentions and URLs. These redundant information were also handled using NLTK and Python pattern matching.

2.2 Fine-Tuned Transformer

The features of this preprocessed data are extracted, then each sentence is tokenized, and these tokens are mapped with their respective word IDs. The following series of steps are followed for all the sentences: a) sentence tokenization, b) prepending of [CLS] token to the start, c) appending of [SEP] token at the end, d) mapping of tokens to their word ID, e) padding or truncation of a sentence depending on the maximum sequence length, and f) mapping of the attention mask. The maximum sequence length used for each case was determined by finding the average length of the text in the dataset. The generated sequence, along with its attention mask, is then encoded. The encoded sentences are processed to yield contextually rich trained embeddings. Afterward, we pass these encodings through the desired transformer models. The transformer models used for the classification tasks were BERT, RoBERTa, ERNIE 2.0, XLM-RoBERTa, Bio Med RoBERTA (Gururangan et al., 2020), and Bio-Clinic BERT (Alsentzer et al., 2019).

2.3 Voting Ensembler

The ensemble is a learning technique in which a collection of neural networks are trained for the same task (Sollich and Krogh, 1995). The generalization ability of a neural network can be significantly enhanced by ensembling a number of neural networks (Hansen and Salamon, 1990). Ensembling involves training many neural networks and combining their predictions. The remarkable performance of this technique has made it popular in both neural network and machine learning techniques (Kohli et al., 2021).

While there are various methods of ensembling neural networks, we used the following technique. BERT, RoBERTa, and ERNIE 2.0 were individually fine-tuned (Section 2.2), after which the labels predicted by each model for each sentence were extracted. A voting system is then applied to these extracted labels, and the label which occurs the maximum number of times is selected as the final label for that sentence, as depicted in Figure 1.

Task	Technique Used	Precision	Recall	F1 Score
Task 1a	**XNLI**	**0.8395**	**0.8201**	**0.8294**
	Ensemble	0.8313	0.7775	0.8003
Task 2a	**RoBERTa**	**0.7414**	**0.7371**	**0.7384**
	Ensemble	0.7186	0.7186	0.7185
Task 2b	**BERT**	**0.7705**	**0.7684**	**0.7694**
	Ensemble	0.763433	0.7641	0.763467
Task 3	**Bio_Med**	**0.6782**	**0.5791**	**0.6028**
	Bio Bert	0.6833	0.5732	0.5967

Task	Technique Used	Precision	Recall	F1 Score
Task 5	**XLM-R**	**0.7612**	**0.7500**	**0.7534**
	XNLI	0.7387	0.7313	0.7349
Task 6	**BERT**	**0.9383**	**0.8419**	**0.8823**
	Ensemble	0.9192	0.8300	0.8723
Task 7	**RoBERTa**	**0.7639**	**0.7337**	**0.7475**
	-	-	-	-
Task 9	**RoBERTa**	**0.9307**	**0.9386**	**0.9345**
	Ensemble	0.924433	0.9327	0.928367

Table 1: Macro-Average Precision, Recall, and F1 Score to perform validation analysis with proposed methodologies

Task	Test Results	Precision	Recall	F1 Score
Task 1a	Submission	0.677	0.297	0.413
	Mean	0.646	0.497	0.562
Task 2a	Submission	-	-	0.501
	Median			0.550
Task 2b	Submission	-	-	0.6213
	Median			0.6472
Task 3	Submission	0.585	0.617	0.557
	Median	0.585	0.617	0.557

Task	Test Result	Precision	Recall	F1 Score
Task 5	Submission	0.83	0.83	0.83
	Median	0.84	0.84	0.84
Task 6	Submission	0.76	0.87	0.68
	Median	0.77	0.9	0.68
Task 7	Submission	0.784	0.689	0.734
	Median	0.790	0.716	0.763
Task 9	**Submission**	**0.896**	**0.941**	**0.918**
	Median	**0.896**	**0.019**	**0.891**

Table 2: Results released by organisers and its comparison with the mean [in Task1] and median scores of the tasks

3 Results and Discussion

3.1 Final Evaluation

Table 2 reports the final results obtained by our best systems and its comparison with the median scores from the task. Our system performed well in Task 9 where it reported a higher f1 score from the median by 0.027. For Task 3a our score was reported as median score. In Task 2a, 2b, 5, 6, 7 our submission was comparable to the arithmetic median scores that were released by the organisers. In task 1a our system is able to outperform the mean score in precision by 0.031. The closeness to the median scores in all the tasks shows that the models could have performed better with accurate hyperparameter tuning. These evaluations helped us in understanding the various shortcomings of our proposed system.

3.2 Validation Study

The hyperparameters were standardised across all of the tasks to allow for experimentation with the suggested techniques. The models were tested on total four checkpoints after being trained on two epochs with two learning rates ($2x10^{-5}$, $3x10^{-5}$). The model that performed the best was chosen for further analysis. The top-performing fine-tuned language model and related Ensemble model for Tasks 1a, 2a, 2b, 6, and 9 have been reported on the Validation Set in Table 1. We used the BIO adjusted versions of RoBERTa and BERT for Task 3a, and only RoBERTa was reported for Task 7. To address the multilingualism challenge in Task 5, we used XLM-RoBERTa.

Table 1 helps us in understanding the validation performance of individual submission.The results reported are macro average Precision, Recall, F1 score since we wanted to give equal contribution to each class. Weighted average was avoided since the imbalance of data resulted in introduction of bias. In general it can be observed that the Ensemble models failed to outperform the Single fine tuned model in Task 1a, 2a, 2b, 6, and 9. This trend highlights the lack of robustness in Ensemble model in the given tasks.This is possible due to lack of performance of 2/3 models in the Voting Ensemble which drags down the overall result. For Task 1a and Task 5 the use of Zero Shot model was also employed to test its performance with imbalanced data.The zero shot technique outperforms in Task 1a but in general the results produced were comparable to single fine tuned language models which also helps us in realising the shortcomings of Meta Learning techniques.For task 2a,7,9 RoBERTa generates best results with F1 reaching 0.7384, 0.6028, 0.9345 respectively. For task 2b,6 BERT became the best performing model with F1 scores of 0.7694 and 0.8823 respectively. The token length was selected by calculating a 25% variation from the mean length of all the text instances available.

4 Error Analysis

This section describes the qualitative analysis of the labels predicted by the proposed architectures, as seen in Table 3. In the first instance the label predicted is noADE which indicates that the model could not understand the semantic of the sentence and focused more on the individual words like

Task	Text	Original Label	Predicted Label
Task 1	This vimpat shit is working but the side effects are hell	ADE	noADE
	i started shaking so i had to eat something :(but now i just had some tylenol pm so hopefully i'll go straight to sleep	noADE	ADE
Task 2a	(stay at home) support our P.M. and promise to follow lockdown. I request Govt to kindly postponed BANK EMJ TILL LOCKDOWN.	FAVOR	NONE
	(stay at home) Stay at home, relax back on your couch and find your dream home. Click here to explore your options.	NONE	AGAINST
Task 2b	Another GREAT perk of wearing a mask is that you can curse at people under your breath in public and they can't read your lips!!	0	1
	My daughter is immune compromised due to a rare genetic disorder, Trump can sacrifice his and the republicans children, but not my baby girl!	1	0
Task 3	Just get cortisone shot neck cool	1	0
	LisainLouKY doc gave ok take one dose excederin migraine today I took it slept woke headache	0	1
Task 5	vengo desde hace días con un dolor que puedo de la espalda alcanzó el coronavirus pero si la escoliosis	non-personal_reports	Self_reports
	El coronavirus empieza con Diarrea Quien sepa pueda decir algo Tengo un familiar con diarrea perdida del olfato gusto fiebre malestar en los huesos	non-personal_reports	Lit-News_mentions
Task 6	The vaccineinduced reduction anxiety starting creep out Good thing I good reason feel anxious cortisol levels spike	Self_reports	Vaccine_chatter
	Just got injection mean injected vaccineCovidVaccine	Vaccine_chatter	Self_reports
Task 7	TeamGivingCom My husband helped friend currently domestically violentabusive relationship We helping get situation	0	1
	johnpavlovitz Deciding testify ex trial domestic violence child abuse He got 35 years crimes BEST decision EVER	1	0
Task 9	DMEK can give you 20/20 but not every time.	0	1
	I wonder it i had it 15yrs ago just dryeye and now im in 40s so it went full blown cuza lak ot estrogen! Ahhh ughh.	1	0

Table 3: Qualitative Testing of data instance of respective shared tasks

"working". The model could not capture the true meaning of the second instance and rather made relations between shaking and tylenol, thus concluding that this was a case of side-effect.
Similarly for Task 2a the model failed to understand the discourse integration and focused on the second sentence only thus making it seem as though the Tweet was a neutral remark about postponing bank emj.In the fifth instance the model focused more on the words like "curse" and "breath", thus misleading it to believe that this statement is an argument. The model has completely failed to understand the semantic meaning of the sixth instance and is hence not labeled it as "0," indicating that there is no argument in this Tweet.

A similar trend can be seen in Task 3, 6, and 7 where the model has focused more on individual words like "dose","I","injected","violence", and "help" rather than understanding the true semantics of the sentence. The model performed relatively well in Task 9. However it failed to understand complex semantic like "now im in 40s" that indirectly indicates the age of the user. The model also puts special focus on numbers since those are commonly used for the indicating age and thus in the fifteenth instance it has made an incorrect prediction.

Our models did not perform well in tasks that comprised of medical discussions because they were not fully able to understand the underlying medical context. In case of Task 3, where Bio BERT and Bio_Med models were used the predictions were yet not very accurate perhaps due to the fact that these models were trained before the COVID-19 period and were thus not able to fully understand the terms that were used.

5 Conclusion

For the SMM4H Tasks we propose an ensemble model that leverages on pretrained representations from BERT, RoBERTa, Ernie 2.0, and a single fine tuned language model as our submission systems. Our system was able to report 91.8% F1 Score in Task 9 and 55.7% in Task 3. For other tasks our results were comparable to the arithmetic median released for all the tasks with F1 reaching 83% in Task 6 and 68% in Task 5. For Task 2a, 2b the scores achieved were 50.1% and 62.13% respectively.

References

Mohammed Ali Al-Garadi, Sangmi Kim, Yuting Guo, Elise Warren, Yuan-Chi Yang, Sahithi Lakamana, and Abeed Sarker. 2022. Natural language model for automatic identification of intimate partner violence reports from twitter. *Array*, 15:100217.

Emily Alsentzer, John R Murphy, Willie Boag, Wei-Hung Weng, Di Jin, Tristan Naumann, and Matthew McDermott. 2019. Publicly available clinical bert embeddings. *arXiv preprint arXiv:1904.03323*.

Alexis Conneau, Kartikay Khandelwal, Naman Goyal, Vishrav Chaudhary, Guillaume Wenzek, Francisco Guzmán, Edouard Grave, Myle Ott, Luke Zettlemoyer, and Veselin Stoyanov. 2019. Unsupervised cross-lingual representation learning at scale. *arXiv preprint arXiv:1911.02116*.

Alexis Conneau, Guillaume Lample, Ruty Rinott, Adina Williams, Samuel R Bowman, Holger Schwenk, and Veselin Stoyanov. 2018. Xnli: Evaluating cross-lingual sentence representations. *arXiv preprint arXiv:1809.05053*.

Luis Gascó Darryl Estrada-Zavala Martin Krallinger Yuting Guo Yao Ge Abeed Sarker Ana Lucia Schmidt Raul Rodriguez-Esteban Mathias Leddin Arjun Magge Juan M. Banda Vera Davydova Elena Tutubalina Graciela Gonzalez-Hernandez. Davy Weissenbacher, Ari Z. Klein. 2022. Overview of the seventh social media mining for health applications smm4h shared tasks at coling 2022. In *Proceedings of the Seventh Social Media Mining for Health (SMM4H) Workshop and Shared Task*.

Jacob Devlin, Ming-Wei Chang, Kenton Lee, and Kristina Toutanova. 2018. Bert: Pre-training of deep bidirectional transformers for language understanding. *arXiv preprint arXiv:1810.04805*.

Suchin Gururangan, Ana Marasović, Swabha Swayamdipta, Kyle Lo, Iz Beltagy, Doug Downey, and Noah A Smith. 2020. Don't stop pretraining: adapt language models to domains and tasks. *arXiv preprint arXiv:2004.10964*.

Lars Kai Hansen and Peter Salamon. 1990. Neural network ensembles. *IEEE transactions on pattern analysis and machine intelligence*, 12(10):993–1001.

Guneet Kohli, Prabsimran Kaur, and Jatin Bedi. 2021. Arguably at comma@ icon: Detection of multilingual aggressive, gender biased, and communally charged tweets using ensemble and fine-tuned indicbert. In *Proceedings of the 18th International Conference on Natural Language Processing: Shared Task on Multilingual Gender Biased and Communal Language Identification*, pages 46–52.

Yinhan Liu, Myle Ott, Naman Goyal, Jingfei Du, Mandar Joshi, Danqi Chen, Omer Levy, Mike Lewis, Luke Zettlemoyer, and Veselin Stoyanov. 2019. Roberta: A robustly optimized bert pretraining approach. *arXiv preprint arXiv:1907.11692*.

Edward Loper and Steven Bird. 2002. Nltk: The natural language toolkit. *arXiv preprint cs/0205028*.

Ana Lucía Schmidt, Raul Rodriguez-Esteban, Juergen Gottowik, and Mathias Leddin. 2022. Applications of quantitative social media listening to patient-centric drug development. *Drug Discovery Today*.

Peter Sollich and Anders Krogh. 1995. Learning with ensembles: How overfitting can be useful. *Advances in neural information processing systems*, 8.

Yu Sun, Shuohuan Wang, Yukun Li, Shikun Feng, Hao Tian, Hua Wu, and Haifeng Wang. 2020. Ernie 2.0: A continual pre-training framework for language understanding. In *Proceedings of the AAAI Conference on Artificial Intelligence*, volume 34, pages 8968–8975.

Ashish Vaswani, Noam Shazeer, Niki Parmar, Jakob Uszkoreit, Llion Jones, Aidan N Gomez, Łukasz Kaiser, and Illia Polosukhin. 2017. Attention is all you need. *Advances in neural information processing systems*, 30.

CASIA@SMM4H'22: A Uniform Health Information Mining System for Multilingual Social Media Texts

Jia Fu[1,2,†], **Sirui Li**[1,3,†], **Minghui Yuan**[1,4],**Zhucong Li**[1,2],**Zhen Gan**[1,4], ,**Yubo Chen**[1,2], **Kang Liu**[1,2], **Jun Zhao**[1,2],**Shengping Liu**[5]

[1] National Laboratory of Pattern Recognition, Institute of Automation, Chinese Academy of Sciences

[2] School of Artificial Intelligence, University of Chinese Academy of Sciences

[3] Department of Computer Science, Emory University, Altlanta

[4] Beijing University of Chemical Technology

[5] Beijing Unisound Information Technology Co., Ltd

{yubo.chen, kliu, jzhao}@nlpr.ia.ac.cn

sirui.li@emory.edu,ganzhen@mail.buct.edu.cn

{fujia2021,lizhucong2019}@ia.ac.cn,liushengping@unisound.com

Abstract

This paper presents a description of our system in SMM4H-2022, where we participated in task 1a, task 4, and task 6 to task 10. There are three main challenges in SMM4H-2022, namely the domain shift problem, the prediction bias due to category imbalance, and the noise in informal text. In this paper, we propose a unified framework for the classification and named entity recognition tasks to solve the challenges, and it can be applied to both English and Spanish scenarios. The results of our system are higher than the median F1-scores for 7 tasks and significantly exceed the F1-scores for 5 tasks. The experimental results demonstrate the effectiveness of our system.

1 Introduction

A large amount of health-related data exists on social media, and this data has attracted a lot of attention in medical applications. The Social Media Mining for Health Applications 2022 (SMM4H-2022) Shared Task(Davy Weissenbacher, 2022) involves natural language processing (NLP) challenges of using social media data for health research, including informal, colloquial expressions and misspellings of clinical concepts, noise, data sparsity, ambiguity, and multilingual posts. We participated in seven English classification tasks (i.e., task 1a, 4, 6, 7, 8, 9) and one Spanish named entity recognition task (i.e., task 10). The classification tasks focus on the classification of tweets mentioning drug changes, tweets indicating self-reported COVID-19 vaccination status, Reddit posts self-reporting exact age, etc. The named entity recognition task is designed to detect the diseases mentioned in the Spanish tweets.

There are three challenges in the SMM4H 2022 shared tasks. The first one is the domain shift problem. When fine-tuning specific downstream tasks using large-scale pre-trained models, the increasing size of the pre-trained models leads to the domain bias problem. To address this problem, we apply continue pre-training (Gururangan et al., 2020) to enhance the adaptive ability of the model in the corresponding domain. Also, we incorporate FGM adversarial training and Child-Tuning (Xu et al., 2021) to improve the robustness of the model. The second challenge is the prediction bias due to category imbalance. The category imbalance in the annotated data affects the focus of the model learning, which eventually leads to prediction bias and also hinders the robustness and generalization of the model. Therefore, we apply oversampling to generate samples from the minority class to achieve data balance. The third challenge is the noise in informal text. When learning from informal social media text data, the text contains a large number of emojis, mentioned usernames, and hyperlinks, which can affect the model's ability to learn the true semantic information in the text. Thus, we apply data cleaning to remove the noise from the data.

[†]First Author and Second Author contribute equally to this work.

Proceedings of the 29th International Conference on Computational Linguistic, pages 143–147
October 12–17, 2022.

Therefore, addressing the three challenges, the three approaches of this paper can be summarized as below:

- Applying continue pre-training and adding FGM adversarial training and Child-Tuning in fine-tuning to address the domain shift problem.
- Applying oversampling to generate more data to address the category imbalance problem.
- Applying data cleaning to solve the noise problem in the corpus.

2 Task Description

2.1 Classification Tasks

We participated in seven English classification tasks including task 1a: Classification of adverse events mentions in tweets;Task 4: Classification of tweets self-reporting exact age; Task 6: Classification of tweets which indicate self-reported COVID-19 vaccination status; Task 7: Classification of self-reported intimate partner violence on Twitter (Al-Garadi et al., 2022); Task 8: Classification of self-reported chronic stress on Twitter; and Task 9: Classification of Reddit posts self-reporting exact age. All the tasks are binary classification and use F1-score for the positive class as the evaluation metric. The training set and evaluation set are released by the organizers, while the testing set is provided afterward.

2.2 Named Entity Recognition Task

Task 10 focuses on detecting drug mentions and entity spans in Spanish tweets(Gasco et al., 2022b). The task aims to use social media as a proxy to better understand the societal perception of disease, from rare immunological and genetic diseases such as cystic fibrosis, highly prevalent conditions such as cancer and diabetes, to often controversial diagnoses such as fibromyalgia and even mental health disorders. Only one class of entities is included in the tweets. The dataset consists of a training set, a validation set and a test set with 5,000, 2,500 and 23,430 data respectively (Gasco et al., 2022a).

3 Methods

3.1 Overall Framework

A uniform framework is used for all the classification and named entity recognition tasks in both English and Spanish. The framework includes

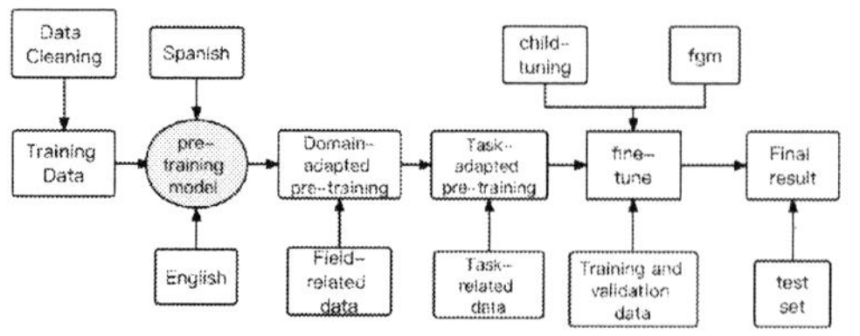

Figure 1: Overall system structure.

three major components: pre-processing social media data, continuing pre-training transformer-based models, and fine-tuning with three strategies.

First, we performed data cleaning to reduce the noise of the tweets and Reddit posts in both languages.

Then, we continued domain-adaptive pre-training using unlabeled data from all the related tasks, aiming to tailor the pre-trained model to the domain of the tasks. Specifically, we continued pre-training RoBERTa_base with data from all the English tasks for 20 rounds, using a smaller learning rate (e.g., 6e-5). We continued pre-training bert_spanish_cased_finetuned_ner (Cañete et al., 2020) using data from all the Spanish tasks, followed by 20 rounds of continued pre-training using large-scale (85k) unlabeled Spanish tweets with mentions of disease.

Finally, for each task, we fine-tuned the model using its training set and validation set. We adopted FGM adversarial training, Child-Tuning, and five-fold cross-validation in this stage to improve the performance. For certain classification tasks, we oversampled the minority class to overcome class imbalance and fine-tuned the model for 24 rounds. For the named entity recognition task, we fuse 15 models with different learning rates and different training epochs to obtain the final results. Our overall system structure is shown in Figure 1.

4 Our Method

4.1 Pre-processing

Data Cleaning: There are a large number of emojis, usernames and hyperlinks in tweets and Reddit posts. For the model to better learn the semantic information in the text, we perform three cleaning operations on the data: replacing all emojis with *emoji*, usernames with @user, and hyperlinks with http.

Oversampling: The annotated data for task 1a,task 6, task 7 are significantly imbalanced where only around 10% of the data are from the positive class (task 1a: 7%,task 6: 11%, task 7: 11%). To overcome the class imbalance problem, we performed oversampling on the training set. Specifically, random samples are duplicated from the minority class until the selected ratio between the positive class and negative class is reached.

4.2 Training Method

Five-fold Cross-voting: Utilizing five-fold cross-validation, we divide the training set into five different datasets, and the inconsistencies of entity labeling in each dataset varies. With the same model structure, we train five models on five training sets and integrate their prediction results on the same test set through majority voting.

Adversarial Training: In order to obtain a model with better robustness, we employ adversarial training to improve the model's stability. Referring to the FGM (Miyato et al., 2016) adversarial training mechanism, we directly impose a small disturbance on the embedding representation of the model and assume the embedding representation of the input text sequence $[v_1, v_2, \ldots, v_T]$ as x. Then the small disturbance r_{adv} applied each time is:

$$r_{adv} = \epsilon \cdot g/\|g\|_2 \tag{1}$$

$$g = \nabla_x L(\theta, x, y) \tag{2}$$

The formulas' meaning is to move the input one step further in the direction of rising loss, which causes the model loss to rise in the fastest direction, generating an attack. In this case, the model must identify more robust parameters in the optimization phase to deal with the attacks against the samples.

Child-Tuning: Nowadays, pre-trained models are becoming increasingly large, and when using pre-trained models to fine-tune on a smaller amount of downstream data, overfitting often occurs, resulting in poor model generalization on downstream tasks. To address the problem, Child-Tuning proposes a new fine-tuning method.in backward propagation, instead of adjusting all the parameters, only a part of them is updated, which is the child network.

Child-Tuning has two algorithms: Child-tuning-F and Child-Tuning-D. The Child-Tuning-F approach is to randomly discard a portion of the gradient when the network parameters are updated in the fine-tuning process.

$$w_{t+1} = w_t - \eta \frac{\partial \zeta(w_t)}{\partial w_t} \odot M_t \tag{3}$$

$$M_t \sim Bernoulli(P_F) \tag{4}$$

Task	AdamW	ChildTuning_F
Task 1a	0.7833	**0.8095**
Task 4	**0.9229**	0.9169
Task 6	**0.7089**	0.6953
Task 7	0.7379	**0.7500**
Task 8	**0.7742**	0.7662
Task 9	**0.8989**	0.8964

Table 1: F1 scores of the classification tasks using different optimizing strategies. Within the same task, all other parameters remain the same.

The Child-Tuning-D approach is to select a child network whose policy can be adaptively adjusted for different downstream tasks, selecting the most relevant and important parameters for the downstream tasks to act as the child network. The Fisher Information Matrix (FIM) (Tu et al., 2016) is introduced to estimate the importance of each parameter to the downstream task.

$$\mathbf{F}^{(i)}(\mathbf{w}) = \frac{1}{|D|} \sum_{j=1}^{|D|} \left(\frac{\partial \log p(y_j|\mathbf{x}_j; \mathbf{w})}{\partial \mathbf{w}^{(i)}} \right)^2 \tag{5}$$

4.3 Continue Pre-training

Due to the increasing size of pre-trained models, domain bias is a problem when fine-tuning for specific downstream tasks using large-scale pre-trained models. This can be well addressed by continuing pre-training. Continue pre-training can be interpreted as the continuation of domain-adaptive or task-adaptive training of a model based on a large-scale pre-trained language model for a specific corpus of downstream NLP tasks. Continued pre-training can be classified into two types:

- Domain-adapted pre-training (DAP), which improves the performance of the model for the corresponding domain-specific task in both low- and high-resource cases
- Task-adapted pre-training (TAP), which is very efficient in improving the performance of the model on specific tasks despite the lack of a small corpus.

5 Experiment

5.1 Classification Tasks

We trained each model with a batch size of 16 for 24 epochs. For RoBERTa_base parameters, we set the learning rate to 3e-5 and L2 normalization's weight to 0.01; for other parameters, we set

Task	TRAIN	TRAIN_1	TRAIN_05
Task 1a	0.7183	0.7343	**0.7368**
Task 6	0.7500	0.7538	**0.7608**
Task 7	0.7015	0.7379	**0.7619**

Table 2: F1 scores of the classification tasks with imbalanced data using oversampling. TRAIN is the training set with the original class distribution. TRAIN_1 is the oversampled set in which the ratio between the positive and negative class is 1 to 1. TRAIN_05 is the other oversampled set in which the ratio between the positive and negative class is 0.5 to 1.

Task	RoBERTa_base	cp-RoBERTa_base
Task 1a	**0.8095**	0.7595
Task 4	**0.9194**	0.7206
Task 6	0.7127	**0.7249**
Task 7	**0.7379**	0.7170
Task 9	0.8973	**0.8989**

Table 3: F1 scores of the classification tasks using continue pre-training. The continue pre-trained RoBERTa_base model is represented by cp-RoBERTa_base.

the learning rate to 3e-4 and L2 normalization's weight to 0. We fine-tuned the all models using five-fold cross-validation on the training set. We experimented on two optimizers, namely AdamW (Loshchilov and Hutter, 2018) and Child-Tuning. The experiments' results are shown in Table 1.

For tasks with significantly imbalanced data (task 1a, 6, 7), we oversampled the minority class with ratios of 1 to 1 and 0.5 to 1 (i.e., the ratio between the number of samples in the minority class and majority class is 0.5 to 1). For other tasks (task 4, 8, 9), we kept the original data and oversampled the minority class with a ratio of 1 to 1. We experimented on these ratios, and the results are shown in Table 2.

Furthermore, we experimented with using the continue pre-trained RoBERTa_base model (Liu et al., 2019). The model is pre-trained on the data from all the English tasks to acquire domain knowledge. The results are shown in Table 3.

For each task, we submitted the predictions of models with the best performances on the validation set. The performances of our framework are above the median and mean for all tasks except task 9. Significantly, our model the mean of task 1a by 7%, exceeds the median of task 4 by 4.7%. exceeds the median of task 7 by 7%, and exceeds the median of task 8 by 3%. The final results are shown in Table 4.

Task	Validation F1	Test F1	Mean F1	Median F1
Task 1a	0.809	0.638	0.562	-
Task 4	0.929	0.916	0.847	0.869
Task 6	0.729	0.78	-	0.77
Task 7	0.780	0.833	-	0.763
Task 8	0.774	0.781	-	0.750
Task 9	0.897	0.885	0.901	0.891

Table 4: Our results, mean results, and median results of each classification task.

Model	Validation F1	Test F1
60	0.9104±0.11	-
20+60	0.9105±0.13	-
20+60+Child-Tuning-F	0.9121±0.21	-
20+60+Child-Tuning-D	0.9064±0.15	-
Final Result	-	0.891

Table 5: Results on the SMM4H-task10 validation set and test set.

5.2 Named Entity Identification Task

For task 10, the BERT model was fine-tuned to 60 epochs on the training set with a learning rate of 1e-5 and a batch size of 12, and we continued to pre-train 10 or 20 epochs on large-scale unlabeled data with a learning rate of 6e-5, followed by adversarial training and Child-Tuning training. The experimental results are shown in the following table. Here, "60" means the BERT model is trained for 60 epochs, "20+60" means the pre-training is continued for 20 epochs and then fine-tuned for 60 epochs, and "20+60+Child-Tuning-F" means adding Child-Tuning-F. The results are the mean value of the training process plus the variation range. The final result is the result of the fusion of 15 models. The results are shown in Table 5.

6 Conclusion and Future Work

The above tasks in the SMM4H 2022 explore social perceptions of health and diseases using social media data, including classification and information extraction tasks. In this paper, we propose a unified system to solve the classification and named entity recognition tasks, which can be applied to both English and Spanish scenarios. Our framework consists of data cleaning, oversampling, continued pre-training, and applying adversarial training and Child-Tuning for fine-tuning. We evaluated different variants of our framework and demonstrated the effectiveness of the framework. For future work, we will introduce knowledge graphs from the medical field to further improve our system.

Acknowledgements

This work is supported by the National Key Research and Development Program of China (No. 2020AAA0106400), the National Natural Science Foundation of China (No.62176257, 61976211). This work is also supported by the CCF-Tencent Open Research Fund and the Youth Innovation Promotion Association CAS.

References

Mohammed Ali Al-Garadi, Sangmi Kim, Yuting Guo, Elise Warren, Yuan-Chi Yang, Sahithi Lakamana, and Abeed Sarker. 2022. Natural language model for automatic identification of intimate partner violence reports from twitter. *Array*, page 100217.

José Cañete, Gabriel Chaperon, Rodrigo Fuentes, Jou-Hui Ho, Hojin Kang, and Jorge Pérez. 2020. Spanish pre-trained bert model and evaluation data. In *PML4DC at ICLR 2020*.

Luis Gascó Darryl Estrada-Zavala Martin Krallinger Yuting Guo Yao Ge Abeed Sarker Ana Lucia Schmidt Raul Rodriguez-Esteban Mathias Leddin Arjun Magge Juan M. Banda Vera Davydova Elena Tutubalina Graciela Gonzalez-Hernandez Davy Weissenbacher, Ari Z. Klein. 2022. Overview of the seventh social media mining for health applications (#smm4h) shared tasks at coling 2022. In *In Proceedings of the Sixth Social Media Mining for Health (#SMM4H) Workshop & Shared Task*.

Luis Gasco, Darryl Estrada, Eulàlia Farré, and Martin Krallinger. 2022a. SocialDisNER corpus: gold standard annotations for detection of disease mentions in Spanish tweets. Funded by the Plan de Impulso de las Tecnologías del Lenguaje (Plan TL).

Luis Gasco, Darryl Estrada-Zavala, Eulàlia Farré-Maduell, Salvador Lima-López, Antonio Miranda-Escalada, and Martin Krallinger. 2022b. Overview of the SocialDisNER shared task on detection of diseases mentions from healthcare related and patient generated social media content: methods, evaluation and corpora. In *Proceedings of the Seventh Social Media Mining for Health (#SMM4H) Workshop and Shared Task*.

Suchin Gururangan, Ana Marasović, Swabha Swayamdipta, Kyle Lo, Iz Beltagy, Doug Downey, and Noah A Smith. 2020. Don't stop pretraining: adapt language models to domains and tasks. *arXiv preprint arXiv:2004.10964*.

Yinhan Liu, Myle Ott, Naman Goyal, Jingfei Du, Mandar Joshi, Danqi Chen, Omer Levy, Mike Lewis, Luke Zettlemoyer, and Veselin Stoyanov. 2019. Roberta: A robustly optimized bert pretraining approach. *arXiv preprint arXiv:1907.11692*.

Ilya Loshchilov and Frank Hutter. 2018. Fixing weight decay regularization in adam.

Takeru Miyato, Andrew M Dai, and Ian Goodfellow. 2016. Adversarial training methods for semi-supervised text classification. *arXiv preprint arXiv:1605.07725*.

Ming Tu, Visar Berisha, Martin Woolf, Jae-sun Seo, and Yu Cao. 2016. Ranking the parameters of deep neural networks using the fisher information. In *2016 IEEE International Conference on Acoustics, Speech and Signal Processing (ICASSP)*, pages 2647–2651. IEEE.

Runxin Xu, Fuli Luo, Zhiyuan Zhang, Chuanqi Tan, Baobao Chang, Songfang Huang, and Fei Huang. 2021. Raise a child in large language model: Towards effective and generalizable fine-tuning. *arXiv preprint arXiv:2109.05687*.

Edinburgh_UCL_Health@SMM4H'22: From Glove to Flair for handling imbalanced healthcare corpora related to Adverse Drug Events, Change in medication and self-reporting vaccination

Imane Guellil University of Edinburgh

Jinge Wu University College London

Honghan Wu University College London

Tony Sun University of Waterloo

Beatrice Alex University of Edinburgh

Abstract

This paper reports on the performance of Edinburgh_UCL_Health's models in the Social Media Mining for Health (SMM4H) 2022 shared tasks. Our team participated in the tasks related to the Identification of Adverse Drug Events (ADEs), the classification of change in medication (change-med) and the classification of self-report of vaccination (self-vaccine). Our best performing models are based on DeepADEMiner (with respective F1= 0.64, 0.62 and 0.39 for ADE identification), on a GloVe model trained on Twitter (with F1=0.11 for the change-med) and finally on a stack embedding including a layer of Glove embedding and two layers of Flair embedding (with F1= 0.77 for self-report).

1 Introduction

Identification of healthcare-related topics from social media is considered meaningful work in society. Therefore, it is attracting attention, particularly from pharmaceutical companies who want to know what people (i.e. patients) think and report about their products. Technically, it requires the ability to apply Natural Language Processing (NLP) techniques for the automatic collection, extraction, representation, analysis and validation of social media data such as Twitter and Reddit posts. This paper summarises our work in the Social Media Mining for Health Applications (#SMM4H) workshop (Davy Weissenbacher, 2022), including identification of ADEs (Task 1), classification of change in medication regimen in tweets (Task 2) and classification of tweets indicating self-reported COVID-19 vaccination. The first task splits into three further sub-tasks of Classification of English tweets reporting ADEs (Task 1a), Extraction of ADE spans in tweets (Task 1b) and Normalization of these colloquial mentions to their standard concept identifiers (IDs) in the MedDRA vocabulary (Task 1c) (Mozzicato, 2009). All of these are explained in detail in Section 2.

2 Task 1: Classification, extraction and normalization of adverse effect mentions

2.1 Data and Pre-processing

The dataset (provided by SMM4H's organizers) includes three parts:

- A training set consisting of 17,385 tweets with 1,711 ADE examples (1,240 examples without duplicates) and 16,145 examples without ADEs (NoADEs).
- A validation dataset containing 915 tweets with 87 ADE examples (65 without repetition) and 850 NoADE examples.
- A test dataset including 10,984 tweets.

The corpus is highly imbalanced as only 14% of the tweets contain ADEs.

We use the *tweet-preprocessor* python API [1] to replace the ambiguous mentions (e.g., typos in sentences) by correcting words and removing URLs and emojis. However, we only use this type of pre-processing for the classification using the first model (Glove_FlairFor_FlairBack) that we detail in Section 2.2.1.

2.2 Sub-task 1a: ADE tweet classification

This sub-task aims to detect tweets that contain an adverse effect (AE), also known as adverse drug effect (ADE), and label them with the tag ADE.

2.2.1 Method

For this binary classification task (containing ADE or not), we compare the performance of two main embeddings, transformers (mainly represented by BERT, RoBERTa) and contextual (mainly represented by FLAIR).

[1] https://pypi.org/project/tweet-preprocessor/

Proceedings of the 29th International Conference on Computational Linguistic, pages 148–152
October 12–17, 2022.

For this purpose, we use two models: 1) Glove_FlairFor_FlairBack, a stack embedding including three types of embeddings (GloVe (Pennington et al., 2014), Flair-Forward and Flair-Backward (Akbik et al., 2018)) and 2) DeepADEMiner_default, a BERT-based model (classifier-bertweet-large) trained on the training data of the shared task , (Magge et al., 2021).[2]

For training the first model (Glove_FlairFor_FlairBack), we use the FLAIR NLP framework (*which enables model training for sequence-labelling-based text classification*). (Akbik et al., 2019). A Long short-term memory network (LSTM) (Hochreiter and Schmidhuber, 1997) was also used over the word embedding in a document (i.e., each tweet) to output the document representation. Document embeddings are different from word embeddings in that they compute one embedding for an entire text, whereas word embeddings produce embeddings for individual words. The Glove_FlairFor_FlairBack model was trained for 30 epochs with a learning rate of 0.05 and the hidden_size variable set to 256. For handling the imbalance in the data, the weight of the ADE label was set to 10, whereas the one for NoADE was kept at 1.

For the second model (classifier-bertweet-large) we used the framework DeepADEMiner and in particular the default model. For training, we also used the embeddings of the large Roberta model *roberta-large*[3]. The DeepADEMiner_default model was trained for 10 epochs with a learning rate of 0.000001. To deal with the imbalanced corpus, the weight loss was fixed to 10 for the ADE class. The corpus was also randomly under-sampled retaining only 20% of the tweets with the label of NoADE. In terms of the evaluation, we directly applied the original model (DeepADEMiner_default) for evaluation without further fine-tuning due to time constraints[4].

2.2.2 Results

During the validation phase, we only trained the Glove_FlairFor_FlairBack model. In the second model (DeepADEMiner_default), we used the default trained model directly on the test dataset. The results regarding both stages (development and testing) are presented in Table 1. We also included the mean results related to the mean score calculated from the results returned by all participants.

Figure 1: The classification results on the dev dataset

Corpus	Model	P	R	F1
Dev	Glove_FlairFor_FlairBack	0.4804	0.7538	0.5868
Test	Glove_FlairFor_FlairBack	0.452	0.505	0.477
	DeepADEMiner_default	0.554	**0.765**	**0.642**
	Mean_results	**0.646**	0.497	0.562

2.2.3 Analysis

By increasing the weight loss to 10 (based on the previous works proposed by (Magge et al., 2021)), we observed that both models classified many false positives. The majority of the tweets that were wrongly classified as ADE include only the name of drugs without any ADEs. Some of them include the reference to *pain or a burn on the skin* which were the initial symptoms and not ADEs related to drugs. In future works, we plan to investigate the relation between weight loss and the results.

2.3 Sub-task 1b: ADE span extraction

2.3.1 Method

We first explored two methods for the span extraction task: 1) The default model used in DeepADEMiner (latest version using *roberta-large*[5]) and 2) A combination of results returned by the default model (*roberta-large*) and a flair-forward model trained using the DeepADEMiner framework.

For the first model, DeepADEMiner_default, we used the default parameters training the model for 10 epochs with a learning rate of 0.000001 and the hidden_size setting of 256. Only 5% of the corpus was used for validating this model.

For the second method, DeepADEMiner_default_FlairFor, we use the union of the two model outputs and we remove any duplicates. The flair-forward model was trained using 100 epochs with all other parameters kept as their defaults. The unique difference is that we used 10% of the corpus for validating this combined model. As a post-processing step, punctuation was automatically removed in case it was detected as an ADE for both approaches.

2.3.2 Results

Table 2 presents the results obtained using DeepADEMiner_default and DeepADEMiner_default_FlairFor on the test dataset. It

[2]https://bitbucket.org/pennhlp/deepademiner/src/master/

[3]https://huggingface.co/roberta-large

[4]https://hlp.ibi.upenn.edu/public_downloads/DeepADEMiner/model/latest/classifier/

[5]https://huggingface.co/roberta-large

also presents the mean scores representing the overlapping and the strict score related to the n precision, recall and f1 score.

Figure 2: The extraction results on the test dataset where _over is for the overlapping results and $_strisstrictresults$

Model	P_over	R_over	F1_over	P_str	R_str	F1_str
DeepADEMiner_default	**0.569**	0.691	**0.624**	**0.427**	**0.526**	**0.471**
DeepADEMiner_default_FlairFor	0.531	**0.709**	0.607	0.352	0.484	0.408
Mean_results	0.539	0.517	0.527	0.344	0.339	0.341

2.3.3 Analysis

Based on the results, we observe that overall DeepADEMiner_default achieves the best performance. DeepADEMiner_default_FlairFor also performs well, especially on the R_over. SemEHR seems to be poor in this task. It has a huge number of false positives. The reason for this is that it is less powerful when facing ambiguous mentions (e.g., "triggers my rapid cycling") and informal language (e.g., "can't sleeeep"). However, the deep learning models could get rid of this due to contextual learning from words. Also, SemEHR isn't designed to just detect ADEs. It detects all types of medical terms. It is mainly for this reason that it performed poorly on our in-domain data.

2.4 Sub-task 1c: ADE resolution

2.4.1 Method

The main goal of the ADE resolution task is to assign each ADE mention found in the text to its corresponding MedDRA code (Link). The file generated during the extraction phase in Sub-task 1b using the best-performing model (i.e., DeepADEMiner_default) was used as input into the resolution sub-task. Different approaches were considered for resolution but the two best-performing models selected were:

1) the default DeepADEMiner model (based on the model *bert-base-uncased*[6], and

2) DeepADEMiner_default+mcn-en-smm4h a pre-trained hugging face model called mcn-en-smm4h was applied[7]. This model was pre-trained using BioBERT and smm4h 2017 data (Sarker et al., 2018). This model was then fine-tuned using a hidden_size of 768, a max_position_embeddings of 512, an attention_probs_dropout_prob of 0.1, and hidden_dropout_prob of 0.1. For each mention, it outputs the linked MedDRA term with the preferred term (PT)/lowest level term (LLT) identifier.

[6]https://huggingface.co/bert-base-uncased

[7]https://huggingface.co/olastor/mcn-en-smm4h

2.4.2 Results

Table 3 lists the results obtained by using the two models just described. One is the default model used in the DeepADEMiner system (DeepADEMiner_default). The other model DeepADEMiner_test+mcn-en-smm4h refers to the normalization results fine-tuned with the *mcn-en-smm4h* model based on the Sub-task 1b results. It also presents the mean scores representing the overlapping and the strict score related to the precision, recall and f1 score.

Figure 3: The normalization results on the test dataset

Model	P_over	R_over	F1_over	P_str	R_str	F1_str
DeepADEMiner_default	0.04	0.04	0.04	0.03	0.03	0.03
DeepADEMiner_default+mcn-en-smm4h	**0.35**	**0.43**	**0.39**	**0.29**	**0.35**	**0.32**
Mean_results	0.120	0.112	0.116	0.085	0.082	0.083

2.4.3 Analysis

For the test dataset, DeepADEMiner_default+mcn-en-smm4h achieves the best performance with an F1_over score of 0.387. This may appear relatively low but presumably, this true positive linked entity can only be achieved if it was extracted properly in the first place. So the resolution performance is affected by the extraction one.

3 Task 3a: Classification of change in medication regimen in tweets

3.1 Data and pre-processing

The organizers provided the dataset of tweets where users self-declare changing their medication treatments, regardless of being advised by a health care professional to do so. This dataset includes 5,898 Tweets for training, 1,572 Tweets for validation and 15,360 for testing, the majority of them being decoy Tweets as only 2,360 Tweets were considered by the organizers for final system evaluation. This dataset is highly imbalanced as only 519 and 139 tweets in the training and the validation datasets, respectively, self-report a change in medication, representing only 9% of the whole dataset. No pre-processing steps were applied to this dataset.

3.2 Method

For this sub-task, we use the Glove Twitter embedding model (pre-trained glove vectors based on 2B tweets, 27B tokens, 1.2M vocab, uncased)[8]. We use the default Glove_Twitter model provided by Flair NLP.The tweets were kept in their original

[8]https://nlp.stanford.edu/projects/glove

form. As we did for classifying ADEs, we also used the flair NLP framework for training (Akbik et al., 2019). Same for the ADE classification, LSTM was also used over the word embedding in a document to output the document representation. The model was trained for 30 epochs with a learning rate of 0.05 and a hidden_size setting of 256. For handling the imbalance, the weight of tweets containing a change in medication (with the label 1) was set to 10 whereas the weights of other tweets were kept at 1.

3.3 Results

Table 4 shows the results of the Glove_Twitter model obtained on the development and test datasets. It also presents the mean scores of all participants.

Figure 4: The change of medication results on the validation dataset

Corpus	Model	P	R	F1
Dev	Glove_Twitter	0.34	0.30	0.32
Test	Glove_Twitter	**0.54**	0.11	0.19
	Mean_results	0.46	**0.54**	**0.46**

3.4 Analysis

The majority of errors are related to the misclassification of tweets including some medications that were prescribed for the first time. To improve the results, the training corpus should be enriched with more samples related to the first-time medication. Also, the chosen model is a fast model to run, however, it is not the best model for taking into account the context. Hence, we plan to use contextual embedding including Flair (Akbik et al., 2019) or Elmo (Peters et al., 2018), in the future, for improving performance.

4 Task 6: Classification of tweets indicating self-reported COVID-19 vaccination

4.1 Data and pre-processing

The shared task organisers provided the dataset of tweets where users self-declare their COVID-19vaccination status. This dataset includes 13,693 tweets for training, 2,784 tweets for development and 5,923 tweets for testing.

4.2 Method

For this sub-task, we use two main models: 1) a Glove Twitter embedding model (the same that we use for Task 3), and 2) a stack embedding including three types of embeddings including Glove, Flair-Forward and Flair-Backward (the same which we use for Sub-task 1a).

With the first, the Glove_Twitter, model the tweets were kept in their original form. As we did for classifying ADEs. This model was trained for 26 epochs. In contrast, we tried three different epochs for training the second, stack embedding model, i.e., 6, 15 and 30 epochs. For training all the models, we also used the flair NLP framework (Akbik et al., 2019). As for the previous tasks, LSTM was also used over the word embedding in a document to output the document representation. Both models were trained with a learning rate of 0.05 and a hidden_size setting of 256. No pre-processing, sampling or increase in weights was applied to the first model. However, a random under-sampling method, where we kept the same number of tweets for both classes, was applied for the second model when the model was trained for 15 epochs. The tweets were also prepossessed using the Twitter-preprocess API when the model was run for 6 and 30 epochs. Also, the weight of tweets self-reporting a vaccination (with label 1) was set to 3 (where the models were trained for 6 and 30 epochs) and 10 (where the model was trained for 15 epochs) whereas the weights of all other tweets were kept 1.

4.3 Results

Table 5 illustrates the results obtained on the test dataset. It also presents the median scores of all participants.

Figure 5: The vaccine self-reporting results on the test dataset

Model	P	R	F1
Glove_Twitter	0.87	0.57	0.69
Glove_FlairFor_FlairBack_epoch6	0.67	0.80	0.73
Glove_FlairFor_FlairBack_epoch15	0.47	**0.93**	0.63
Glove_FlairFor_FlairBack_epoch30	0.69	0.86	**0.77**
Median_results	**0.9**	0.77	0.68

4.4 Analysis

As the majority of people reporting a vaccination are using other terms such as "COVID-19", when these terms are used in a vaccine chatter context, they are misclassified. Also, the tweets that are not directly referencing vaccination or the sarcastic tweets are misclassified.

5 Conclusion

In this work, we explore the use of different models associated with different techniques for dealing with imbalanced healthcare-related social media corpora. We observed that the best technique relied on the use of deep learning models such as *Roberta or Flair*, under-sampling and the adjustment of the weight loss. In the future, we plan to explore more techniques for handling imbalanced datasets, augmenting the training corpus and relying on more embedding models.

6 Acknowledgement

This study/project is funded by the National Institute for Health Research (NIHR) Artificial Intelligence and Multimorbidity: Clustering in Individuals, Space and Clinical Context (AIM-CISC) grant NIHR202639. The views expressed are those of the authors and not necessarily those of the NIHR or the Department of Health and Social Care.

References

Alan Akbik, Tanja Bergmann, Duncan Blythe, Kashif Rasul, Stefan Schweter, and Roland Vollgraf. 2019. Flair: An easy-to-use framework for state-of-the-art nlp. In *Proceedings of the 2019 conference of the North American chapter of the association for computational linguistics (demonstrations)*, pages 54–59.

Alan Akbik, Duncan Blythe, and Roland Vollgraf. 2018. Contextual string embeddings for sequence labeling. In *COLING 2018, 27th International Conference on Computational Linguistics*, pages 1638–1649.

Luis Gascó Darryl Estrada-Zavala Martin Krallinger Yuting Guo Yao Ge Abeed Sarker Ana Lucia Schmidt Raul Rodriguez-Esteban Mathias Leddin Arjun Magge Juan M. Banda Vera Davydova Elena Tutubalina Graciela Gonzalez-Hernandez Davy Weissenbacher, Ari Z. Klein. 2022. Overview of the seventh social media mining for health applications smm4h shared tasks at coling 2022. *In Proceedings of the Seventh Social Media Mining for Health (SMM4H) Workshop and Shared Task.*

Sepp Hochreiter and Jürgen Schmidhuber. 1997. Long short-term memory. *Neural computation*, 9(8):1735–1780.

Arjun Magge, Elena Tutubalina, Zulfat Miftahutdinov, Ilseyar Alimova, Anne Dirkson, Suzan Verberne, Davy Weissenbacher, and Graciela Gonzalez-Hernandez. 2021. Deepademiner: a deep learning pharmacovigilance pipeline for extraction and normalization of adverse drug event mentions on twitter. *Journal of the American Medical Informatics Association*, 28(10):2184–2192.

Patricia Mozzicato. 2009. Meddra. *Pharmaceutical Medicine*, 23(2):65–75.

Jeffrey Pennington, Richard Socher, and Christopher D Manning. 2014. Glove: Global vectors for word representation. In *Proceedings of the 2014 conference on empirical methods in natural language processing (EMNLP)*, pages 1532–1543.

Matthew E Peters, Mark Neumann, Mohit Iyyer, Matt Gardner, Christopher Clark, Kenton Lee, and Luke Zettlemoyer. 2018. Deep contextualized word representations. corr abs/1802.05365 (2018). *arXiv preprint arXiv:1802.05365*.

Abeed Sarker, Maksim Belousov, Jasper Friedrichs, Kai Hakala, Svetlana Kiritchenko, Farrokh Mehryary, Sifei Han, Tung Tran, Anthony Rios, Ramakanth Kavuluru, et al. 2018. Data and systems for medication-related text classification and concept normalization from twitter: insights from the social media mining for health (smm4h)-2017 shared task. *Journal of the American Medical Informatics Association*, 25(10):1274–1283.

KUL@SMM4H'22: Template Augmented Adaptive Pre-training for Tweet Classification

Sumam Francis and **Marie-Francine Moens**
KU Leuven, Belgium

Abstract

This paper describes models developed for the Social Media Mining for Health 2022 shared tasks. Our team participated in the first subtask that classifies tweets with Adverse Drug Effect mentions. Our best-performing model comprises of a template augmented task adaptive pre-training and further fine-tuning on target task data. Augmentation with random prompt templates increases the amount of task-specific data to generalize the pretrained language model to the target task domain. We explore 2 pre-training strategies: masked language modeling and simple contrastive pre-training and the impact of adding template augmentations with these pre-training strategies. Our system achieves an F1 score of 0.433 on the test set without using supplementary resources and medical dictionaries.

1 Introduction

The goal of the shared task as part of the Social Media Mining for Health (SMM4H) - 2022 (Weissenbacher et al., 2022) involves detecting tweets that have Adverse Drug Effect (ADE) mentions. Organizers of SMM4H Task 1 provided datasets of English tweets with annotations ADE and noADE indicating the presence or absence of ADE mentions in the tweet. Recent work (Lee et al., 2020; Gururangan et al., 2020; Beltagy et al., 2019) showed that downstream performance can be improved by further adapting a general pre-trained model by continued pre-training on more relevant set of downstream tasks. The representations learned in the task adaptive pre-training (TAPT) (Gururangan et al., 2020) involves pre-training using an unsupervised objective on not just the similar domain but the actual end-task and dataset itself. It has shown to improve the performance of the model for the target task and is less computationally demanding. We explore task adaptive pre-training strategies together with fine-tuning for the classification problem. We further explore the use templates as augmentations to improve the task adaptive pre-training.

2 Data

The dataset (Magge et al., 2021) comprises of a training set ($18,000$ tweets), validation set (915 tweets), and test set ($10,000$ tweets). The dataset is very imbalanced with only around 7% of the tweets consisting of ADE mentions. We use oversampling to deal with this class imbalance.

As part of the pre-processing on the dataset, we removed URL, retweets, mentions, extra space, non-ascii words and characters. Further we lower-cased and striped off white spaces at both ends. We inserted space between punctuation marks.

3 Model

We first apply TAPT (Gururangan et al., 2020) on the pre-trained model to expose it to the task domain sentences and increase the overlap between the language model domain and the target task domain.

The two pre-training strategies we explored in this system setup are: 1) **Simple Contrastive pre-training (SimSCE)** (Gao et al., 2021): The idea is to encode the same sentence twice. Due to the dropout used in the transformer models, both sentence embeddings will be at slightly different positions in the vector space. The distance between these two embeddings will be minimized, while the distance to other embeddings of the other sentences in the same batch will be maximized which serve as negative examples.

2) **Masked language modelling (MLM) (Devlin et al., 2019)** : MLM is an efficient pre-training strategy for learning sentence embeddings. It is trained with a masked language modeling objective (i.e., cross-entropy loss on predicting randomly masked tokens).

Proceedings of the 29th International Conference on Computational Linguistic, pages 153–155
October 12–17, 2022.

Given that amount of task data is insufficient for TAPT,we further augment the task data and generate multiple sentences with various prompt templates appended at the end of each sentence. The templates are randomly assigned from a set of predefined templates based on label information. Examples of templates include: "It contains an Adverse Drug Effect.","It contains an ADE","ADE is present","It mentions an ADE". We thus expose the pre-trained language model (LM) to syntactically diverse template prompts to either cluster the similar representations (SimSCE_aug) or probe the pre-trained LM using masked tokens (mlm_aug). We further fine-tune the continually task pre-trained model on the task data. The supervised TAPT with augmentations phase helps the pre-trained LM encapsulate a broader range of task distribution.

We use BERTweet (Nguyen et al., 2020) as the base encoder to compute the sentence representations. To tackle class imbalance, we experiment with oversampling. For oversampling approach, we randomly sampled positive examples with replacement until ADE class contained 10,000 tweets.

4 Experiments

For the classification task, each model is fine-tuned for 10 epochs with a learning rate of $5e-5$ using Adam optimizer (Loshchilov and Hutter, 2019). We set the batch size to 32 and the maximum sequence length to 128. We utilize PyTorch (Paszke et al., 2017) and HuggingFace library [1] for training BERT using cross-entropy loss. We save model_checkpoint every 200 steps against the validation set using the F1-score of the ADE class for evaluation. For adaptive pre-training with MLM, we train for 10 epochs with masking probability set as 0.15. For adaptive contrastive pre-training, we train for 10 epochs with batch size of 64 and the maximum sequence length 100.

5 Discussion

It is evident from Table 1 that template augmented TAPT yields improved classification accuracy in detecting ADE mentions in tweets. It demonstrates that augmenting with prompt templates can better generalize LM to target task distribution compared to conventional adaptive pre-training method. MLM pre-training with template augmentations (mlm_aug) outperforms MLM only pre-training quite significantly on the validation set. This difference can be attributed to the lack of sufficient task-related data in MLM only pre-training compared to template augmented pre-training. The results from Table 1 indicate that oversampling the ADE class (BERTweet) improved the class imbalance compared to not oversampling (BERTweet_n) and is clearly indicative in the F1-scores.

Table 1: Micro-average Precision (P), Recall (R) and F1 scores (F1) on the validation set of the SMM4H 2022 Task 1a.

TAPT	FT	P	R	F1
-	BERTweet_n	0.6912	0.7231	0.7068
-	BERTweet	0.7065	0.7223	0.7143
mlm	BERTweet	0.7042	0.7692	0.7353
mlm_aug	BERTweet	**0.7605**	**0.8307**	**0.7941**
simsce	BERTweet	0.7344	0.7231	0.7287
simsce_aug	BERTweet	0.7385	0.75	0.7441

Table 2: Micro-average Precision (P), Recall (R) and F1 scores (F1) on the test set of the SMM4H 2022 Task 1a.

model (BERTweet)	P	R	F1
mlm_aug	0.614	0.334	0.433
mlm_aug+ (post-eval)	**0.776**	0.4680	**0.584**
Mean_results	0.646	0.497	0.562

Table 2 shows the performance of the best performing model on the test set of SMM4H 2022 Task1a. Our model's performance is relatively less on the test set compared to the validation set, which can be attributed to over-fitting. Further the performance was improved in post-eval by augmenting data with more templates and achieved better F1 scores than mean_results. To further improve the performance, back translations can be used in addition to template augmentations.

6 Conclusion

In this work, we explore and propose the use of template based TAPT for improving the downstream task performance of detection ADE entity mentions in English tweets. We demonstrate the significance of the template augmentation in comparison to traditional adaptive pre-training strategies. Experiments have shown that our model with template augmentations with TAPT has achieved an F1-score of 0.43, precision of 0.61, and recall of 0.33 on the test set. The future directions would be to use supplementary data and back-translation together with template augmentations.

[1] https://huggingface.co/models

References

Iz Beltagy, Kyle Lo, and Arman Cohan. 2019. Scibert: A pretrained language model for scientific text. In *Proceedings of the 2019 Conference on Empirical Methods in Natural Language Processing and the 9th International Joint Conference on Natural Language Processing, EMNLP-IJCNLP 2019, Hong Kong, China, November 3-7, 2019*, pages 3613–3618. Association for Computational Linguistics.

Jacob Devlin, Ming-Wei Chang, Kenton Lee, and Kristina Toutanova. 2019. BERT: pre-training of deep bidirectional transformers for language understanding. In *Proceedings of the 2019 Conference of the North American Chapter of the Association for Computational Linguistics: Human Language Technologies, NAACL-HLT*. Association for Computational Linguistics.

Tianyu Gao, Xingcheng Yao, and Danqi Chen. 2021. Simcse: Simple contrastive learning of sentence embeddings. In *Proceedings of the 2021 Conference on Empirical Methods in Natural Language Processing, EMNLP 2021, Virtual Event / Punta Cana, Dominican Republic, 7-11 November, 2021*, pages 6894–6910. Association for Computational Linguistics.

Suchin Gururangan, Ana Marasovic, Swabha Swayamdipta, Kyle Lo, Iz Beltagy, Doug Downey, and Noah A. Smith. 2020. Don't stop pretraining: Adapt language models to domains and tasks. In *Proceedings of the 58th Annual Meeting of the Association for Computational Linguistics, ACL 2020, Online, July 5-10, 2020*, pages 8342–8360. Association for Computational Linguistics.

Jinhyuk Lee, Wonjin Yoon, Sungdong Kim, Donghyeon Kim, Sunkyu Kim, Chan Ho So, and Jaewoo Kang. 2020. Biobert: a pre-trained biomedical language representation model for biomedical text mining. *Bioinform.*, 36(4):1234–1240.

Ilya Loshchilov and Frank Hutter. 2019. Decoupled weight decay regularization. In *7th International Conference on Learning Representations, ICLR 2019, New Orleans, LA, USA, May 6-9, 2019*. OpenReview.net.

Arjun Magge, Elena Tutubalina, Zulfat Miftahutdinov, Ilseyar Alimova, Anne Dirkson, Suzan Verberne, Davy Weissenbacher, and Graciela Gonzalez-Hernandez. 2021. Deepademiner: a deep learning pharmacovigilance pipeline for extraction and normalization of adverse drug event mentions on twitter. *J. Am. Medical Informatics Assoc.*, 28(10):2184–2192.

Dat Quoc Nguyen, Thanh Vu, and Anh Tuan Nguyen. 2020. Bertweet: A pre-trained language model for english tweets. In *Proceedings of the 2020 Conference on Empirical Methods in Natural Language Processing: System Demonstrations, EMNLP 2020 - Demos, Online, November 16-20, 2020*, pages 9–14. Association for Computational Linguistics.

Adam Paszke, Sam Gross, Soumith Chintala, Gregory Chanan, Edward Yang, Zachary DeVito, Zeming Lin, Alban Desmaison, Luca Antiga, and Adam Lerer. 2017. Automatic differentiation in pytorch. In *NIPS-W*.

Davy Weissenbacher, Ari Z. Kleinand Luis Gascó, Darryl Estrada-Zavala, Martin Krallinger, Yuting Guo, Yao Ge, Abeed Sarker, Ana Lucia Schmidt, Raul Rodriguez-Esteban, Mathias Leddin, Arjun Magge, Juan M. Banda, Vera Davydova, Elena Tutubalina, and Graciela Gonzalez-Hernandez. 2022. Overview of the seventh social media mining for health applications smm4h shared tasks at coling 2022. In *In Proceedings of the Seventh Social Media Mining for Health (SMM4H) Workshop and Shared Task.*

Enolp musk@SMM4H'22 : Leveraging Pre-trained Language Models for Stance And Premise Classification

Millon Madhur Das, Archit Mangrulkar, Ishan Manchanda
Manav Nitin Kapadnis, Sohan Patnaik
Indian Institute of Technology Kharagpur, India
{millonmadhurdas, archit.mang, ishanmanchanda}@kgpian.iitkgp.ac.in,
{iammanavk, sohanpatnaik106}@iitkgp.ac.in

Abstract

This paper covers our approaches for the Social Media Mining for Health (SMM4H) Shared Tasks 2a and 2b. Apart from the baseline architectures, we experiment with Parts of Speech (PoS), dependency parsing, and Tf-Idf features. Additionally, we perform contrastive pretraining on our best models using a supervised contrastive loss function. In both the tasks, we outperformed the mean and median scores and ranked first on the validation set. For stance classification, we achieved an F1-score of **0.636** using the CovidTwitterBERT model, while for premise classification, we achieved an F1-score of **0.664** using BART-base model on test dataset. Additionally, all the code and data to generate reproducible results is available on Github[1].

1 Introduction

With the growth of social media platforms such as Twitter and Facebook, people can express their views on any topic. This lets them reach a vast audience, as these platforms have millions of users, and helps them get recognition for their issues online. In the case of trending events on social media, many people post their opinions around the same time, thus rapidly shaping the public opinion on those events. Misinformation may be spread unknowingly or knowingly by malicious parties, creating the wrong public opinion among social media users.

Generally, when governments introduce new schemes or policies, they need to know the people's reviews, suggestions, and complaints. Social Media sites like Twitter are essential in this aspect as they let users post microblog tweets of up to 280 characters and allow them to use hashtags to link the tweet to other tweets of the same topic that use this hashtag. During the COVID crisis, many public mandates were imposed regarding the closure of public places, lockdown in many places, and social distancing norms. This unusual change made many people post their views on these topics. To derive the public opinion on these topics, we must know the text's claim i.e. towards what entity is this argument targeted, the user's stance associated with this claim i.e. whether the argument is in favor or against the claim, and the premise associated with this claim i.e whether the text has a convincing argument or not.

Similar to the work carried out by (Glandt et al., 2021), we perform stance classification for health mandate-related tweets by employing pre-trained transformer models and then trying to improve classification results by using additional features as described in the section 3.2.

2 Task & Dataset Description

Task 2 of the shared tasks for Social Media Mining for Health (Weissenbacher et al., 2022) has the motive for stance and premise classification of tweets related to the health mandates - "Stay at Home Orders", "Face Masks" and "School Closures". The two subtasks are:

- Task 2a: Classification of stance in tweets about health mandates related to COVID-19 (in English)
- Task 2b: Classification of premise in tweets about health mandates related to COVID-19 (in English)

The dataset (Davydova and Tutubalina, 2022) consists of views of the general public on various issues and government mandates for COVID-19 in the form of Twitter Comments. It is manually labeled for claims, stance, and premise. The claims are of three categories "Stay at Home Orders", "Face Masks" and "School Closures". Each of the annotated entries in the dataset consists of stance which is of one of the three classes- "FAVOR", "AGAINST" or "NEITHER", and the

[1] https://github.com/architmang/enolp_musk-SMM4H_COLING2022

Proceedings of the 29th International Conference on Computational Linguistic, pages 156–159
October 12–17, 2022.

premise which contains the binary annotations 0 (if the tweet does not have any premise) or 1 (if the tweet contains a premise). The training, validation, and test sets are of size 3556, 600, and 2000 tweets, respectively.

3 Implementation Details

Section 3.1 describes the baseline architectures used for our experiments, Section 3.2 and Section 3.3 describe the techniques tried to improve the models using additional features and contrastive learning respectively. We discuss the details of our experimental setting in Section 3.4.

3.1 Baseline Architectures

We adopt pre-trained transformer models (Vaswani et al., 2017) from HuggingFace and add linear layers over the 786 or 1024 dimensional sentence representation from the models (Figure 1) and finetune them for our use case. We use monolingual pretrained language models as the data corpus consists of English comments only.

The details of exact architectural details is discussed in this section.

- BERT[2](Devlin et al., 2018) is a transformers model pretrained on a large corpus of English data in a self-supervised fashion with the Masked Language Modelling objective. We use the uncased version of the model for our experiments.

- RoBERTa[3] (Liu et al., 2019), pretrained on a large corpus of English data in a semi-supervised setting with dynamic masked language modelling objective.

- DeBERTa-V3[4] (He et al., 2021), is an improvement over the original DeBERTa model that uses ELECTRA-style pre-training with Gradient Disentangled Embedding Sharing. DeBERTa improves the BERT and RoBERTa models using disentangled attention and enhanced mask decoder.

- BART[5] (Lewis et al., 2019) is a encoder-decoder transformer model with a BERT like bidirectional encoder and auto-regressive decoder like GPT (Radford and Narasimhan, 2018).

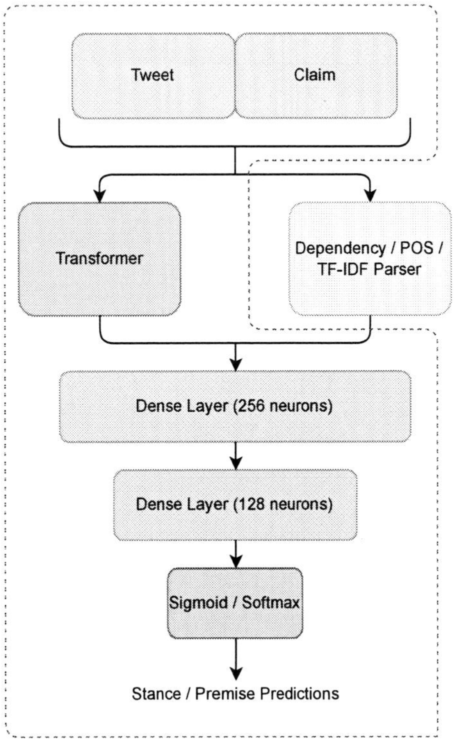

Figure 1: Model Architectures
The part of the figure inside the blue dotted line is the baseline architecture.

- CovidTwitterBERT-V2[6](Müller et al., 2020), is a BERT$_{large}$ model pretrained on large corpus of Tweets on COVID-19.

3.2 Additional Features

As suggested in (Kapadnis et al., 2021), we pass additional information to the model in the form of Parts of Speech features (POS), Dependency Parsing features and term frequency-inverse document frequency (Tf-Idf) features. The claims corresponding to the tweets are appended to the text before computing the feature vectors.

We find the POS label for each lexicon and then label encode it according to the descending order of occurrences. This feature vector is then merged with the pooled layer output and passed to the next set of dense layers.
Similarly, we do it for dependency parsing features and Tf-idf bi-gram features generation.

3.3 Contrastive Pretraining

We try to improve the representations in the embedding space of our best model (CovidTwitterBERT

[2]huggingface.co/bert-base-uncased
[3]huggingface.co/roberta-large
[4]huggingface.co/microsoft/deberta-v3-large
[5]huggingface.co/facebook/bart-large

[6] huggingface.co/digitalepidemiologylab/covid-twitter-bert-v2

& BART-base) by using a supervised contrastive loss function (Khosla et al., 2020), instead of the usual cross-entropy loss. This can leverage label information better and morph the embedding space by pulling data points from the same class closer and pushing apart data points from other classes. The loss is defined as follows,

$$\mathcal{L}^{sup}_{out} = \sum_{i \in I} \mathcal{L}^{sup}_{out,i} = \sum_{i \in I} \frac{-1}{|P(i)|} \sum_{p \in P(i)} \log \frac{exp\left(z_i z_p/\tau\right)}{\sum\limits_{a \in A(i)} exp\left(z_i z_a/\tau\right)} \tag{1}$$

The dataset for contrastive training was created by adding positive and negative labels to the samples present in the task dataset. We try three different strategies, finetuning with contrastive loss only, pretraining with contrastive loss then finetuning with cross-entropy loss and finetuning with a weighted loss with 0.7 weight given to cross-entropy and 0.3 to contrastive loss. The weights were chosen considering the performance when finetuning with cross-entropy and contrastive loss. We achieved an F1-score of 0.289, 0.835, and 0.802 with the above mentioned strategies respectively.

3.4 Experimentation Details

Preliminary analysis of the training set reveals that the dataset is balanced among the classes for stances. However, this is not the case for the premise prediction task. The most frequent tri-grams derived from the tweets that favor the claim are *'#closetheschools #closetheschools #closetheschools'*, *'wear damm mask'* & *'people wearing masks'*; the same for against *'#schoolreopenindia #schoolreopenindia #schoolreopenindia'*, *'open. don't think'* & *'don't wear mask'*. The none class has tri-grams from random topics like *'liveon #ourapponplaystore #websiteonbio'*.

We preprocess the dataset before passing it through the transformer models. The whitespaces, URLs, HTML tags, and emoji patterns are removed from the tweet string. The processed tweets and claims are passed to the model tokenizer, which returns the encoded input which consists of input ids and attention masks. The stance values are mapped to numerical values using a label binarizer.

All the models are trained with a batch size of 16, except CovidTwitterBERT which is trained with a batch size of 8 due to computational constraints. A very low learning rate of $10e^{-5}$ is used to finetune the transformer models for 10 epochs. Weight decay and learning rate scheduling is used to prevent overfitting. The input sequence length after tokenization is set at 512 to get the full advantage of these large language models.

MODELS	base	large	large + noun	large + dependency	Contrastive
BERT	0.606	0.602	0.475	0.537	-
RoBERTa	0.645	0.731	0.744	0.743	-
DeBERTa-V3	0.630	0.773	0.784	0.784	-
BART	0.479	0.691	0.728	0.708	-
CovidTwitterBERT	-	**0.855**	0.843	0.827	0.835

Table 1: Performance on Stance Prediction

MODELS	base	large	large + noun	large + dependency	Contrastive
BERT	0.803	0.785	0.753	0.655	-
RoBERTa	0.823	0.817	0.741	0.675	-
DeBERTa-V3	0.819	0.812	0.760	0.729	-
BART	**0.857**	0.803	0.715	0.738	0.704
CovidTwitterBERT	-	0.781	0.764	0.734	-

Table 2: Performance on Premise Prediction

4 Evaluation Metrics

The F1-scores for stance and premise are calculated as follows,

$$F_1 = 1/n \sum_{c \in C} F_{1rel,c}$$

where, C is the set of claims, n is the number of unique claims and F_{1rel} is the macro F_1-score averaged over first two relevance classes (the class "neither" is excluded).

5 Summary of Results

We note the crucial observations for Stance & Premise prediction on the development dataset in Table 1 & Table 2[7] respectively. The first column mentions the model name, and the rest of the columns are the performance while using a specific model size or other improvement strategies described earlier.

We observe that, in general the transformer models pre-trained on in-domain datasets perform better than the base models. Furthermore, we also notice a jump in performance gained with CovidTwitterBERT by pretraining the baseline BERT_{large} model with in-domain data.

We performed contrastive training only for the best model for each subtask. However, the performance didn't improve with this strategy. Passing more contextual information in the form of additional features gives a considerable boost to the performance of baseline models for stance prediction but fails for premise prediction.

[7]Note that the results for Task 2b are according to the evaluation metric described in the Task Description and doesn't match with Codalab results.

Surprisingly, BART-base outperforms the large and in-domain pre-trained models for Task 2b.

References

Vera Davydova and Elena Tutubalina. 2022. Smm4h 2022 task 2: Dataset for stance and premise detection in tweets about health mandates related to covid-19. In *Proceedings of the Seventh Social Media Mining for Health Applications (#SMM4H) Workshop Shared Task*, pages –.

Jacob Devlin, Ming-Wei Chang, Kenton Lee, and Kristina Toutanova. 2018. Bert: Pre-training of deep bidirectional transformers for language understanding. *arXiv preprint arXiv:1810.04805*.

Kyle Glandt, Sarthak Khanal, Yingjie Li, Doina Caragea, and Cornelia Caragea. 2021. Stance detection in COVID-19 tweets. In *Proceedings of the 59th Annual Meeting of the Association for Computational Linguistics and the 11th International Joint Conference on Natural Language Processing (Volume 1: Long Papers)*, pages 1596–1611, Online. Association for Computational Linguistics.

Pengcheng He, Jianfeng Gao, and Weizhu Chen. 2021. Debertav3: Improving deberta using electra-style pretraining with gradient-disentangled embedding sharing. *arXiv preprint arXiv:2111.09543*.

Manav Kapadnis, Sohan Patnaik, Siba Panigrahi, Varun Madhavan, and Abhilash Nandy. 2021. Team enigma at ArgMining-EMNLP 2021: Leveraging pre-trained language models for key point matching. In *Proceedings of the 8th Workshop on Argument Mining*, pages 200–205, Punta Cana, Dominican Republic. Association for Computational Linguistics.

Prannay Khosla, Piotr Teterwak, Chen Wang, Aaron Sarna, Yonglong Tian, Phillip Isola, Aaron Maschinot, Ce Liu, and Dilip Krishnan. 2020. Supervised contrastive learning. *Advances in Neural Information Processing Systems*, 33:18661–18673.

Mike Lewis, Yinhan Liu, Naman Goyal, Marjan Ghazvininejad, Abdelrahman Mohamed, Omer Levy, Ves Stoyanov, and Luke Zettlemoyer. 2019. Bart: Denoising sequence-to-sequence pre-training for natural language generation, translation, and comprehension. *arXiv preprint arXiv:1910.13461*.

Yinhan Liu, Myle Ott, Naman Goyal, Jingfei Du, Mandar Joshi, Danqi Chen, Omer Levy, Mike Lewis, Luke Zettlemoyer, and Veselin Stoyanov. 2019. Roberta: A robustly optimized bert pretraining approach. *arXiv preprint arXiv:1907.11692*.

Martin Müller, Marcel Salathé, and Per E Kummervold. 2020. Covid-twitter-bert: A natural language processing model to analyse covid-19 content on twitter. *arXiv preprint arXiv:2005.07503*.

Alec Radford and Karthik Narasimhan. 2018. Improving language understanding by generative pretraining.

Ashish Vaswani, Noam Shazeer, Niki Parmar, Jakob Uszkoreit, Llion Jones, Aidan N Gomez, Łukasz Kaiser, and Illia Polosukhin. 2017. Attention is all you need. *Advances in neural information processing systems*, 30.

Davy Weissenbacher, Ari Z. Klein, Luis Gascó, Darryl Estrada-Zavala, Martin Krallinger, Yuting Guo, Yao Ge, Abeed Sarker, Ana Lucia Schmidt, Raul Rodriguez-Esteban, Mathias Leddin, Arjun Magge, Juan M. Banda, Vera Davydova, Elena Tutubalina, and Graciela Gonzalez-Hernandez. 2022. Overview of the seventh social media mining for health applications smm4h shared tasks at coling 2022. In *Proceedings of the Seventh Social Media Mining for Health Applications (#SMM4H) Workshop Shared Task*, pages –.

AIR-JPMC@SMM4H'22: Identifying Self-Reported Spanish COVID-19 Symptom Tweets Through Multiple-Model Ensembling

Adrian Garcia Hernandez, Leung Wai Liu, Akshat Gupta, Vineeth Ravi, Saheed O. Obitayo, Xiaomo Liu, Sameena Shah
J.P. Morgan AI Research, New York, NY, USA
ag4482@columbia.edu, leungwai@wustl.edu,
{akshat.x.gupta, vineeth.ravi, saheed.o.obitayo, xiaomo.liu, sameena.shah}@jpmchase.com

Abstract

We present our response to Task 5 of the Social Media Mining for Health Applications (SMM4H) 2022 competition. We share our approach into classifying whether a tweet in Spanish about COVID-19 symptoms pertain to themselves, others, or not at all. Using a combination of BERT based models, we were able to achieve results that were higher than the median result of the competition.

1 Introduction

The Social Media Mining for Health Applications (SMM4H) 2022 Shared Tasks competition (Weissenbacher et al., 2022) aim to encourage the use of Natural Language Processing for health research. This paper will describe our group's response to Task 5, which deals with the classification of Spanish-language tweets containing self-reported COVID-19 symptoms. The motivation for this task is the need to further develop the volume of natural language processing research on languages other than English. As Joshi et al. (2020) point out, only a small number of the world's 7000 languages are represented in Natural Language Processing technologies and related conferences. Task 5 identifies a need to increase the amount of NLP and social media research with regards to COVID-19 in all languages other than English.

Task 5 is of a three-way classification problem where a Spanish-language tweet has to be identified as one of three possible classes: **Self_reports**, **non_personal_reports**, and **Lit-News_Mentions**. What follows is a description the data-sets provided, various experiments performed, and our final submission results.

2 Dataset Information

We were presented with three data-sets consisting of labeled tweets to be used for training, an unlabeled validation set to check the final performance of our models, and an unlabeled test set to submit our final predictions. As we did not originally have the labels of the validation set during the validation stage, two approaches were applied to split the labeled training data into training and validation sets: a 80%-20% split and a 50%-50% split.

Dataset	Self-reports	Non-personal-reports	Lit-news-Mentions	Total
Training	1654	2413	5984	10051
Validation	572	859	2146	3577
Test	—	—	—	6850

Table 1: Distribution of Spanish-language tweets of each class in their respective data-sets. Since the test set is unlabeled, the distribution is unknown.

The motivation for the 80%-20% split was to feed as much data as possible when training while allowing us to measure the performance of our models. However, since the distribution of the training set skews towards more **Lit-News_mentions** tweets, the 50%-50% split approach is used to provide a better performance metric. Table 1 shows the distribution of the training, validation, and test sets.

3 Methods

3.1 Pre-processing

To explore the effects of the unbalanced class distribution in the original training set, three training sets were created on the training data from the 80%-20% split: an 'Equalized' training set in which the minority labels were randomly over-sampled to match the number of samples for the **Lit-News** label, an 'Over-Under' training set in which the minority labels were randomly over-sampled to contain 70% of the **Lit-News** label's samples while the **Lit-News** label was then randomly under-sampled to contain just 80% of its previous size, and a 'Reversed' training split set in which the positive label (self-reports) was randomly over-sampled to be the

Proceedings of the 29th International Conference on Computational Linguistic, pages 160–162
October 12–17, 2022.

size of the **Lit-News** label while the **Lit-News** label was then randomly under-sampled to be the size of the **Self_reports** label. No pre-processing was performed for the 50%-50% split training and validation data sets.

3.2 Models

To address this task, we utilized various versions of BERT (Devlin et al., 2019), as it is known to produce state-of-the-art results while being adaptable for many different applications. We used the cased and uncased versions of BERT-base-multilingual (Devlin et al., 2019) and Spanish-BERT (Cañete et al., 2020) as well as XLM-RoBERTa (Conneau et al., 2019).

3.3 Model Ensembling Post-processing

For the 80%-20% split, models were trained on the modified training data-sets for 5 iterations of 10 epochs each. The epoch with the best F1-score for the positive label (**Self_reports**) was kept per iteration. At the end of each model's training period, the best performing iteration was selected. Then, a majority-vote ensemble predictor was created by combining the predictions of each individual model and used to submit the predictions on our unlabeled test set.

A similar approach was used for the 50%-50% data-set split. However, models were trained for 5 iterations of 15 epochs each. Then, multiple model ensembling methods were tested: majority-vote of all 25 models, unweighted average and weighted averages, and a combination of separating per model and taking the top 5/10/15/20 models.

This equation was used to calculate the unweighted average: $\lfloor(\Sigma x)/25\rceil$, with x being the numeric value of the prediction, and the result rounded to nearest digit. This equation was used to calculate the weighted average: $\lfloor(\Sigma xf)/c\rceil$, with f being the F1-score of each particular model, and $c = 17.8153$ being the sum of all the F1-scores of all 25 models, with the result rounded to the nearest digit. Ultimately, the majority vote of all 25 models were used, as it produced the highest results.

Table 2 shows the baseline results for all models, while Table 3 shows the validation results after model ensembling. Ensembling was used because it has been shown to have a better performance than individual models (Jayanthi and Gupta, 2021).

Method \ Model	BERT$_{BASE}$-multilingual	BETO	XLM-RoBERTa
Original (80-20)	0.761 (0.756)	0.743 (0.737)	0.748
Over-Under (80-20)	0.751 (0.752)	0.751	—
Reversed (80-20)	0.758	—	—
Equalized (80-20)	0.747	—	—
Original (50-50)	0.750 (0.739)	0.731 (0.736)	0.740
Ensemble Results	80%-20%: **0.770** | 50%-50%: **0.755**		

Table 2: F1-Scores for the positive label (Self_reports) for various pre-processed data-sets. NOTE: Values in parenthesis refer the cased model. Not all models were tested on each of the different data-sets.

Submission \ Metric	F1-Score	Precision	Recall
80-20 Split Ensemble	0.749 (0.839)	0.649 (0.839)	0.883 (0.839)
50-50 Split Ensemble	0.753 (0.843)	0.657 (0.843)	0.881 (0.843)

Table 3: Final ensembling performance on validation data. NOTE: The top value is F1-Score for the positive label used during our testing, while the bottom value in parenthesis is the micro-average F1-Score used by CodaLab.

Classifier \ Metric	F1-Score	Precision	Recall
Baseline	0.90	0.90	0.90
Median	0.84	0.84	0.84
50-50 Split Ensemble	**0.85**	**0.85**	**0.85**
80-20 Split Ensemble	0.84	0.84	0.84

Table 4: Micro-averaged performance on test data.

4 Results

The results of our ensemble predictions for the test data compared with the Baseline and Median results are shown in Task 4. The two submissions have a different F1-scores when experimenting compared to the CodaLab submission because when experimenting, the F1-Score with respect to the **Self_report** label was used while the micro-averaged F1 Score was used in the CodaLab submission, which is akin to accuracy with a multi-class classification.

The 50%-50% split ensemble performed better than the 80%-20% split ensemble with regards to the micro-average F1-Score and vice-versa when considering the F1-Score for the positive label only. We hypothesize this is because the 80%-20% split ensemble was optimized to predict tweets belonging to the positive class. This led to an overall lowering of its accuracy, which is shown in its micro-averaged metrics. However, both the 80%-

20% split and the 50%-50% performed similar to the median score.

5 Conclusion

As demonstrated in our results section, the best performance was achieved through a majority-vote ensemble of all models. Possible opportunities for improvement would be exploring the usage of data augmentation techniques such as back-translation to address the class imbalance present in the training data-set. Moreover, we could explore different pre-processing techniques to remove possible noise from our data such as user mentions, hyperlinks or hashtags.

References

José Cañete, Gabriel Chaperon, Rodrigo Fuentes, Jou-Hui Ho, Hojin Kang, and Jorge Pérez. 2020. Spanish pre-trained bert model and evaluation data. In *PML4DC at ICLR 2020*.

Alexis Conneau, Kartikay Khandelwal, Naman Goyal, Vishrav Chaudhary, Guillaume Wenzek, Francisco Guzmán, Edouard Grave, Myle Ott, Luke Zettlemoyer, and Veselin Stoyanov. 2019. Unsupervised cross-lingual representation learning at scale. *CoRR*, abs/1911.02116.

Jacob Devlin, Ming-Wei Chang, Kenton Lee, and Kristina Toutanova. 2019. BERT: Pre-training of deep bidirectional transformers for language understanding. In *Proceedings of the 2019 Conference of the North American Chapter of the Association for Computational Linguistics: Human Language Technologies, Volume 1 (Long and Short Papers)*, pages 4171–4186, Minneapolis, Minnesota. Association for Computational Linguistics.

Sai Muralidhar Jayanthi and Akshat Gupta. 2021. SJ_AJ@DravidianLangTech-EACL2021: Task-adaptive pre-training of multilingual BERT models for offensive language identification. In *Proceedings of the First Workshop on Speech and Language Technologies for Dravidian Languages*, pages 307–312, Kyiv. Association for Computational Linguistics.

Pratik Joshi, Sebastin Santy, Amar Budhiraja, Kalika Bali, and Monojit Choudhury. 2020. The state and fate of linguistic diversity and inclusion in the NLP world. In *Proceedings of the 58th Annual Meeting of the Association for Computational Linguistics*, pages 6282–6293, Online. Association for Computational Linguistics.

Davy Weissenbacher, Ari Z. Klein, Luis Gascó, Darryl Estrada-Zavala, Martin Krallinger, Yuting Guo, Yao Ge, Abeed Sarker, Ana Lucia Schmidt, Raul Rodriguez-Esteban, Mathias Leddin, Arjun Magge, Juan M. Banda, Vera Davydova, Elena Tutubalina, and Graciela Gonzalez-Hernandez. 2022. Overview of the Seventh Social Media Mining for Health Applications (#SMM4H) shared tasks at COLING 2022. In *Proceedings of the Seventh Social Media Mining for Health (#SMM4H) Workshop and Shared Task*, pages –.

AIR-JPMC@SMM4H'22: BERT + Ensembling = Too Cool: Using Multiple BERT Models Together for Various COVID-19 Tweet Identification Tasks

Leung Wai Liu, Akshat Gupta, Saheed O. Obitayo, Xiaomo Liu, Sameena Shah
J.P. Morgan AI Research, New York, NY, USA
leungwai@wustl.edu, {akshat.x.gupta, saheed.o.obitayo, xiaomo.liu, sameena.shah}@jpmchase.com

Abstract

This paper presents my submission for Tasks 1 and 2 for the Social Media Mining of Health (SMM4H) 2022 Shared Tasks competition. I first describe the background behind each of these tasks, followed by the descriptions of the various subtasks of Tasks 1 and 2, then present the methodology. Through model ensembling, this methodology was able to achieve higher results than the mean and median of the competition for the classification tasks.

1 Introduction

Social media has grown to be an essential communication platform. Massive platforms like Facebook have billions of users (Heath, 2022), enabling it to gauge public sentiment about a plethora of topics, two of them being pharmacovigilance and health mandates. Pharmacovigilance is the assessment of adverse effects from medical drugs (WHO, 2022). In the past, pharmacovigilance information were gathered from marketing clinical trials, or post-marketing reporting by physicians, which pose limitations on information quantity. (O'Connor et al., 2014). Platforms like Twitter are also a key tool to gauge sentiment about public events, such as health mandates during a public health crisis. Understanding public sentiment is pertinent to get everyone on the same page.

The issue with Twitter data, however, is its lack of uniformity, making it hard to automate sentiment classification. This is where Natural Language Processing (NLP) and this competition comes into play. Hosting for the seventh time (Weissenbacher et al., 2022), this Shared Tasks competition, like in previous years, tackles many social media data related tasks, including pharmacovigilance (Magge et al., 2021a). This system builds upon state-of-the-art models to further improve results for classification tasks.

1.1 Task Description

Task 1 is a pharmacoviligiance task consisting of three subtasks, building upon one another:

- **Task 1a - ADE Classification:** Classify whether a tweet is about an Adverse Event of a drug (**ADE**) or not (**noADE**).
- **Task 1b - ADE Span Detection:** Given an **ADE** classified tweet, detect the span that pertains to such ADE.
- **Task 1c - ADE Entity Normalization:** Given an **ADE** span prediction, map that to its respective MedDRA concept ID Label.

Task 2 is a health mandate related task. Given three health mandate labels: **face masks**, **stay at home orders**, and **school closures**, the two subtasks are:

- **Task 2a - Stance Detection:** Classify whether the tweet of a particular label is in **FAVOR**, **AGAINST**, or **NONE** opinion of the mandate.
- **Task 2b - Premise Classification:** Detect whether the tweet has a premise pertaining to the label (labeled **1**), or not (labeled **0**).

Tasks 1a, 2a, and 2b are classification tasks, sharing similar methodology, while Tasks 1b (span detection), and 1c (entity normalization) take a different approach.

2 Datasets

The dataset used for Task 1 is Version 2 of the DeepADEMiner dataset from Magge et al. (2021b), while the dataset used for Task 2 is the stance detection dataset from Davydova and Tutubalina (2022). The training and validation sets have labels for training and validation, respectively, while the testing sets did not. Tables 1 and 2 shows the distribution of tweets for Tasks 1 and 2, respectively.

Proceedings of the 29th International Conference on Computational Linguistic, pages 163–167
October 12–17, 2022.

Dataset	ADE	noADE	Total
Training	1214	15960	17174
Validation	65	844	909
Test	—	—	10969

Table 1: Distribution of Task 1's ADE and noADE classes. Since the test set is unlabeled, the distribution is unknown.

Dataset	AGAINST	FAVOR	NONE	1	0	Total
Training	874	1346	1336	1331	2225	3556
Validation	158	244	198	220	380	600
Test	—	—	—	—	—	2000

Table 2: Distribution of Task 2's stance and premise classes. Since the test set is unlabeled, the distribution is unknown.

Task 1's distribution of labels are unbalanced, skewing towards more **noADE** class tweets. Task 2 has roughly equal distribution across its stance and premise classes.

3 Methodology

3.1 Pre-Processing

No pre-processing was performed besides merging the dataset file of the Tweet IDs and its labels together for training and validation for the classification tasks. For Task 1b, a different transformer model from the HuggingFace library was used (BERTForTokenClassification instead of AutoModelForSequenceClassification), and the label tokens were set up as 0/1/2: 0 being an irrelevant word in the sentence, 1 being the relevant span in the sentence, and 2 being padding for outside the sentence. This approach was adapted from Batcha (2021)'s Kaggle guide.

3.2 Models Used

For all tasks, the BERT language model (Devlin et al., 2019) was used, as it is a state-of-the-art language model that can be fine-tuned for many different tasks. The classification tasks were also trained with the RoBERTa language model (Liu et al., 2019), which improves upon BERT with tweaks in its training method. **BERT$_{\mathbf{BASE}}$-uncased** and **RoBERTa$_{\mathbf{BASE}}$** were used to get a baseline performance metric. After that, the datasets were trained on **BERT$_{\mathbf{LARGE}}$-uncased** and **RoBERTa$_{\mathbf{LARGE}}$**, as the larger models improve on accuracy than the base models.

The base models were trained on the training datasets for 5 iterations and 15 epochs per iteration across all classification tasks. The large models were trained on the training datasets for 3 iterations for Task 1a and 5 iterations for Tasks 2a and 2b, all with 15 epochs per iteration.

The methods to calculate the F1, precision, and recall metrics were different across subtasks as well. Task 1a requires the F1-score, precision, and recall for the **ADE** class only, while Task 2 requires the macro F1 score for the **AGAINST** and **FAVOR** classes only for Task 2a, and for both classes **1** (premise) and **0** (no premise) for Task 2b. In addition, Task 2 uses this formula to calculate the F1-score: $F1_{total} = \frac{1}{n} * \sum_{c \in C} F1_{relc}$, with C being the different labels (**face masks, stay at home orders, school closures**), and $n = 3$ being the size of set C. This means that after generating the predictions of the validation set, the results were split by label to find their individual F1 scores first, then combined together to find the total F1-score. Table 3 shows the baseline results of the F1, precision and recall of the classification tasks.

For Task 1b, only **BERT$_{\mathbf{LARGE}}$-uncased** was used, as specific parsing of its tokens was required during post-processing. The model was trained on the training datasets for 5 iterations and 20 epochs per iteration, keeping the last epoch per iteration. No metrics were measured for Tasks 1b and 1c due to time constraints.

Task \ F1	BERT-base-uncased	RoBERTa-base	BERT-large-uncased	RoBERTa-large
Task 1a	0.587	0.667	0.742*	0.803*
Task 2a	0.629	0.665	0.682	0.746
Task 2b	0.765	0.769	0.651	0.647

Table 3: Performance of BERT and RoBERTa models on validation data for all classification tasks. The F1 scores is the average F1-score among all five models for each category (with the exception of certain Task 1a results marked with a (*), which takes the average of 3 models).

3.3 Span Detection Post-Processing

Although the classification tasks did not require post-processing before model ensembling, it was necessary for the span detection task to perform post-processing, to convert the predicted tokens to a predicted span string with beginning and end indices for where it is in the original sentence. Here is an explanation of how the post-processing code works:

1. Convert the original and predicted token ids

into word tokens.

2. If predicted span list IS empty, then predicted span string = "", begin and end indices = 0.
3. ELSE if the predicted span list is NOT empty:
   (a) Process original tweet, removing all punctuation and capitalization.
   (b) Retrieve the first and last word in predicted span list
   (c) Search for that word (removing ## if it is a partial word) in the full processed tweet using `regex`. This returns the beginning of the span.
   (d) If the beginning of the span is empty, then start index = 0, otherwise retrieve start index.
   (e) If predicted span list has ONLY one token: retrieve end index of that begin span.
   (f) ELSE if predicted token token list has more than one token:
      i. Reduce the processed full tweet to start at the begin span's start index.
      ii. Find first word that matches token (removing ## if it is a partial word) , then retrieve end index of that span.

Once a predicted span string, begin and end indices have been generated, model ensembling occurs.

3.4 Model Ensembling

After the models have been trained, a series of ensembling methods were conducted to improve the accuracy of the predictions. This includes: majority-vote, unweighted average, and weighted average of all 10 large models, and separately by model. The unweighted average was calculated with this equation: $\lfloor(\Sigma x)/n\rceil$, with x being the prediction and n being the number of models used in the ensemble, rounded to the nearest digit. The weighted average was calculated with this equation: $\lfloor(\Sigma xf)/c\rceil$, with f being the F1 score of the particular model, and c being the sum of all the models' F1-score used in the ensemble, rounded to the nearest digit.

Ultimately, a majority ensemble using **RoBERTa$_{\text{LARGE}}$** models only was used for Tasks 1a and 2a, while a weighted average ensemble with **BERT$_{\text{LARGE}}$-uncased** models was used for Task 2b. Table 4 shows the F1, precision, and recall metrics after ensembling for the classification tasks.

Metric / Task	F1-Score	Precision	Recall
Task 1a	0.838	0.803	0.877
Task 2a	0.771	—	—
Task 2b	0.816	—	—

Table 4: Performance metric on model ensembles for classification tasks. NOTE: Tasks 2a and 2b does not have the overall precision and recall scores calculated, as only the overall F1-score is used.

For Task 1b, the span with the median span length out of all 5 models is chosen. This enables the filling an answer for any tweets a particular model was unable to predict. If after ensembling there is no prediction, the entire tweet will be used as the prediction, as it helps with the overlapping performance metric.

The model ensembling method is used because multiple models generating a prediction together is proven to be effective in further increasing accuracy of predictions compared to a singular model (Jayanthi and Gupta, 2021).

3.5 Task 1c Post-Processing

Due to time constraints, entity normalization was not performed for Task 1c. Using the predicted span strings from Task 1b, the NLTK framework was used to remove the stop words from the span. Then, each remaining word in the span is searched in the MedDRA library, and the first match found for the span is submitted as the prediction.

4 Results

Table 5 on the next page shows the metrics on test data for Tasks 1 and 2. The classification tasks performed better than the mean and median scores of the competition across all metrics.

However, Task 1b (span prediction) performed at or above the mean in overlapping metrics while worse in strict metrics. This concurs with the submitted system, as it only takes the start index of the first word and last index of the last word in the predicted token list to build the predicted span, meaning that any gaps in between the predicted tokens are included in the final span. This was done to account for multiple spans in the same tweet in the training data.

For Task 1c (entity normalization), the first-term approach, as expected, performed worse than the mean metrics across the board.

TASK 1						
	Mean F1	**Submitted F1**	**Mean Precision**	**Submitted Precision**	**Mean Recall**	**Submitted Recall**
1a	0.562	**0.693**	0.646	**0.772**	0.497	**0.629**
1b	0.527 **(0.341)**	**0.568** (0.091)	0.539 **(0.344)**	**0.671** (0.114)	**0.517** **(0.339)**	0.492 (0.076)
1c	**0.116** **(0.083)**	0.070 (0.011)	**0.120** **(0.085)**	0.087 (0.014)	**0.112** **(0.082)**	0.058 (0.009)
TASK 2						
	Mean F1-score		**Median F1-score**		**Submitted F1-score**	
2a	0.491		0.550		**0.577**	
2b	0.574		0.647		**0.701**	

Table 5: Performance Metrics on Test Data for Tasks 1 and 2. NOTE: For Tasks 1b and 1c, the values on the left are overlapping metrics while the values in the right in parenthesis are strict metrics. The values in bold are the higher value in the specific category and/or subtask.

5 Conclusion

As shown in the results section, better results were achieved than the baseline through model ensembling. This resulted in our system performing better than the mean and median metrics of the competition for all of the classification tasks and on-par with the overlapping metrics for Task 1b.

Further work can be done to improve this system. For the classification tasks, pre-processing of the dataset, such as removing extraneous punctuation or hashtags, may generate better results. For Task 1b, keeping the best epoch based on F1-score in addition to getting rid of the gaps in the span prediction may help especially in the strict metrics. Lastly, having the opportunity to conduct entity normalization properly may help improve Task 1c metrics as well.

References

Thanish Batcha. 2021. Bert for Token Classification (NER) - Tutorial.

Vera Davydova and Elena Tutubalina. 2022. SMM4H 2022 Task 2: Dataset for stance and premise detection in tweets about health mandates related to COVID-19. In *Proceedings of the Seventh Social Media Mining for Health (#SMM4H) Workshop and Shared Task*, pages –.

Jacob Devlin, Ming-Wei Chang, Kenton Lee, and Kristina Toutanova. 2019. BERT: Pre-training of Deep Bidirectional Transformers for Language Understanding. In *Proceedings of the 2019 Conference of the North American Chapter of the Association for Computational Linguistics: Human Language Technologies, Volume 1 (Long and Short Papers)*, pages 4171–4186, Minneapolis, Minnesota. Association for Computational Linguistics.

Alex Heath. 2022. Facebook lost daily users for the first time ever last quarter.

Sai Muralidhar Jayanthi and Akshat Gupta. 2021. SJ_AJ@DravidianLangTech-EACL2021: Task-Adaptive Pre-Training of Multilingual BERT models for Offensive Language Identification. In *Proceedings of the First Workshop on Speech and Language Technologies for Dravidian Languages*, pages 307–312, Kyiv. Association for Computational Linguistics.

Yinhan Liu, Myle Ott, Naman Goyal, Jingfei Du, Mandar Joshi, Danqi Chen, Omer Levy, Mike Lewis, Luke Zettlemoyer, and Veselin Stoyanov. 2019. RoBERTa: A Robustly Optimized BERT Pretraining Approach.

Arjun Magge, Ari Klein, Antonio Miranda-Escalada, Mohammed Ali Al-Garadi, Ilseyar Alimova, Zulfat Miftahutdinov, Eulalia Farre, Salvador Lima López, Ivan Flores, Karen O'Connor, Davy Weissenbacher, Elena Tutubalina, Abeed Sarker, Juan Banda, Martin Krallinger, and Graciela Gonzalez-Hernandez. 2021a. Overview of the Sixth Social Media Mining for Health Applications (#SMM4H) shared tasks at NAACL 2021. In *Proceedings of the Sixth Social Media Mining for Health (#SMM4H) Workshop and Shared Task*, pages 21–32, Mexico City, Mexico. Association for Computational Linguistics.

Arjun Magge, Elena Tutubalina, Zulfat Miftahutdinov, Ilseyar Alimova, Anne Dirkson, Suzan Verberne, Davy Weissenbacher, and Graciela Gonzalez-Hernandez. 2021b. DeepADEMiner: a deep learning pharmacovigilance pipeline for extraction and normalization of adverse drug event mentions on Twitter. In *Journal of the American Medical Informatics Association*, pages 2184–2192. American Medical Informatics Association.

Karen O'Connor, Pranoti Pimpalkhute, Azadeh Nikfarjam, Rachel Ginn, Karen L Smith, and Graciela Gonzalez. 2014. Pharmacovigilance on Twitter? Mining Tweets for Adverse Drug Reactions. In *Pharmacovigilance on Twitter? Mining Tweets for Ad-*

verse Drug Reactions, pages 924–933, United States. American Medical Informatics Association.

Davy Weissenbacher, Ari Z. Klein, Luis Gascó, Darryl Estrada-Zavala, Martin Krallinger, Yuting Guo, Yao Ge, Abeed Sarker, Ana Lucia Schmidt, Raul Rodriguez-Esteban, Mathias Leddin, Arjun Magge, Juan M. Banda, Vera Davydova, Elena Tutubalina, and Graciela Gonzalez-Hernandez. 2022. Overview of the Seventh Social Media Mining for Health Applications (#SMM4H) shared tasks at COLING 2022. In *Proceedings of the Seventh Social Media Mining for Health (#SMM4H) Workshop and Shared Task*, pages –.

WHO. 2022. What is Pharmacovigilance?

CAISA@SMM4H'22: Robust Cross-Lingual Detection of Disease Mentions on Social Media with Adversarial Methods

Akbar Karimi and **Lucie Flek**
Conversational AI and Social Analytics (CAISA) Lab
Department of Mathematics and Computer Science, University of Marburg
http://caisa-lab.github.io

Abstract

We propose adversarial methods for increasing the robustness of disease mention detection on social media. Our method applies adversarial data augmentation on the input and the embedding spaces to the English BioBERT model. We evaluate our method in the SocialDisNER challenge at SMM4H'22 on an annotated dataset of disease mentions in Spanish tweets. We find that both methods outperform a heuristic vocabulary-based baseline by a large margin. Additionally, utilizing the English BioBERT model shows a strong performance and outperforms the data augmentation methods even when applied to the Spanish dataset, which has a large amount of data, while augmentation methods show a significant advantage in a low-data setting.

1 Motivation

Social media are a rich source of discussion about diseases in various contexts, including newly emerging cases, experienced symptoms, drug effects, or social consequences. Such content can be used to make more informed decisions regarding the societal perception of these diseases, as well as to augment existing medical knowledge.

Recently, pre-trained models such as BERT (Devlin et al., 2019) in the general domain and BioBERT (Lee et al., 2020) in the medical domain have become the go-to solutions for semantic annotation tasks, including previously held similar shared tasks (Klein et al., 2020; Magge et al., 2021). However, most participants have opted for a model trained on the same language (Spanish) as the task language Magge et al. (2021); Fidalgo et al. (2021); Canete et al. (2020); Ruas et al. (2021). Our research questions in this work are: (1) Can we use a domain-specific model pre-trained in one language (English) to produce reasonable performance in another? (2) Can we still further improve the robustness of such pre-trained models with methods effective on small datasets?

To address these questions, we explore the effects of adversarial and augmentation techniques, namely **adversarial training** (Goodfellow et al., 2014) and **AEDA** (Karimi et al., 2021b), on the pre-trained English BioBERT model for detecting disease mentions in Spanish tweets.

2 System Description

Our system is an ensemble of BioBERT, a bag-of-words model, improved BERT model, and adversarial data augmentation techniques:

BioBERT (Lee et al., 2020) is a BERT-based (Devlin et al., 2019) model with the same architecture trained on the general domain (Wikipedia and Book Corpus) and the medical domain (PubMed abstracts and PMC full-text articles).

BoW (Karimi et al., 2021c) or Bag-of-Words model creates a dictionary from the training words and calculates a ratio of their occurrences as a disease word. Based on this ratio (0.7 in our work), words in the test set are labeled as diseases. This is a fast and strong baseline.

Adversarial Training (AT) (Miyato et al., 2016; Karimi et al., 2021a) is a method which creates adversarial examples in the embedding space and uses them in the training process in order to make a neural network model more robust to small perturbations.

PSUM (Karimi et al., 2020) or parallel aggregation is a module that is added on top of the BERT model to improve its performance. It applies an extra BERT layer to the last four layers of the BioBERT model and aggregates the losses.

AEDA (Karimi et al., 2021b) is an adversarial data augmentation technique which inserts punctuation marks into a sentence randomly creating more data for training.

In addition to each individual method above, we also pair BioBERT, adversarial training, and PSUM models with the AEDA data augmentation technique. We acquire our final results by combining

Proceedings of the 29th International Conference on Computational Linguistic, pages 168–170
October 12–17, 2022.

labels produced for each token in the sequence using majority voting.

3 Dataset Statistics

The dataset was provided as part of the shared task of detecting disease mentions (Task 10) at the Social Media Mining for Health 2022 (SMM4H'22) (Gasco et al., 2022). The initial 5K tweets were later expanded by the organizers to make a 90K training set. The dev and test sets contained 2.5K and 23.5K tweets. Covid19 was by far the most discussed disease. The 10 most frequent disease words include *covid* (13478), *covid19* (11661), *cáncer* (7705), *covid-19* (7394), *ansiedad* (4921), *diabetes* (4253), *discapacidad* (3284), *enfermedades* (2567), *enfermedad* (2455), and *depresión* (2229).

4 Results

Table 1 summarizes our results. In order for our models to handle the labels, we convert them from the character span format into the BIO notation (with only two labels of B and O) and back.We report the performance of the models both on the BIO formatted data and the converted span match format, as this conversion process caused performance loss in our predictions due to multiple spacing, extra punctuation marks etc.

As can be seen, English BioBERT achieves the best recall. Our qualitative analysis indicates that there are already many similar words in English and Spanish in the medical entities domain. In addition, the large amount of Spanish data used for fine-tuning injects a sufficient amount of extra Spanish knowledge into the model. BioBERT also outperforms the data augmentation methods since the 90k training set seems to be large enough to neglect the augmentation benefits. However, these models seem to be more promising in low-data regimes as we can see from Table 2. While they improve the models marginally with 1000 training samples, they show significant improvements (up to nearly 10 percent) when we consider 100 tweets for training.

5 Error Analysis

A confusion matrix of the two classes indicates that the number of false positives was almost triple that of false negatives (549 vs. 192). Examining the top misclassified tokens (Table 3), we can see that the top false positives are not disease mentions, but rather prepositions that were part of the (wrongly) annotated disease spans in the training data.

	BIO			**Spans**		
Model	P	R	F1	P	R	F1
BioBERT	96.4	92.0	**94.7**	90.7	80.2	85.1
PSUM	96.8	91.4	94.0	90.8	79.9	85.0
AT	96.8	91.5	94.1	91.2	80.2	**85.4**
AEDA	96.7	91.0	93.8	90.8	79.9	85.0
PSUM+AEDA	97.0	91.2	94.0	91.1	79.9	85.1
AT+AEDA	96.9	91.1	93.9	90.8	79.5	84.8
BoW	-	-	-	76.8	57.3	65.6
Voting	-	-	-	91.4	80.1	**85.4**

Table 1: Performance on the dev set in BIO format and matching spans

	100			**1000**		
Model	P	R	F1	P	R	F1
BioBERT	61.1	73.6	66.8	86.3	89.4	87.8
PSUM	74.9	68.7	71.7	87.4	88.4	87.9
AT	62.9	72.2	67.2	86.8	89.2	**88.0**
AEDA	74.3	77.5	75.8	86.3	89.4	87.8
PSUM+AEDA	76.2	76.3	**76.2**	87.4	88.4	87.9
AT+AEDA	70.8	78.3	74.4	86.8	89.2	**88.0**

Table 2: Performance on the dev set in BIO format with 100 and 1000 training tweets

Word	de	la	#	en	y
Error frequency	41	27	25	17	13

Table 3: Top 5 errors

6 Conclusion

We introduced and successfully evaluated adversarial and augmentation strategies, which proved to be significantly influential in a low-data regime, to be combined with the BioBERT model in order to address the problem of detecting disease mentions in Spanish tweets. We also showed that the English BioBERT can have a strong performance on the same dataset; this behavior can be attributed to the dataset domain (medical entities), where the two languages share many similar words. We believe that this can be applied to other languages as well, as long as there are some similarities between them.

References

José Canete, Gabriel Chaperon, Rodrigo Fuentes, Jou-Hui Ho, Hojin Kang, and Jorge Pérez. 2020. Spanish pre-trained bert model and evaluation data. *Pml4dc at iclr*, 2020:1–10.

Jacob Devlin, Ming-Wei Chang, Kenton Lee, and Kristina Toutanova. 2019. Bert: Pre-training of deep bidirectional transformers for language understanding. In *Proceedings of the 2019 Conference of the North American Chapter of the Association for Computational Linguistics: Human Language Technologies, Volume 1 (Long and Short Papers)*, pages 4171–4186.

David Carreto Fidalgo, Daniel Vila-Suero, Francisco Aranda Montes, and Ignacio Talavera Cepeda. 2021. System description for profner-smmh: Optimized finetuning of a pretrained transformer and word vectors. In *Proceedings of the Sixth Social Media Mining for Health (# SMM4H) Workshop and Shared Task*, pages 69–73.

Luis Gasco, Darryl Estrada-Zavala, Eulàlia Farré-Maduell, Salvador Lima-López, Antonio Miranda-Escalada, and Martin Krallinger. 2022. Overview of the SocialDisNER shared task on detection of diseases mentions from healthcare related and patient generated social media content: methods, evaluation and corpora. In *Proceedings of the Seventh Social Media Mining for Health (#SMM4H) Workshop and Shared Task.*

Ian J Goodfellow, Jonathon Shlens, and Christian Szegedy. 2014. Explaining and harnessing adversarial examples. *arXiv preprint arXiv:1412.6572.*

Akbar Karimi, Leonardo Rossi, and Andrea Prati. 2020. Improving bert performance for aspect-based sentiment analysis. *arXiv preprint arXiv:2010.11731.*

Akbar Karimi, Leonardo Rossi, and Andrea Prati. 2021a. Adversarial training for aspect-based sentiment analysis with bert. In *2020 25th International Conference on Pattern Recognition (ICPR)*, pages 8797–8803. IEEE.

Akbar Karimi, Leonardo Rossi, and Andrea Prati. 2021b. Aeda: An easier data augmentation technique for text classification. *arXiv preprint arXiv:2108.13230.*

Akbar Karimi, Leonardo Rossi, and Andrea Prati. 2021c. Uniparma at semeval-2021 task 5: Toxic spans detection using characterbert and bag-of-words model. *arXiv preprint arXiv:2103.09645.*

Ari Klein, Ilseyar Alimova, Ivan Flores, Arjun Magge, Zulfat Miftahutdinov, Anne-Lyse Minard, Karen O'connor, Abeed Sarker, Elena Tutubalina, Davy Weissenbacher, et al. 2020. Overview of the fifth social media mining for health applications (# smm4h) shared tasks at coling 2020. In *Proceedings of the Fifth Social Media Mining for Health Applications Workshop & Shared Task*, pages 27–36.

Jinhyuk Lee, Wonjin Yoon, Sungdong Kim, Donghyeon Kim, Sunkyu Kim, Chan Ho So, and Jaewoo Kang. 2020. Biobert: a pre-trained biomedical language representation model for biomedical text mining. *Bioinformatics*, 36(4):1234–1240.

Arjun Magge, Ari Klein, Antonio Miranda-Escalada, Mohammed Ali Al-Garadi, Ilseyar Alimova, Zulfat Miftahutdinov, Eulalia Farre, Salvador Lima-López, Ivan Flores, Karen O'Connor, et al. 2021. Overview of the sixth social media mining for health applications (# smm4h) shared tasks at naacl 2021. In *Proceedings of the Sixth Social Media Mining for Health (# SMM4H) Workshop and Shared Task*, pages 21–32.

Takeru Miyato, Andrew M Dai, and Ian Goodfellow. 2016. Adversarial training methods for semi-supervised text classification. *arXiv preprint arXiv:1605.07725.*

Pedro Ruas, Vitor Andrade, and Francisco M Couto. 2021. Lasige-biotm at profner: Bilstm-crf and contextual spanish embeddings for named entity recognition and tweet binary classification. In *Proceedings of the Sixth Social Media Mining for Health (# SMM4H) Workshop and Shared Task*, pages 108–111.

OFU@SMM4H'22: Mining Advent Drug Events Using Pretrained Language Models

Omar Adjali
Université Paris-Saclay, CEA
Laboratoire Analyse
Sémantique Texte et Image,
Gif-sur-Yvette, F-91191 France
omar.adjali@cea.fr

Fréjus A. A. Laleye,
Opscidia, Paris, France
frejus.laleye@opscidia.com

Umang Aggarwal
Eco-Compteur, France
umang.aggarwal@eco-counter.com

Abstract

We describe in this paper our proposed systems for the Social Media Mining for Health 2022 shared task 1. In particular, we participated in the three sub-tasks, tasks that aim at extracting and processing Adverse Drug Events. We investigate different transformer-based pretrained models we fine-tuned on each task and proposed some improvement on the task of entity normalization.

1 Introduction

Social media platforms became a compulsory source of information thanks to its immediacy and accessibility. The recent COVID-19 global crisis has demonstrated the necessity to increase the research effort toward processing health-related information with the aim of infoveillance, pharmacovigilance (Chew and Eysenbach, 2010) and fighting missinformation (Zarocostas, 2020). Indeed, the abundance of data provided by social media poses new challenges which encouraged the Natural Language Processing (NLP) community to design and adapt suitable text processing approaches. The Social Media Mining for Health (SMM4H) shared tasks (Weissenbacher et al., 2022) follow this initiative and propose several NLP tasks for health information monitoring. In particular, SMM4H task 1 proposes to tackle the problem of mining Adverse Drug Events (ADEs) in tweets using the annotated dataset proposed in (Magge et al., 2021). We describe in this paper the systems that solve the three subtasks we participated in: 1) Binary ADEs tweets classification. 2) ADE span detection in tweets and 3) ADE mention normalization.

2 Motivation

Active research effort have been spent on learning language representation models from large general-purpose corpora. In particular, Transformers-based models such as BERT (Devlin et al., 2018), XL-Net (Yang et al., 2019) and RoBERTa (Liu et al., 2019) greatly contributed to boosting the performance of many natural language processing (NLP) applications. Although, such models have strong generalization abilities, they need an additional domain-adaptation training step in order to be employed in supervised settings for downstream tasks. For example, in the biomedical domain, pretrained language models (LMs) such as clinicalBert (Huang et al., 2019), BioBERT (Lee et al., 2020) have been proposed. While these domain-adapted models unarguably outperform general ones, the performance gain may vary depending on a number of factors, including the nature of the downstream task, the proximity of the task domain to the domain-adaptation training corpus and the task dataset size (Gururangan et al., 2020). Furthermore, the characteristics of social media posts may adversely affect the general performance, as they are often in the form of short noisy texts with an informal writing style that involves abbreviations and misspelled words. In this work, we evaluate different pretrained language models on the three aforementioned tasks, investigating how well they transfer to different information extraction tasks in the context of social media and pharmacovigilance.

3 Task1a: Tweet Classification

In Task1a, the objective is to classify tweets as ADE or noADE. The training dataset is imbalanced and contains 1711 tweets reporting ADE and 15674 NoADE examples. Previous work such as (Vydiswaran et al., 2019; Cortes-Tejada et al., 2019), proposed a text processing pipeline and bidirectional LSTM to classify ADE tweets. (Rawal et al., 2019) proposed a lexicon-based approach that identified keywords used as markers of whether a tweet contains an ADE. (Miftahutdinov et al., 2019) relied on a bert architecture for tweet classification, while (Mahata et al., 2019) additionally showed that ULMFiT(Howard and Ruder, 2018) produces descent performance.

Proceedings of the 29th International Conference on Computational Linguistic, pages 171–175
October 12–17, 2022.

Following (Miftahutdinov et al., 2019), we first preprocessed tweets using the tweet-preprocessor package[1] to remove re-tweets, Hashtags, emojis and URLs. We investigate several Transformer-based pretrained language models (Vaswani et al., 2017) which we fine-tuned on ADE tweet classification. For each model, we take the pooled output ([CLS] token representation) followed by a linear layer and a softmax function for classification. For all tasks, we used the following models:

BERT: We used the "base-uncased" BERT pretrained on general-domain text corpora.

BioBERT (Lee et al., 2020): It is pretrained on biomedical domain corpora (PubMed abstracts + PMC full-text articles). We experimented with the BioBERT-Large v1.1 (PubMed 1M), a version based on BERT-large-Cased with a custom 30k vocabulary.

XLNet: With a different architecture, training objectives and without sequence length limit, XLNet performs autoregressive (AR) and permutation-based Language Modelling (PLM) to encode bidirectional contexts. (Wang et al., 2018) showed XLNet performance improvements over BERT on several benchmark datasets. We used the XLNet-Base pretrained model version.

Roberta (Liu et al., 2019): similar to BERT, RoBERTa is trained on larger general-domain corpora without the next sentence prediction objective. It has been shown that it outperforms BERT on several downstream tasks.

3.1 Experiments Setup

For all experiments, our model implementations are based on the huggingface library (Wolf et al., 2019) and pytorch (Paszke et al., 2019). Texts are tokenized with respect to each architecture tokenization scheme and special tokens. We fine-tuned the pretrained models on the down stream tasks for a maximum of 100 epochs with a batch size of 32 for BERT and RoBERTa models, and 64 for XLnet models. Early stopping is applied to stop learning after no improvement on the validation loss for 10 epochs. Model selection is based on the best validation performance on the development set, which represents 20% of the training set. Optimization is done using Adam optimizer with a learning rate of 3e-5 and a weight decay of 0.1. The performance was measured on the F1-score of the ADE classe.

[1]https://pypi.org/project/tweet-preprocessor/

Meth.	F1
BERT	0.692
BioBERT	0.634
XLNet	0.683
Roberta	**0.715**

Table 1: Classification results on the validation set

3.2 Discussion

Table 1 shows the classification results on the validation set. Results show that RoBERTa is the best performing model, while BERT performs slightly better than XLNet. Without surprise, RoBERTa being trained on larger corpora and with longer sequences, it confirms superiority over BERT and the XLNet architecture. Althought, BioBERT is a domain-specific model pretrained on biomedical corpora, it achieved the worse performance. We suggest that general-domain pretrained models have better generalization capabilities on social media posts even when their texts involve biomedical-related terms. Obviously, extensive experiments are necessary to support such an assumption. We conducted additional experiments by oversampling the minority class (ADE) in the training set to overcome the imbalance nature of the dataset, however experiments with BERT showed no improvements.

4 Task1b: ADE span detection in Tweets

Given a tweet, the goal is to detect a span that refers to an expressed ADE. Similar to (Rawal et al., 2019; Mahata et al., 2019; Miftahutdinov et al., 2019), we formulated the span detection task as a sequence labeling problem. We automatically annotated the dataset with labels following the BIO tagging scheme. Hence, the beginning of ADE mention is tagged with B-ADE, the inside tokens of an ADE mention are tagged with I-ADE, and the rest of tokens are tagged with O (outside ADE mention). Our model relies mainly on a sequence encoder (BERT, RoBERTa, BioBERT), followed by a Conditional Random Field (CRF) layer to predict output labels. We additionally implement a BiLSTM + CRF as a strong baseline for sequence labeling. We used the same hyper-parameter settings as for the classification task, and computed the F1 score of the correct detected spans. In this task, we did not experiment with XLNet because of implementation incompatibilities with the CRF layer.

Meth.	F1
BiLSTM + CRF	0.455
BERT + CRF	0.574
RoBERTa + CRF	**0.591**
BioBERT + CRF	0.526

Table 2: Span detection results on the validation set

4.1 Discussion

Table 2 shows the results obtained with different sequence encoders it terms of strict F1 score. The results are in line with the those obtained in task1a. Obviously, Transformers-based models outperformed the BiLSTM+CRF basline. RoBERTa produced the best performance while BioBERT failed to produce satisfying results. We suggest that pretraining on well formatted domain-specific documents poorly transfers on downstream tasks that involve short noisy texts like tweets. In particular, the variability of ADE span forms in tweets makes the detection very challenging.

5 Task1c: ADE mention normalization

In task1c, we map ADE mentions to their standard concept IDs in the Medical Dictionary for Regulatory Activities (MedDRA) vocabulary. Like (Vydiswaran et al., 2019), we used MedDRA dictionary from SIDER, the database of drugs and side effects (Kuhn et al., 2016).

We followed a metric learning approach to address the entity normalization task, where mentions are encoded using a neural network, and entity labels are represented using pretrained continuous vector representations. During training, these representations are optimized with the objective of minimizing the embedding distance between mention representations and their normalized entity representation according to a given metric.

In this work, span mention representations are obtained by encoding their texts using one of the previously used encoders (BERT, BioBERT, XLNet, RoBERTa), while entities are encoded using Word2Vec (Mikolov et al., 2013) representations obtained after training word embeddings on PubMed and PMC texts (Chiu et al., 2016). We therefore average the word embeddings of the words composing entity labels to build vector representations of all entities in the MedDRA dictionary.

During inference, we computed the cosine similarity between the vector representations of each detected span mention and all the entity vector representations in the MedDRA dictionary. Finally, we normalize mentions following two retrieval methods: 1) the nearest neighbor (NN) retrieval method which consists in selecting the closest entity representation in the embedding space, according to the cosine similarity. 2) the Cross-Domain Similarity Local Scaling (CSLS) (Conneau et al., 2017), which is a similarity measure that computes the cosine similarity normalized with the mean cosine similarity of each entity vector representation with its K entity neighbors (see (Conneau et al., 2017) for more details).

5.1 Distant supervision

The dateset being small sized, we offset by following a distant supervision strategy to synthetically augment the training data. We therefore used the the MedDRA dictionary labels as training examples of ADE mentions, where each entity label is normalized with its entity ID resulting in a training set of relatively large size (90k examples).

5.2 Experiments

We run several experiments to evaluate the two retrieval methods and the impact of distant supervision. We trained the encoders used in the previous tasks with the same hyper-parameter settings. Evaluation is done using the strict F1 metric. We assume that all span were correctly detected, as we used the ground truth span in the validation set to compute F1 scores.

5.3 Discussion

Table 4 shows the different configuration results. We first observe that the CSLS metric slightly improves the performance over the NN retrieval except for XLNet models. The CSLS metric achieved significant improvement in word translation retrieval by alleviating the problem of hubness in high-dimensional vector spaces (Conneau et al., 2017). We note a substantial performance gain for all distant supervision settings, with a 4 points performance gain which demonstrates the benefit of data augmentation on such small datasets. Moreover, we note that the performance are close when comparing the different encoders. BERT slightly outperforms RoBERTa, While XLNet has the worse performance on this task.

Task	$Ol-P$	$Ol-R$	$Ol-F1$	$Strict-P$	$Strict-R$	$Strict-F1$
1a	-	-	-	0.235	0.409	0.299
1b	0.178	0.288	0.22	0.097	0.158	0.12
1c	0.094	0.152	0.116	0.067	0.108	0.082

Table 3: Final results on the Test set. (OL:Overlapping, P:precision, R:recall, F1: F1 score).

Meth.	F1
BERT-NN	0.609
BERT-CSLS	0.616
BERT-CSLS+ Distant superv.	**0.656**
XLNet-NN	0.594
XLNet-CSLS	0.592
XLNet-CSLS + Distant superv.	0.638
RoBERTa-NN	0.612
RoBERTa-CSLS	0.614
RoBERTa-CSLS+ Distant superv.	0.651
BioBERT-NN	0.608
BioBERT-CSLS	0.612
BioBERT-CSLS+ Distant superv.	0.626

Table 4: Normalization results on the validation set

6 Conclusion

In this work, we explored different transformer-based pretrained models in a transfer learning setting for several information extraction tasks. Our approaches proposed to solve the three subtasks of SMM4H Shared Task 1, by exploring the best transfer pretrained model. The results we obtained are different from our expectations since domain-specific pretrained models such as BioBERT did not outperform general-domain on domain-specific downstream tasks. We suggested that general-domain pretrained models have better generalization abilities to extract ADE on social media texts. Further experiment and analysis are planned in order to draw more meaningful conclusions. Our results on the test set are showed in table 3. For each task, the run corresponds to the best system we obtained on the validation set. Unfortunately, results on the test are not in line with those obtained on the validation set. Only our entity normalization results reached the mean score of all task1c submissions. As Task 1 requires an end-to-end pipeline, the results on each subtask are affected by the performance of its upstram subtasks. We will further investigate our implementations used for test runs.

References

Cynthia Chew and Gunther Eysenbach. 2010. Pandemics in the age of twitter: content analysis of tweets during the 2009 h1n1 outbreak. *PloS one*, 5(11):e14118.

Billy Chiu, Gamal Crichton, Anna Korhonen, and Sampo Pyysalo. 2016. How to train good word embeddings for biomedical nlp. In *Proceedings of the 15th workshop on biomedical natural language processing*, pages 166–174.

Alexis Conneau, Guillaume Lample, Marc'Aurelio Ranzato, Ludovic Denoyer, and Hervé Jégou. 2017. Word translation without parallel data. *arXiv preprint arXiv:1710.04087.*

Javier Cortes-Tejada, Juan Martinez-Romo, and Lourdes Araujo. 2019. Nlp@ uned at smm4h 2019: neural networks applied to automatic classifications of adverse effects mentions in tweets. In *Proceedings of the fourth social media mining for health applications (# SMM4H) workshop & shared task*, pages 93–95.

Jacob Devlin, Ming-Wei Chang, Kenton Lee, and Kristina Toutanova. 2018. Bert: Pre-training of deep bidirectional transformers for language understanding. *arXiv preprint arXiv:1810.04805.*

Suchin Gururangan, Ana Marasović, Swabha Swayamdipta, Kyle Lo, Iz Beltagy, Doug Downey, and Noah A Smith. 2020. Don't stop pretraining: adapt language models to domains and tasks. *arXiv preprint arXiv:2004.10964.*

Jeremy Howard and Sebastian Ruder. 2018. Universal language model fine-tuning for text classification. *arXiv preprint arXiv:1801.06146.*

Kexin Huang, Jaan Altosaar, and Rajesh Ranganath. 2019. Clinicalbert: Modeling clinical notes and predicting hospital readmission. *arXiv preprint arXiv:1904.05342.*

Michael Kuhn, Ivica Letunic, Lars Juhl Jensen, and Peer Bork. 2016. The sider database of drugs and side effects. *Nucleic acids research*, 44(D1):D1075–D1079.

Jinhyuk Lee, Wonjin Yoon, Sungdong Kim, Donghyeon Kim, Sunkyu Kim, Chan Ho So, and Jaewoo Kang. 2020. Biobert: a pre-trained biomedical language representation model for biomedical text mining. *Bioinformatics*, 36(4):1234–1240.

Yinhan Liu, Myle Ott, Naman Goyal, Jingfei Du, Mandar Joshi, Danqi Chen, Omer Levy, Mike Lewis, Luke Zettlemoyer, and Veselin Stoyanov. 2019. Roberta: A robustly optimized bert pretraining approach. *arXiv preprint arXiv:1907.11692*.

Arjun Magge, Elena Tutubalina, Zulfat Miftahutdinov, Ilseyar Alimova, Anne Dirkson, Suzan Verberne, Davy Weissenbacher, and Graciela Gonzalez-Hernandez. 2021. Deepademiner: a deep learning pharmacovigilance pipeline for extraction and normalization of adverse drug event mentions on twitter. *Journal of the American Medical Informatics Association*, 28(10):2184–2192.

Debanjan Mahata, Sarthak Anand, Haimin Zhang, Simra Shahid, Laiba Mehnaz, Yaman Kumar, and Rajiv Shah. 2019. Midas@ smm4h-2019: identifying adverse drug reactions and personal health experience mentions from twitter. In *Proceedings of the fourth social media mining for health applications (# SMM4H) workshop & shared task*, pages 127–132.

Zulfat Miftahutdinov, Ilseyar Alimova, and Elena Tutubalina. 2019. Kfu nlp team at smm4h 2019 tasks: Want to extract adverse drugs reactions from tweets? bert to the rescue. In *Proceedings of the fourth social media mining for health applications (# SMM4H) workshop & shared task*, pages 52–57.

Tomas Mikolov, Kai Chen, Greg Corrado, and Jeffrey Dean. 2013. Efficient estimation of word representations in vector space. *arXiv preprint arXiv:1301.3781*.

Adam Paszke, Sam Gross, Francisco Massa, Adam Lerer, James Bradbury, Gregory Chanan, Trevor Killeen, Zeming Lin, Natalia Gimelshein, Luca Antiga, Alban Desmaison, Andreas Kopf, Edward Yang, Zachary DeVito, Martin Raison, Alykhan Tejani, Sasank Chilamkurthy, Benoit Steiner, Lu Fang, Junjie Bai, and Soumith Chintala. 2019. Pytorch: An imperative style, high-performance deep learning library. In *Advances in Neural Information Processing Systems 32*, pages 8024–8035. Curran Associates, Inc.

Samarth Rawal, Siddharth Rawal, Saadat Anwar, and Chitta Baral. 2019. Identification of adverse drug reaction mentions in tweets–smm4h shared task 2019. In *Proceedings of the Fourth Social Media Mining for Health Applications (# SMM4H) Workshop & Shared Task*, pages 136–137.

Ashish Vaswani, Noam Shazeer, Niki Parmar, Jakob Uszkoreit, Llion Jones, Aidan N Gomez, Łukasz Kaiser, and Illia Polosukhin. 2017. Attention is all you need. *Advances in neural information processing systems*, 30.

VG Vinod Vydiswaran, Grace Ganzel, Bryan Romas, Deahan Yu, Amy Austin, Neha Bhomia, Socheatha Chan, Stephanie Hall, Van Le, Aaron Miller, et al. 2019. Towards text processing pipelines to identify adverse drug events-related tweets: university of michigan@ smm4h 2019 task 1. In *Proceedings of the fourth social media mining for health applications (# SMM4H) workshop & shared task*, pages 107–109.

Alex Wang, Amanpreet Singh, Julian Michael, Felix Hill, Omer Levy, and Samuel R Bowman. 2018. Glue: A multi-task benchmark and analysis platform for natural language understanding. *arXiv preprint arXiv:1804.07461*.

Davy Weissenbacher, Ari Z. Klein, Luis Gascó, Darryl Estrada-Zavala, Martin Krallinger, Yuting Guo, Yao Ge, Abeed Sarker, Ana Lucia Schmidt, Raul Rodriguez-Esteban, Mathias Leddin, Arjun Magge, Juan M. Banda, Vera Davydova, Elena Tutubalina, and Graciela Gonzalez-Hernandez. 2022. Overview of the seventh social media mining for health applications #smm4h shared tasks at coling 2022. In *In Proceedings of the Seventh Social Media Mining for Health (#SMM4H) Workshop and Shared Task*.

Thomas Wolf, Lysandre Debut, Victor Sanh, Julien Chaumond, Clement Delangue, Anthony Moi, Pierric Cistac, Tim Rault, Rémi Louf, Morgan Funtowicz, et al. 2019. Huggingface's transformers: State-of-the-art natural language processing. *arXiv preprint arXiv:1910.03771*.

Zhilin Yang, Zihang Dai, Yiming Yang, Jaime Carbonell, Russ R Salakhutdinov, and Quoc V Le. 2019. Xlnet: Generalized autoregressive pretraining for language understanding. *Advances in neural information processing systems*, 32.

John Zarocostas. 2020. How to fight an infodemic. *The lancet*, 395(10225):676.

CompLx@SMM4H'22: In-domain pretrained language models for detection of adverse drug reaction mentions in English tweets

Orest Xherija
Independent Researcher
xherija.orest@gmail.com

Hojoon Choi
Nielsen
hojoon.choi@nielsen.com

Abstract

The paper describes the system that team CompLx developed for sub-task 1a of the Social Media Mining for Health 2022 (#SMM4H) Shared Task. We finetune a RoBERTa model, a pretrained, transformer-based language model, on a provided dataset to classify English tweets for mentions of Adverse Drug Reactions (ADRs), i.e. negative side effects related to medication intake. With only a simple finetuning, our approach achieves competitive results, significantly outperforming the average score across submitted systems. We make the model checkpoints[1] and code[2] publicly available. We also create a web application[3] to provide a user-friendly, readily accessible interface for anyone interested in exploring the model's capabilities.

1 Introduction

The Shared Task (Weissenbacher et al., 2022) of the 2022 Social Media Mining for Health Applications (#SMM4H) workshop proposed ten sub-tasks in the domain of social media mining for health monitoring and surveillance. From the perspective of Natural Language Processing (NLP), these tasks present a considerable challenge since the nature of social media posts requires dealing with both a significant level of language variation (informal and colloquial expressions, ambiguity, multilingual posts) and data sparsity, as well as a widespread presence of noise such as misspellings of clinical concepts and syntactic errors.

In the 2022 instantiation of the #SMM4H Shared Task, our team participated in: (i) sub-task 1a, the classification of English tweets containing mentions of Adverse Drug Reactions (ADRs) (Magge et al., 2021), (ii) sub-task 3, the classification of English tweets (3a) and WebMD reviews (3b) containing mentions of changes in medication treatments, and (iii) sub-task 8, the classification of English tweets self-reporting chronic stress. In this paper we primarily describe our approach for task 1a, as that constituted the major focus of our efforts.

To address these challenges, we finetune a variant of a RoBERTa (Liu et al., 2019) model, a transformer-based (Vaswani et al., 2017) language model pretrained on approximately 128 million tweets (Loureiro et al., 2022) on each sub-task's provided dataset. Without any domain adaptation efforts (apart from standard finetuning on the downstream task) or hyperparameter optimizations, the model outperforms the average of all submissions for sub-task 1a by a 9% absolute difference in F1-score.

In the following sections, we introduce the sub-tasks' datasets, describe the model architecture and training setup, report our results, and conclude with a discussion of related research and potential avenues for future work.

2 Datasets

In Section [1] we provided a brief summary of each sub-task in which we participated. For each of them, participants were given access to a labeled training and validation set, as well as an unlabeled evaluation set that was used to determined the final performance of the submitted systems. Table [1] summarizes the number of samples per dataset per task. Additionally, Table [2] provides representative samples from sub-task 1a. As can be noted upon quick inspection, merely depending

	1a	3a	3b	8
training	17174	5898	10378	2936
validation	909	1572	1297	420
evaluation	10969	15360	13132	839

Table 1: Number of samples per split per task.

[1]https://huggingface.co/orestxherija/roberta-base-adr-smm4h2022

[2]https://github.com/orestxherija/CompLx-SMM4H2022

[3]https://huggingface.co/spaces/orestxherija/adr-mention-classifier

Proceedings of the 29th International Conference on Computational Linguistic, pages 176–181
October 12–17, 2022.

Sentence	Label
vyvanse make me so hyper and creative and i think of so many tweets	ADR
feed an ocd vyvanse and cover him in crayons	No ADR
trazodone has screwed up my sleep schedule. its helping tho.	ADR

Table 2: Selection of samples from training set of sub-task 1a.

on medication-related keywords for label assignment is going to be problematic: both the first and the second example contain the medication term "vyvanse" but they have been assigned different labels, "ADR" and "No ADR" respectively. This motivates the use of a modeling approach that leverages the overall semantic content of the sentence, rather than keyword matching with individual constituents.

3 Modeling Approach

3.1 Model Architecture

The establishment of language modeling as the pretraining step in the transfer learning pipeline revolutionized modern NLP with models such as ULMFiT (Howard and Ruder, 2018), ELMo (Peters et al., 2018) and, most notably, transformer-based language models such as GPT (Radford et al., 2018) and BERT (Devlin et al., 2019). In recent years, there have been intensive efforts in the research community to produce ever-larger transformer-based pretrained language models that are trained using a variety of datasets, transformer-model architectures, training objectives and optimization techniques. This should come as no surprise, since such language models have dominated virtually all NLP leaderboards, most notably GLUE (Wang et al., 2018) and SuperGLUE (Wang et al., 2019).

Considering this overwhelming success, we opt for a RoBERTa (Liu et al., 2019) model[4] that has been trained on approximately 128 million tweets (Loureiro et al., 2022). Our exact modeling approach is depicted in Figure [1]. We opt for a model that has been trained on an in-domain corpus, namely tweets, as transfer learning has been shown to yield improved results when there is in-domain pretraining (Gururangan et al., 2020). We do not use any text normalization steps.

[4]https://huggingface.co/cardiffnlp/twitter-roberta-base-mar2022

3.2 Training Regime

We train the model to minimize the negative log-likelihood loss using back-propagation with stochastic gradient descent and a mini-batch size of 16. To monitor model performance, we use the train/validation split provided by the organizers. For optimization, we use the AdamW optimizer (Loshchilov and Hutter, 2019) with gradient clipping (Pascanu et al., 2013) and a linear scheduler with no warm-up. We use FP-16 mixed precision (Micikevicius et al., 2018) training (and inference) in order to afford a larger batch size and increased training speed. To optimize GPU use by minimizing the amount of memory allocated for padding tokens, we use dynamic padding and length-based batching in the sense of (Skinner, 2018). Finally, we employ label smoothing (Szegedy et al., 2016) with a smoothing factor of 0.1.

3.3 Hyperparameters

As mentioned in Section [1], we do not experiment with hyperparameter tuning but rather keep the default parameters of the `Trainer` API in the Hugging Face `transformers` library. More specifically, we use $\beta_1 = 0.9, \beta_2 = 0.999$ and $\epsilon = 10^{-8}$ for the AdamW optimizer parameter values and a learning rate of 0.00005. We train the models for a maximum of 25 epochs with an early stopping patience level set to 0.001 for 3 epochs. Finally, we set a maximum sequence length of 128 since input sentences are generally short and we would like to avoid consuming GPU memory for padding tokens.

4 Experiments and Results

In this section, we give a brief description of the system we used to conduct our experiments, share our results and provide a brief discussion.

4.1 Setup

The model was developed using the PyTorch (Paszke et al., 2019) implementation of the Hugging Face `transformers` (Wolf et al., 2020) library. The experiments were executed on a

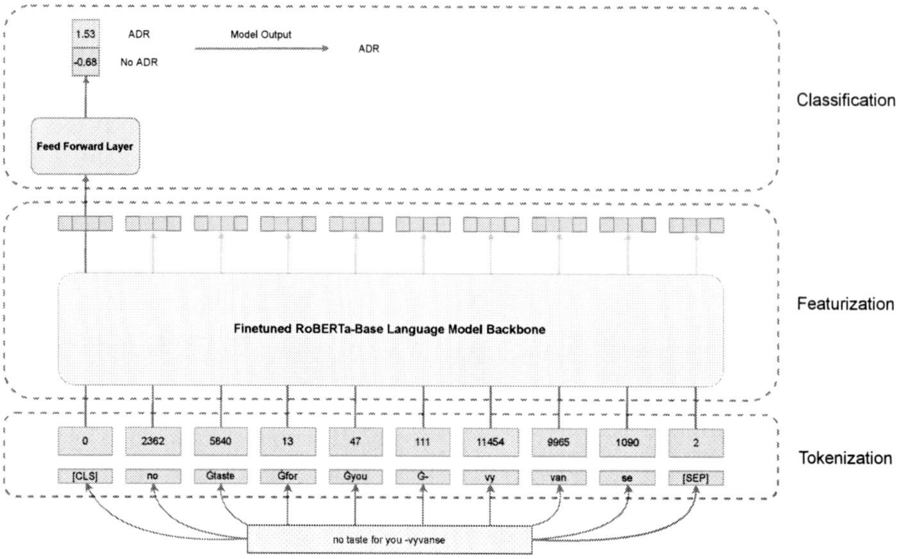

Figure 1: Illustration of modeling approach (inference step): the string input is tokenized and the tokens are passed through the language model backbone so as to obtain contextualized (vector) representations of the tokens. The vector associated with the [CLS] token is passed through a feed-forward layer and the logit outputs are used to decide the sample's class label, "ADR" or "No ADR".

machine with an Intel Core i9-9820X CPU @ 3.30GHz and a NVIDIA GeForce RTX 2080 Ti GPU with 11GB of memory.

4.2 Results

Table [3] summarizes the performance of our approach in the validation set for each sub-task. Note that in this set of experiments, the validation set was used both during training (e.g. for early stopping or selection of batch size) as well as for the reporting of the systems' performance. Table [4] summarizes the performance of our approach in the evaluation set for each sub-task. The organizers chose to disclose to each team only their respective score along with the average score of all submitted systems. Our system performed considerably better than the average in sub-task 1a and surpassed the existing state-of-the-art F1-score of 0.63 reported in (Magge et al., 2021). Performance was considerably poorer for sub-tasks 3 and 8. As mentioned in Section [1], our main efforts were dedicated to sub-task 1a and the system developed did not transfer well to the remaining sub-tasks.

	P	**R**	**F1**
1a	0.769	0.769	0.769
3a	0.030	0.312	0.055
3b	0.571	0.995	0.725
8	0.372	1.0	0.543

Table 3: Validation set results for sub-tasks 1a, 3 and 8.

	P	**R**	**F1**
Subtask-1a	**0.737**	**0.585**	**0.652**
	(0.646)	(0.497)	(0.562)
Subtask-3a	0.034	0.341	0.061
	(0.535)	(0.458)	(0.456)
Subtask-3b	0.567	**1.0**	0.723
	(0.778)	(0.888)	(0.818)
Subtask-8	0.372	**1.0**	0.542
	(0.720)	(0.760)	(0.750)

Table 4: Results on the evaluation set for sub-tasks 1a, 3 and 8. Average score of all participating systems in parentheses. Metric is F1-score for class 1.

5 Conclusion and Future Directions

We demonstrated that a RoBERTa model (Liu et al., 2019) pretrained on approximately 128 million tweets performs very competitively when finetuned on English tweet classification for ADRs. Using only a standard finetuning approach, our model obtained competitive results, outperforming the average of all submissions for sub-task 1 by a 9% absolute difference in F1-score. This constitutes yet another testament of the fact that large pre-trained language models have rightfully become the default approach in virtually all NLP tasks.

With respect to potential future work, there is a large collection of available options. Text classification and, more generally, binary classification is one of the oldest and most widely researched topics in NLP. Most approaches aiming to improve performance of classification models can be broadly categorized into three groups, depending on the segment of the machine learning workflow that they are targeting. Data augmentation methods typically target the initial part of the workflow, the data, aiming to increase the quantity, quality and diversity of the training dataset to ensure that model performance is robust to small syntactic or semantic perturbations in the inputs. Transformations acting directly on strings, such random token insertions or deletions, synonym/antonym replacements and related techniques (Wei and Zou, 2019; Karimi et al., 2021, inter alia) have shown significant performance improvements, especially in low-resource scenarios much like the one in this shared task.

A second approach, evidently a natural extension of the previous technique, would be to target vector encodings of the tokens and/or documents that are produced by the various layers of the neural networks. We can distinguish two different approaches here: (i) improve the language model backbone during the pretraining phase, or (ii) improve the weights of the language model backbone during finetuning. The research community has devoted intensive efforts in the former approach, as can be observed by the ever-increasing list of transformer-based pretrained language models (Devlin et al., 2019; Joshi et al., 2020; Kitaev et al., 2020; Raffel et al., 2020; Brown et al., 2020, inter multi alia) released. Model size, in terms of total number of trainable parameters, has been consistently shown to correlate strongly with downstream performance, so opting for a larger pretrained model would be a reasonable first steps towards more transferable vector representations (and hence improved performance) in the downstream task. The latter approach would include domain adaptation techniques, such as continued self-supervised pretraining followed by supervised finetuning, which has been shown (Gururangan et al., 2020) to consistently lead to superior results relative to direct finetuning.

Finally, one could aim to improve performance by modifying aspects of the objective function. (Hui and Belkin, 2021), in an extensive series of experiments, show that the established practice of using a cross-entropy loss for classification is not well-founded and show through a variety of diverse experiments that a square loss can, in many cases, significantly improve performance.

References

Tom Brown, Benjamin Mann, Nick Ryder, Melanie Subbiah, Jared D Kaplan, Prafulla Dhariwal, Arvind Neelakantan, Pranav Shyam, Girish Sastry, Amanda Askell, Sandhini Agarwal, Ariel Herbert-Voss, Gretchen Krueger, Tom Henighan, Rewon Child, Aditya Ramesh, Daniel Ziegler, Jeffrey Wu, Clemens Winter, Chris Hesse, Mark Chen, Eric Sigler, Mateusz Litwin, Scott Gray, Benjamin Chess, Jack Clark, Christopher Berner, Sam McCandlish, Alec Radford, Ilya Sutskever, and Dario Amodei. 2020. Language Models are Few-Shot Learners. In *Advances in Neural Information Processing Systems*, volume 33, pages 1877–1901. Curran Associates, Inc.

Jacob Devlin, Ming-Wei Chang, Kenton Lee, and Kristina Toutanova. 2019. BERT: Pre-training of Deep Bidirectional Transformers for Language Understanding. In *Proceedings of the 2019 Conference of the North American Chapter of the Association for Computational Linguistics: Human Language Technologies, Volume 1 (Long and Short Papers)*, pages 4171–4186, Minneapolis, Minnesota. Association for Computational Linguistics.

Suchin Gururangan, Ana Marasović, Swabha Swayamdipta, Kyle Lo, Iz Beltagy, Doug Downey, and Noah A. Smith. 2020. Don't Stop Pretraining: Adapt Language Models to Domains and Tasks. In *Proceedings of the 58th Annual Meeting of the Association for Computational Linguistics*, pages 8342–8360, Online. Association for Computational Linguistics.

Jeremy Howard and Sebastian Ruder. 2018. Universal Language Model Fine-tuning for Text Classification. In *Proceedings of the 56th Annual Meeting of the Association for Computational Linguistics (Volume 1: Long Papers)*, pages 328–339, Melbourne, Australia. Association for Computational Linguistics.

Like Hui and Mikhail Belkin. 2021. Evaluation of neural architectures trained with square loss vs cross-entropy in classification. In *International Conference on Learning Representations*.

Mandar Joshi, Danqi Chen, Yinhan Liu, Daniel S. Weld, Luke Zettlemoyer, and Omer Levy. 2020. SpanBERT: Improving Pre-training by Representing and Predicting Spans. *Transactions of the Association for Computational Linguistics*, 8:64–77.

Akbar Karimi, Leonardo Rossi, and Andrea Prati. 2021. AEDA: An Easier Data Augmentation Technique for

Text Classification. In *Findings of the Association for Computational Linguistics: EMNLP 2021*, pages 2748–2754, Punta Cana, Dominican Republic. Association for Computational Linguistics.

Nikita Kitaev, Lukasz Kaiser, and Anselm Levskaya. 2020. Reformer: The Efficient Transformer. In *International Conference on Learning Representations*.

Yinhan Liu, Myle Ott, Naman Goyal, Jingfei Du, Mandar Joshi, Danqi Chen, Omer Levy, Mike Lewis, Luke Zettlemoyer, and Veselin Stoyanov. 2019. RoBERTa: A Robustly Optimized BERT Pretraining Approach. *CoRR*, abs/1907.11692.

Ilya Loshchilov and Frank Hutter. 2019. Decoupled Weight Decay Regularization. In *International Conference on Learning Representations*.

Daniel Loureiro, Francesco Barbieri, Leonardo Neves, Luis Espinosa Anke, and Jose Camacho-collados. 2022. TimeLMs: Diachronic Language Models from Twitter. In *Proceedings of the 60th Annual Meeting of the Association for Computational Linguistics: System Demonstrations*, pages 251–260, Dublin, Ireland. Association for Computational Linguistics.

Arjun Magge, Elena Tutubalina, Zulfat Miftahutdinov, Ilseyar Alimova, Anne Dirkson, Suzan Verberne, Davy Weissenbacher, and Graciela Gonzalez-Hernandez. 2021. DeepADEMiner: a deep learning pharmacovigilance pipeline for extraction and normalization of adverse drug event mentions on twitter. *Journal of the American Medical Informatics Association*, 28(10):2184–2192.

Paulius Micikevicius, Sharan Narang, Jonah Alben, Gregory Diamos, Erich Elsen, David Garcia, Boris Ginsburg, Michael Houston, Oleksii Kuchaiev, Ganesh Venkatesh, and Hao Wu. 2018. Mixed Precision Training. In *International Conference on Learning Representations*.

Razvan Pascanu, Tomas Mikolov, and Yoshua Bengio. 2013. On the Difficulty of Training Recurrent Neural Networks. In *Proceedings of the 30th International Conference on Machine Learning*, volume 28 of *Proceedings of Machine Learning Research*, pages 1310–1318, Atlanta, GA, USA. PMLR.

Adam Paszke, Sam Gross, Francisco Massa, Adam Lerer, James Bradbury, Gregory Chanan, Trevor Killeen, Zeming Lin, Natalia Gimelshein, Luca Antiga, Alban Desmaison, Andreas Kopf, Edward Yang, Zachary DeVito, Martin Raison, Alykhan Tejani, Sasank Chilamkurthy, Benoit Steiner, Lu Fang, Junjie Bai, and Soumith Chintala. 2019. PyTorch: An Imperative Style, High-Performance Deep Learning Library. In *Advances in Neural Information Processing Systems*, volume 32. Curran Associates, Inc.

Matthew E. Peters, Mark Neumann, Mohit Iyyer, Matt Gardner, Christopher Clark, Kenton Lee, and Luke Zettlemoyer. 2018. Deep Contextualized Word Representations. In *Proceedings of the 2018 Conference of the North American Chapter of the Association for Computational Linguistics: Human Language Technologies, Volume 1 (Long Papers)*, pages 2227–2237, New Orleans, Louisiana. Association for Computational Linguistics.

Alec Radford, Karthik Narasimha, Tim Salimans, and Ilya Sutskever. 2018. Improving Language Understanding by Generative Pre-Training. Blog post, OpenAI.

Colin Raffel, Noam Shazeer, Adam Roberts, Katherine Lee, Sharan Narang, Michael Matena, Yanqi Zhou, Wei Li, and Peter J. Liu. 2020. Exploring the Limits of Transfer Learning with a Unified Text-to-Text Transformer. *Journal of Machine Learning Research*, 21(140):1–67.

Michael Skinner. 2018. Product Categorization with LSTMs and Balanced Pooling Views. In *SIGIR 2018 Workshop on eCommerce (ECOM 18)*, SIGIR '18, Ann Arbor, MI, USA. ACM.

Christian Szegedy, Vincent Vanhoucke, Sergey Ioffe, Jon Shlens, and Zbigniew Wojna. 2016. Rethinking the Inception Architecture for Computer Vision. In *2016 IEEE Conference on Computer Vision and Pattern Recognition (CVPR)*, pages 2818–2826.

Ashish Vaswani, Noam Shazeer, Niki Parmar, Jakob Uszkoreit, Llion Jones, Aidan N Gomez, Ł ukasz Kaiser, and Illia Polosukhin. 2017. Attention is All you Need. In *Advances in Neural Information Processing Systems*, volume 30. Curran Associates, Inc.

Alex Wang, Yada Pruksachatkun, Nikita Nangia, Amanpreet Singh, Julian Michael, Felix Hill, Omer Levy, and Samuel Bowman. 2019. SuperGLUE: A Stickier Benchmark for General-Purpose Language Understanding Systems. In *Advances in Neural Information Processing Systems*, volume 32. Curran Associates, Inc.

Alex Wang, Amanpreet Singh, Julian Michael, Felix Hill, Omer Levy, and Samuel Bowman. 2018. GLUE: A Multi-Task Benchmark and Analysis Platform for Natural Language Understanding. In *Proceedings of the 2018 EMNLP Workshop BlackboxNLP: Analyzing and Interpreting Neural Networks for NLP*, pages 353–355, Brussels, Belgium. Association for Computational Linguistics.

Jason Wei and Kai Zou. 2019. EDA: Easy Data Augmentation Techniques for Boosting Performance on Text Classification Tasks. In *Proceedings of the 2019 Conference on Empirical Methods in Natural Language Processing and the 9th International Joint Conference on Natural Language Processing (EMNLP-IJCNLP)*, pages 6382–6388, Hong Kong, China. Association for Computational Linguistics.

Davy Weissenbacher, Ari Z. Klein, Luis Gascó, Darryl Estrada-Zavala, Martin Krallinger, Yuting Guo, Yao Ge, Abeed Sarker, Ana Lucia Schmidt, Raul Rodriguez-Esteban, Mathias Leddin, Arjun Magge,

Juan M. Banda, Vera Davydova, Elena Tutubalina, and Graciela Gonzalez-Hernandez. 2022. Overview of the Seventh Social Media Mining for Health Applications (#SMM4H) Shared Tasks at COLING 2022. In *Proceedings of the Seventh Social Media Mining for Health (#SMM4H) Workshop and Shared Task*, Gyeongju, Republic of Korea. Association for Computational Linguistics.

Thomas Wolf, Lysandre Debut, Victor Sanh, Julien Chaumond, Clement Delangue, Anthony Moi, Pierric Cistac, Tim Rault, Rémi Louf, Morgan Funtowicz, Joe Davison, Sam Shleifer, Patrick von Platen, Clara Ma, Yacine Jernite, Julien Plu, Canwen Xu, Teven Le Scao, Sylvain Gugger, Mariama Drame, Quentin Lhoest, and Alexander M. Rush. 2020. Transformers: State-of-the-Art Natural Language Processing. In *Proceedings of the 2020 Conference on Empirical Methods in Natural Language Processing: System Demonstrations*, pages 38–45, Online. Association for Computational Linguistics.

The SocialDisNER shared task on detection of disease mentions in health-relevant content from social media: methods, evaluation, guidelines and corpora

Luis Gascó†, Darryl Estrada-Zavala, Eulàlia Farré-Maduell, Salvador Lima-López, Antonio Miranda-Escalada, Martin Krallinger
Barcelona Supercomputing Center (BSC), Barcelona, Spain
†Corresponding author: `luis.gasco@bsc.es`

Abstract

There is a pressing need to exploit health-related content from social media, a global source of data where key health information is posted directly by citizens, patients and other healthcare stakeholders. Use cases of disease-related social media mining include disease outbreak/surveillance, mental health and pharmacovigilance. Current efforts address the exploitation of social media beyond English. The SocialDisNER task, organized as part of the SMM4H 2022 initiative (Weissenbacher et al., 2022), has applied the LINKAGE methodology to select and annotate a Gold Standard corpus of 9,500 tweets in Spanish enriched with disease mentions generated by patients and medical professionals. As a complementary resource for teams participating in the SocialDisNER track, we have also created a large-scale corpus of 85,000 tweets, where in addition to disease mentions, other medical entities of relevance (e.g., medications, symptoms and procedures, among others) have been automatically labelled. Using these large-scale datasets, co-mention networks or knowledge graphs were released for each entity pair type. Out of the 47 teams registered for the task, 17 teams uploaded a total of 32 runs. The top-performing team achieved a very competitive 0.891 f-score, with a system trained following a continue pre-training strategy. We anticipate that the corpus and systems resulting from the SocialDisNER track might further foster health-related text mining of social media content in Spanish and inspire disease detection strategies in other languages. Corpus: `https://doi.org/10.5281/zenodo.6359365`

1 Introduction

With more than 4.2 billion users worldwide, social media have become the most widely used digital platform for interacting with peers as well as accessing information relevant to specific groups (Kemp, 2021). Specifically Twitter, an online micro-blogging social network (OSN), has been widely used to extract information about people: from opinions and effects of environmental pollution (Gasco et al., 2019; Otero et al., 2021) to biomedical aspects such as adverse drug reactions (OConnor et al., 2014; MacKinlay et al., 2017), public health (Collier et al., 2008), and the psychological effects of a pandemic on the population (Aiello et al., 2021).

Most analysis and information extraction studies are carried out in English, the main language used by the platform's users. However, other major languages such as Spanish have generated a large amount of potentially usable data for analysis (Alshaabi et al., 2021), which increases the impact of NLP systems able to extract information in this language.

To date, the resources for working with Twitter data in Spanish have focused on corpora to extract information related to emotions (Plaza-del Arco et al., 2020; Martinez-Camara et al., 2015) and professions (Miranda-Escalada et al., 2021). Recently, language models trained with Spanish tweets have been developed to improve the performance of NLP tasks applied to data obtained from this OSN (Huertas-Tato et al., 2022). However, there is a clear lack of corpora to train systems capable of extracting biomedical information from Spanish content that could be used for real-time mining of health information posted by patients, which is of growing interest for different scenarios such as screening for rare diseases (Miller et al., 2021).

One of the main entities to effectively recognise health-related content is diseases. Shared-task such as DisTEMIST (Miranda-Escalada et al., 2022; Nentidis et al., 2022), which focused on the detection of disease mentions in clinical cases, fostered the creation of many tools for this purpose. These systems focus on extracting information from technical and scientific texts, but their performance is limited to the more lay language used in social networks. In the SocialDisNER shared task, we have applied the knowledge and experience from clinical cases acquired in DisTEMIST to detect disease mentions in Spanish tweets written by patients and

Proceedings of the 29th International Conference on Computational Linguistic, pages 182–189
October 12–17, 2022.

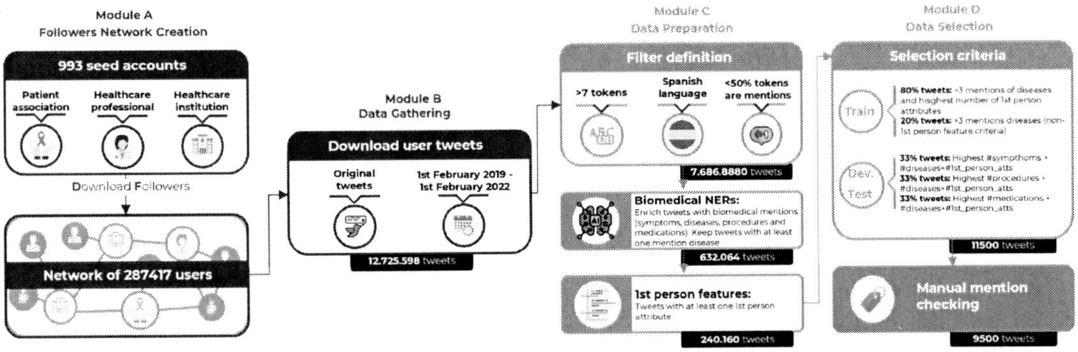

Figure 1: LINKAGE (reLevant socIal NetworK dAta GathEring) methodology for twitter data selection.

medical professionals.

2 Task Description

Shared Task Goal. SocialDisNER focuses on the recognition of disease mentions in Twitter posts. Tweets originate from the following: patients' accounts, with firsthand health reports; friends, support network and relatives, who share the difficulties faced by patients; and medical professionals, who disseminate reliable information about diseases. Tweets in this task include information on rheumatic diseases such as lupus erythematosus, highly prevalent diseases such as cancer, diabetes, obesity and mental disorders, fibromyalgia and autism spectrum conditions.

Shared Task Setting. The track comprised a single challenge in which participants were required to train NLP systems capable of detecting disease mentions automatically. We conducted the task using CodaLab in a three-stage scenario. In the *practice phase*, the training and evaluation set were released so that participants could build their systems and evaluate their models on the validation set. For the *evaluation phase*, the test and background set were released without annotations for the participant teams to compute and submit their predictions. Teams were only evaluated on the test set, using the background data to avoid possible manual corrections and to evaluate the scalability of the systems. In this phase, the participants were only allowed to upload 2 runs. In order to further develop disease mention recognition systems in tweets, the competition has been kept open in the *post-evaluation phase*, so that researchers continue to measure and compare the performance of their systems.

Evaluation metrics. Teams were evaluated and ranked using micro-average F1 score. In addition to this metric, micro-averaged precision and recall were computed. The evaluation script is available on Github[1]. We also compared the systems versus a baseline model following a lexical search approach (TeMU-BSC, 2022).

3 Corpus and resources

3.1 SocialDisNER Gold Standard

Gold Standard selection methodology. SocialDisNER has compiled a set of 9,500 tweets written in Spanish containing patient and family members experiences about diseases and relevant content written by clinical professionals. The LINKAGE methodology was created to obtain a larger number of tweets posted by patients themselves and relevant professionals. This methodology has been designed to avoid noisy health twitter content and biases associated with keyword-based selection (Kowald and Lex, 2018). The procedure, shown in the Figure 1, consists of 4 stages:

1. **Followers Network Creation**: First, a user network with a common interest in the health field is downloaded. For SocialDisNER, a total of 993 seed Twitter accounts of patient associations, health professionals and healthcare institutions were selected and manually curated by task organizers. We obtained the followers from this pool of accounts, resulting in a community of 287,417 users.

2. **Data Gathering**: Second, the content published by this community of users was downloaded using the Twitter API. In the task, only the original tweets written by users were

[1]https://github.com/TeMU-BSC/socialdisner_evaluation_script

downloaded, ignoring retweets and replies to other tweets. Tweets posted between February 2019 and February 2022 were downloaded to prevent having data only about the COVID-19 pandemic, as we have tweets prior to the beginning of it. After this downloading process, more than 12.7 million tweets were obtained.

3. **Data Preparation**: Third, rules were applied to filter out irrelevant content as follows: a) tweets written in a language other than Spanish according the Twitter API; b) tweets with less than 7 tokens, as they might not contain substantial information about diseases; and c) documents in which more than 50% of tokens were mentions. A data enrichment process was applied to the resulting set to produce the final candidates. On the one hand, biomedical mentions such as symptoms, procedures, diseases and drugs were extracted using NER systems developed in previous works (Gonzalez-Agirre et al., 2019). Only tweets with at least one disease mention were selected. First-person characteristics, such as the presence of first-person pronouns, were also calculated, considering only tweets that at least had a first-person attribute. After these constraints, a set of more than 240k tweets were obtained.

4. **Data Selection**: Finally, criteria were determined for selecting candidate tweets to be annotated by experts. Selection strategies were applied so that there was content with several mentions of diseases written in the first person and also had mentions of other biomedical diseases such as symptoms, procedures and medications. The selection criteria for the development and test sets were the same. At the end of this phase, 11,500 tweets were selected for annotation.

Gold Standard statistics. Tweets were annotated by a medical expert during 3 months using an adaptation of the DisteMIST annotation guidelines, whose agreement was tested among medical and linguistic experts with a IAA score of 0.823. A total of 9,500 tweets were finally selected for the task. This set of tweets was divided into a training, a development and a test set. Table 1 shows the distribution and statistics of the corpus,

	Corpus name	Documents	Annotations	Tokens
Gold Stand.	**Training**	5,000	15,173	211,555
	Development	2,500	4,252	84,478
	Test	2,000	3,859	70,244
	Total	9,500	23,284	366,277
Silver Stand.	**Socialdisner-Diseases**	85,077	116,260	3,236,411
	Socialdisner-Symptoms	12,624	12,896	521,503
	Socialdisner-Procedures	11,464	10,080	467,059
	Socialdisner-Pharma	1,759	1,029	68,269
	Socialdisner-Morphology_neoplasms	8,518	8,943	332,539
	Socialdisner-Professions	15,831	18,590	660,071
	Socialdisnerv-Person	41,033	58,007	1,689,479
	Socialdisner-Species	12,118	14,014	486,249

Table 1: SocialDisNER corpora summary.

3.2 SocialDisNER Large Scale corpus

A set of 85,000 tweets was selected to generate large-scale corpora of Spanish tweets with several biomedical mentions. These datasets were published as Silver Standard and were generated using NER systems trained on data previously published by our team. Since those NERs were not trained with tweets, programmatic cleanups were performed by eliminating mentions containing URLs and more than one twitter mention. Additionally, we conducted a manual review of the most recurrent mentions to eliminate false positives. The statistics for each large-scale corpus are shown in Table 1. Due to the selection process of the SocialDisNER data, which focused on diseases, there is a higher presence of content with this entity.

3.3 SocialDisNER co-mention networks

Inspired by (Hope et al., 2020), several co-occurrence matrices of mentions present in the large-scale corpus have been made available to the community. On the one hand, we have published a disease co-occurrence matrix, which could be used to analyze the comorbidity of diseases among users. Co-mention matrices between diseases and other entities have also been published, providing interesting associations between diseases and symptoms, professions and medicines, among others. To the best of our knowledge, this is the first graph of social network biomedical mentions in Spanish.

3.4 SocialDisNER guidelines

We have also published the SocialDisNER guidelines. This document shows the annotation criteria used by medical experts when creating the corpus and ensure its quality and replicability. The guidelines were created by adapting the DisTEMIST annotation rules to the special features of social me-

Team	Country	A/I	Tool	P	R	F1	Ref
CASIA	China	A	-	**0,906**	<u>0,876</u>	**0,891**	(Fu et al., 2022)
READ-BioMed	Australia	A	(READ-BioMed, 2022)	0,868	0,875	<u>0,871</u>	(Yepes and Verspoor, 2022)
Clac	Canada	A	-	0,851	**0,888**	0,869	(Verma et al., 2022)
PLN CMM	Chile	A	(PLN-CMM, 2022)	<u>0,882</u>	0,843	0,862	(Rojas et al., 2022)
NLP-CIC-WFU	México	A	(Tamayo, 2022)	0,842	0,860	0,851	(Tamayo et al., 2022)
dezzai	Spain	I	-	0,828	0,845	0,836	(Ortega-Martín et al., 2022)
RACAI	Romania	A	(RACAI, 2022)	0,868	0,779	0,821	(Avram et al., 2022)
KU_EDI	Korea	A	-	0,809	0,798	0,803	(Lain et al., 2022)
SINAI	Spain	A	(SINAI, 2022)	0,756	0,795	0,775	(Chizhikova et al., 2022)
ITAINNOVA	Spain	I	(ITAINNOVA, 2022)	0,779	0,769	0,774	(Montañés-Salas et al., 2022)
FRE	Spain	I	-	0,680	0,805	0,738	(Cetina and García-Santa, 2022)
baseline			(TeMU-BSC, 2022)	0,776	0,701	0,737	
HIBA	Argentina	I	-	0,759	0,644	0,697	(Castano et al., 2022)[a]
TEAM IAI	France	A	-	0,640	0,655	0,647	(Sinha et al., 2022)
CAISA[b]	Germany	A	-	0,836	0,494	0,621	(Karimi and Flek, 2022)
ResearchX	India	A	-	0,505	0,625	0,559	-
AILAB Udine	Italy	A	-	0,504	0,461	0,481	(Portelli et al., 2022)
JSL[c]	USA	I	-	0,004	0,004	0,004	(Kocaman et al., 2022)

[a]Same system that the one used in DisTEMIST.
[b]The best system of this team was a post-workshop evaluation.
[c]This team had problems in the evaluation phase due to the format used in the submission.

Table 2: SocialDisNER ranking with the best submission per team. Best result bolded, second best underlined. A/I stands for Academy/Industry.

dia. The final version contains 56 annotation and restriction rules in relation to disease concepts. The guidelines are freely available at Zenodo (Farré-Maduell et al., 2022).

4 Results

Participation. SocialDisNER achieved a considerable impact in the scientific-technical community. A total of 47 teams were registered, out of which 17 submitted a total of 32 runs for evaluation. Although most of the teams came from academic environments, a significant number (4) came from industry. Interestingly, 10 teams came from non-Spanish-speaking countries, which indicates the community interest in developing systems in languages other than English.

Results. Table 2 shows the results of SocialDisNER. Eleven teams achieved better performance than the baseline of the task. The best performing system was developed by the CASIA team, with a micro-average F1-score of 0.891. The Clac team obtained the best performance in terms of Recall, with a value of 0.888. The top-performing team developed a Unified Named Entity Recognition system by following a continual pre-training strategy based on transformers architecture. This system was able to correctly predict ≈93% (2871/3083) of disease mentions from the test set that also appeared in the training. Out of 776 mentions that the model had not previously seen, it was able to properly predict 511 (≈66%).

Error analysis. In common with similar biomedical entity recognition tasks, the longer the mentions, the more difficult is for models to predict them correctly (Augenstein et al., 2017). In SocialDisNER we have found a correlation of -0.24 between the prediction errors of the systems and the length of the mentions, with a tendency to error when the length of the mention increases. Regarding specific issues of detection, we have detected 4 common detection problems in several participating systems:

1. *Difficulty recognizing capitalized mentions*: Systems are able to detect a lowercase mention, but not its uppercase version if they have not seen it before in the training set.
2. *Mentions containing punctuation marks and/or special Twitter characters*: Mentions with internal punctuation marks are difficult to be correctly extracted by systems. They are usually detected as several independent mentions or detecting only one of the segments.
3. *Composite mentions*: Mentions that refer to more than one disease using conjunctions or prepositions are noted as a single mention, but participating systems tend to split such mentions when extracted.

4. *Detection of mention boundaries*:The systems developed in the task predict longer mentions than expected when there are symbols such as hashtag "#" or emojis before and/or after the mention.

Some examples of these errors can be found in Table 3 in appendix A. The systems also fail to properly detect mentions semantically similar to those seen in the training phase, but expressed with another verbal construction. For example, the models are able to detect "suicide" but not "thinking about suicide"

5 Discussion

SocialDisNER is the first task focused on extracting diseases from social media content written in Spanish. The first Gold Standard corpus of tweets with diseases annotated by medical experts has been built specially for the task. The corpus documents were selected to contain first-person patient experiences and relevant biomedical information written by experts.

We also published additional resources such a large-scale Silver Standard corpus annotated with additional biomedical entities that might be used to train systems to detect entities in Spanish tweets that previously could not be detected due to lack of resources. This large-scale corpus made it possible to generate co-mention networks that can be used toward Knowledge Graph mining to replicate studies on adverse drug effects in Spanish-speaking people (Nikfarjam et al., 2015). A knowledge graph similar to the one in Figure 2 might allow descriptive analysis of disease co-morbidities, to detect new symptoms in rare diseases, to discover diseases associated with specific professions, and even to discover adverse effects of biomaterials and prosthetics.

SocialDisNER has been very well received by the community. There have been participants from the academy and the industry, probably enticed by the methodology followed to collect content, which ensured that a significant percentage of the documents were of high relevance. Nevertheless, the modularity with which the methodology has been defined enables its improvement for future shared-tasks and projects. For example, more sophisticated first-person detection systems such as those developed in Al-Garadi et al. (2020) could be used, based on manual annotation of previous data. NER systems could also be trained with the large-scale corpus to improve the data enrichment process. This data selection methodology to retrieve relevant information could also be easily transferred to other languages as it relies on the presence of patient associations on Twitter, which is relatively common.

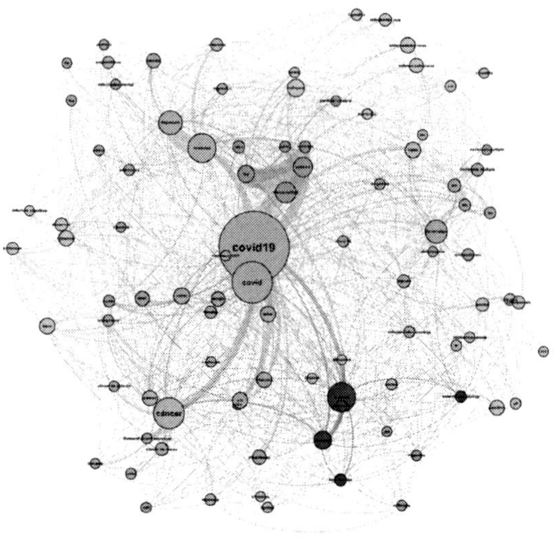

Figure 2: Simplified SocialDisNER co-morbility network,

When organizing the task, we contacted several patient associations with Twitter accounts. These groups showed interest in SocialDisNER, its results and how the output of the task may benefit the patients. The interest shown by associations, which in many cases helped to disseminate the event, shows the importance of involving all relevant stakeholders during the development and organization phases in order to increase the impact and use cases of our work.

Acknowledgements

We acknowledge the Encargo of PlanTL (SEDIA) to BSC for supporting SocialDisNER and the scientific committee for their valuable guidance and help. Due to the relevance of detecting disease content and symptoms from social media to analyze adverse effects of prosthetics and biomaterials, this work has been supported by BIOMATDB, a European Unions Horizon Europe Coordination Support Action under Grant Agreement No 101058779. Due to the relevance of the detection of diseases related to work environments, this project has also been supported by the AI4ProfHealth project (PID2020-119266RA-I00). Icons from Figure 1 created by Flaticon.

References

Luca Maria Aiello, Daniele Quercia, Ke Zhou, Marios Constantinides, Sanja Šćepanović, and Sagar Joglekar. 2021. How epidemic psychology works on twitter: Evolution of responses to the covid-19 pandemic in the us. *Humanities and Social Sciences Communications*, 8(1):1–15.

Mohammed Ali Al-Garadi, Yuan-Chi Yang, Sahithi Lakamana, and Abeed Sarker. 2020. A text classification approach for the automatic detection of twitter posts containing self-reported covid-19 symptoms.

Thayer Alshaabi, David Rushing Dewhurst, Joshua R Minot, Michael V Arnold, Jane L Adams, Christopher M Danforth, and Peter Sheridan Dodds. 2021. The growing amplification of social media: Measuring temporal and social contagion dynamics for over 150 languages on twitter for 2009–2020. *EPJ data science*, 10(1):15.

Isabelle Augenstein, Mrinal Das, Sebastian Riedel, Lakshmi Vikraman, and Andrew McCallum. 2017. Semeval 2017 task 10: Scienceie-extracting keyphrases and relations from scientific publications. *arXiv preprint arXiv:1704.02853*.

Andrei-Marius Avram, Vasile Pais, and Maria Mitrofan. 2022. Racai@smm4h'22: Tweets disease mention detection using a neural lateral inhibitory mechanism. In *Proceedings of the Seventh Social Media Mining for Health Applications (#SMM4H) Workshop & Shared Task*, pages 1–3.

Jose Castano, Maria Laura Gambarte, Carlos Otero, and Daniel Luna. 2022. A simple terminology-based approach to clinical entity recognition.

Kendrick Cetina and Nuria García-Santa. 2022. Fre at socialdisner: Joint learning of language models for named entity recognition. In *Proceedings of the Seventh Social Media Mining for Health Applications (#SMM4H) Workshop & Shared Task*, pages 68–70.

Mariia Chizhikova, Pilar López-Úbeda, Manuel C. Díaz-Galiano, L. Alfonso Ureña-López, and M. Teresa Martín-Valdivia. 2022. Sinai@smm4h'22: Transformers for biomedical social media text mining in spanish. In *Proceedings of the Seventh Social Media Mining for Health Applications (#SMM4H) Workshop & Shared Task*, pages 27–30.

Nigel Collier, Son Doan, Ai Kawazoe, Reiko Matsuda Goodwin, Mike Conway, Yoshio Tateno, Quoc-Hung Ngo, Dinh Dien, Asanee Kawtrakul, Koichi Takeuchi, et al. 2008. Biocaster: detecting public health rumors with a web-based text mining system. *Bioinformatics*, 24(24):2940–2941.

Eulàlia Farré-Maduell, Salvador Lima, Luis Gascó, Antonio Miranda-Escalada, and Martin Krallinger. 2022. SocialDisNER Guidelines: detection of disease mentions in spanish social media content. Funded by the Plan de Impulso de las Tecnologías del Lenguaje (Plan TL).

Jia Fu, Sirui Li, Hui Ming Yuan, Zhucong Li, Zhen Gan, Yubo Chen, Kang Liu, Jun Zhao, and Shengping Liu. 2022. Casia@smm4h'22: A uniform health information mining system for multilingual social media texts. In *Proceedings of the Seventh Social Media Mining for Health Applications (# SMM4H) Workshop & Shared Task*, pages 143–147.

Luis Gasco, Chloé Clavel, Cesar Asensio, and Guillermo de Arcas. 2019. Beyond sound level monitoring: Exploitation of social media to gather citizens subjective response to noise. *Science of The Total Environment*, 658:69–79.

Aitor Gonzalez-Agirre, Montserrat Marimon, Ander Intxaurrondo, Obdulia Rabal, Marta Villegas, and Martin Krallinger. 2019. Pharmaconer: Pharmacological substances, compounds and proteins named entity recognition track. In *Proceedings of The 5th Workshop on BioNLP Open Shared Tasks*, pages 1–10.

Tom Hope, Jason Portenoy, Kishore Vasan, Jonathan Borchardt, Eric Horvitz, Daniel S Weld, Marti A Hearst, and Jevin West. 2020. Scisight: Combining faceted navigation and research group detection for covid-19 exploratory scientific search. *arXiv preprint arXiv:2005.12668*.

Javier Huertas-Tato, Alejandro Martin, and David Camacho. 2022. Bertuit: Understanding spanish language in twitter through a native transformer. *arXiv preprint arXiv:2204.03465*.

ITAINNOVA. 2022. Socialdisner code. `https://github.com/ITAINNOVA/SocialDisNER`.

Akbar Karimi and Lucie Flek. 2022. Caisa@smm4h'22: Robust cross-lingual detection of disease mentions on social media with adversarial methods. In *Proceedings of the Seventh Social Media Mining for Health Applications (#SMM4H) Workshop & Shared Task*, pages 168–170.

Simon Kemp. 2021. [link].

Veysel Kocaman, Cabir Celik, Damla Gurbaz, Gursev Pirge, Bunyamin Polat, Halil Saglamlar, Meryem Vildan Sarikaya, Gokhan Turer, and David Talby. 2022. John_snow_labs@smm4h'22: Social media mining for health (#smm4h) with spark nlp. In *Proceedings of the Seventh Social Media Mining for Health Applications (#SMM4H) Workshop & Shared Task*, pages 44–47.

Dominik Kowald and Elisabeth Lex. 2018. Studying confirmation bias in hashtag usage on twitter. *arXiv preprint arXiv:1809.03203*.

Antoine Lain, Wonjin Yoon, Hyunjae Kim, Jaewoo Kang, and Ian Simpson. 2022. Ku_ed at socialdisner: Extracting disease mentions in tweets written in spanish. In *Proceedings of the Seventh Social Media Mining for Health Applications (#SMM4H) Workshop & Shared Task*, pages –.

Andrew MacKinlay, Hafsah Aamer, and Antonio Jimeno Yepes. 2017. Detection of adverse drug reactions using medical named entities on twitter. In *AMIA Annual Symposium Proceedings*, volume 2017, page 1215. American Medical Informatics Association.

Eugenio Martinez-Camara, M Teresa Martín-Valdivia, L Alfonso Urena-Lopez, and Ruslan Mitkov. 2015. Polarity classification for spanish tweets using the cost corpus. *Journal of Information Science*, 41(3):263–272.

Emily G Miller, Amanda L Woodward, Grace Flinchum, Jennifer L Young, Holly K Tabor, and Meghan C Halley. 2021. Opportunities and pitfalls of social media research in rare genetic diseases: a systematic review. *Genetics in Medicine*, 23(12):2250–2259.

Antonio Miranda-Escalada, Eulàlia Farré-Maduell, Salvador Lima-López, Luis Gascó, Vicent Briva-Iglesias, Marvin Agüero-Torales, and Martin Krallinger. 2021. The profner shared task on automatic recognition of occupation mentions in social media: systems, evaluation, guidelines, embeddings and corpora. In *Proceedings of the Sixth Social Media Mining for Health (# SMM4H) Workshop and Shared Task*, pages 13–20.

Antonio Miranda-Escalada, Luis Gascó, Salvador Lima-López, Eulàlia Farré-Maduell, Darryl Estrada, Anastasios Nentidis, Anastasia Krithara, Georgios Katsimpras, Georgios Paliouras, and Martin Krallinger. 2022. Overview of distemist at bioasq: Automatic detection and normalization of diseases from clinical texts: results, methods, evaluation and multilingual resources.

Rosa M. Montañés-Salas, Irene López-Bosque, Luis García-Garcés, and Rafael del Hoyo-Alonso. 2022. Itainnova at socialdisner: A transformers cocktail for disease identification in social media in spanish. In *Proceedings of the Seventh Social Media Mining for Health Applications (#SMM4H) Workshop & Shared Task*, pages 71–74.

Anastasios Nentidis, Georgios Katsimpras, Eirini Vandorou, Anastasia Krithara, Antonio Miranda-Escalada, Luis Gasco, Martin Krallinger, and Georgios Paliouras. 2022. Overview of bioasq 2022: The tenth bioasq challenge on large-scale biomedical semantic indexing and question answering. In *Experimental IR Meets Multilinguality, Multimodality, and Interaction*, pages 337–361, Cham. Springer International Publishing.

Azadeh Nikfarjam, Abeed Sarker, Karen Oconnor, Rachel Ginn, and Graciela Gonzalez. 2015. Pharmacovigilance from social media: mining adverse drug reaction mentions using sequence labeling with word embedding cluster features. *Journal of the American Medical Informatics Association*, 22(3):671–681.

Miguel Ortega-Martín, Alfonso Ardoiz, Jorge Álvarez, Oscar García-Sierra, and Adrián Alonso. 2022. dezzai@smm4h'22: Tasks 5 10 - hybrid models everywhere. In *Proceedings of the Seventh Social Media Mining for Health Applications (#SMM4H) Workshop & Shared Task*, pages –.

P Otero, J Gago, and P Quintas. 2021. Twitter data analysis to assess the interest of citizens on the impact of marine plastic pollution. *Marine Pollution Bulletin*, 170:112620.

Karen OConnor, Pranoti Pimpalkhute, Azadeh Nikfarjam, Rachel Ginn, Karen L Smith, and Graciela Gonzalez. 2014. Pharmacovigilance on twitter? mining tweets for adverse drug reactions. In *AMIA annual symposium proceedings*, volume 2014, page 924. American Medical Informatics Association.

Flor Miriam Plaza-del Arco, Carlo Strapparava, L Alfonso Urena Lopez, and M Teresa Martín-Valdivia. 2020. Emoevent: A multilingual emotion corpus based on different events. In *Proceedings of the 12th Language Resources and Evaluation Conference*, pages 1492–1498.

PLN-CMM. 2022. Socialdisner. `https://github.com/plncmm/socialdisner`.

Beatrice Portelli, Simone Scaboro, Emmanuele Chersoni, Enrico Santus, and Giuseppe Serra. 2022. Ailab-udine@smm4h'22: Limits of transformers and bert ensembles. In *Proceedings of the Seventh Social Media Mining for Health Applications (#SMM4H) Workshop & Shared Task*, pages 130–134.

RACAI. 2022. Rner. `https://github.com/racai-ai/RNER`.

READ-BioMed. 2022. socialdisner-2022. `https://github.com/READ-BioMed/socialdisner-2022`.

Matias Rojas, Jose Barros, Kinan R. Martin, Mauricio Araneda-Hernandez, and Jocelyn Dunstan. 2022. Pln cmm at socialdisner: Improving detection of disease mentions in tweets by using document-level features. In *Proceedings of the Seventh Social Media Mining for Health Applications (#SMM4H) Workshop & Shared Task*, pages 52–54.

SINAI. 2022. Spanish_disease_finder. `https://huggingface.co/chizhikchi/Spanish_disease_finder`.

Aman Sinha, Cristina Garcia Holgado, Marianne Clausel, and Matthieu Constant. 2022. Iai @ socialdisner : Catch me if you can! capturing complex disease mentions in tweets. In *Proceedings of the Seventh Social Media Mining for Health Applications (#SMM4H) Workshop & Shared Task*, pages 85–89.

Antonio Tamayo. 2022. Nlp-cic-wfu at socialdisner: Disease mention extraction in spanish tweets using transfer learning and search by propagation. `https://github.com/ajtamayoh/NLP-CIC-WFU-Contribution-to-SocialDisNER-shared-task-2022`.

Antonio Tamayo, Diego Burgos, and Alexander Gelbukh. 2022. Nlp-cic-wfu at socialdisner: Disease mention extraction in spanish tweets using transfer learning and search by propagation. In *Proceedings of the Seventh Social Media Mining for Health Applications (#SMM4H) Workshop & Shared Task*, pages 19–22.

TeMU-BSC. 2022. Socialdisner-st baseline 1 - lookup. `https://github.com/TeMU-BSC/socialdisner_baseline_lookup`.

Harsh Verma, Parsa Bagherzadeh, and Sabine Bergler. 2022. Claclab at socialdisner: Using medical gazetteers for named-entity recognition of disease mentions in spanish tweets. In *Proceedings of the Seventh Social Media Mining for Health Applications (#SMM4H) Workshop & Shared Task*, pages 55–57.

Davy Weissenbacher, Juan M. Banda, Vera Davydova, Darryl Estrada-Zavala, Luis Gascó, Yao Ge, Yuting Guo, Ari Z. Klein, Martin Krallinger, Mathias Leddin, Argun Magge, Raul Rodriguez-Esteban, Abeed Sarker, Ana Lucía Schmidt, Elena Tutubalina, and Graciela Gonzalez-Hernandez. 2022. Overview of the seventh social media mining for health applications (#SMM4H) shared tasks at coling 2022. In *Proceedings of the Seventh Social Media Mining for Health Applications (#SMM4H) Workshop & Shared Task*.

Antonio Jimeno Yepes and Karin Verspoor. 2022. Readbiomed@socialdisner: Adaptation of an annotation system to spanish tweets. In *Proceedings of the Seventh Social Media Mining for Health Applications (#SMM4H) Workshop & Shared Task*, pages 48–51.

A Appendix prediction errors

Name	Tweet with mention	Example of participants extractions
Capitalized mentions	LAS CICATRICES No hay cicatriz, (...)	Systems predict the mention "cicatrices", but not "CICATRICES", only present in the test set.
	(...) NO SOMOS DEPRESIVAS, TENEMOS DEPRESIÓN! Sabías (...)	Systems predict "depresión", but not "DEPRESIÓN"
Mentions with punctuation marks and/or special characters	visibilidad para esta enfermedad crónica asociada al #dolor	*enfermedad crónica* *enfermedad crónica* and *dolor*]
	(...)teniendo una enfermedad crónica (Crohn) y en tratamiento(...)	*enfermedad crónica* *enferemedad crónica (Crohn* *enfermedad cronica* and *Crohn*
	(...)Antraciclinas en Her2+ en ca de mama temprano(...)	*ca de mama* *ca de mama temprano* *her2* and *ca de mama*
Composite mentions	(...)debido a una malformación o disfunción de los órganos que(...)	*malformación* *disfunción de los órganos* *malformación* and *disfunción de los órganos* *malformación o disfunción de los órganos*
	(...)cuidar a su marido con cáncer y metástasis. No(...)	*cancer* and *metástasis*
Detection of mention boundaries	(...)y el consiguiente daño NEURONAL🧠...)	*daño NEURONAL*🧠
	(...)lo agradecemos 👏👏👏😂✌ #ConferenciaCovid19 (...)	*ConferenciaCovid19*

Table 3: Example of prediction errors of SocialDisNER systems

Romanian micro-blogging named entity recognition including health-related entities

Vasile Păiș, Verginica Barbu Mititelu, Elena Irimia, Maria Mitrofan, Carol Luca Gasan, Roxana Micu
Research Institute for Artificial Intelligence "Mihai Drăgănescu", Romanian Academy
vasile,vergi,elena,maria@racai.ro

Abstract

This paper introduces a manually annotated dataset for named entity recognition (NER) in micro-blogging text for the Romanian language. It contains gold annotations for 9 entity classes and expressions: persons, locations, organizations, time expressions, legal references, disorders, chemicals, medical devices and anatomical parts. Furthermore, word embedding models computed on a larger micro-blogging corpus are made available. Finally, several NER models are trained and their performance is evaluated against the newly introduced corpus.

1 Introduction

Social media networks can present an alternative to formal data sources, and prove to be a reliable resource for different natural language processing (NLP) tasks. Since users often submit domain-specific information in a wide variety of social networking resources, such as specific communities, blogs, microblogs, public news websites, or forums, there has been a significant interest in extracting information from these resources. Micro-blogging platforms, such as Twitter, Reddit or Gab, as a source of such comments, in general, contain relevant information that can be used to develop NLP resources for both general and specific domains, such as health (dos Santos et al., 2020) and legal aspects (Altoaimy, 2018). Furthermore, mining and studying health-related information can ultimately be used to improve public health (Magge et al., 2021).

Our contribution is threefold. First, we introduce the MicroBloggingNERo corpus (Păiș et al., 2022a), comprised of micro-blogging specific messages, manually annotated with 9 named entity (NE) classes and expressions, including 4 health-related classes. This is the largest available Romanian micro-blogging NE corpus to date and the only one containing health-related entities. Second, we release word embeddings computed on a larger micro-blogging corpus. Finally, we train several NER models using the newly released corpus and we make these available for the research community.

This paper is organized as follows: Section 2 presents related work, Section 3 describes the manually annotated dataset, and Section 4 introduces the experiments performed, including the generated word embedding representations and NER models. Finally, Section 5 gives conclusions.

2 Related work

Biomedical corpora have been collected for languages with different extent of technological support: Hmong (White, 2022), Romanian (Mitrofan et al., 2019), Spanish (Carrino et al., 2021) and others. There are even parallel medical corpora available[1]. Many of them are annotated with UMLS medical entities (Mohan and Li, 2019; Ohta et al., 2002).[2] Most of the biomedical corpora were collected from abstracts (Ohta et al., 2002), but also from full articles (Alex et al., 2008; Bada et al., 2012); others combine various sources (Mitrofan et al., 2019).

The interest in gathering corpora made up of micro-blogging texts characterizes the research dedicated to various languages: Arabic (Zaatari et al., 2016), Chinese (Wang et al., 2012), English (Sharma et al., 2020), French (Mazoyer et al., 2020), German and Danish (Bick, 2020), Italian (Sanguinetti et al., 2018), Romanian (Manolescu and Çöltekin, 2021; Ciobotaru et al., 2022), Turkish (Çöltekin, 2020), etc.

Micro-blogging corpora are used for various tasks: sentiment analysis application development (Sharma et al., 2020; Cieliebak et al., 2017), emotion annotation (Roberts et al., 2012), credibility

[1] https://ufal.mff.cuni.cz/ufal_medical_corpus

[2] Such a collection is available at https://github.com/BaderLab/Biomedical-Corpora.

Proceedings of the 29th International Conference on Computational Linguistic, pages 190–196
October 12–17, 2022.

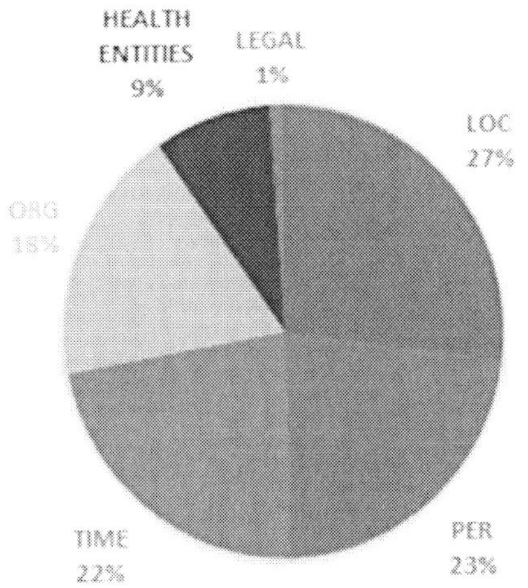

Figure 1: Label distribution of entity classes.

analysis (Zaatari et al., 2016), event detection (Mazoyer et al., 2020), hate speech detection (Bick, 2020; Çöltekin, 2020; Manolescu and Çöltekin, 2021; Sanguinetti et al., 2018) and others. Another vivid line of research is identifying certain types of entity names in these corpora, such as occupations (Miranda-Escalada et al., 2021; Santamaría Carrasco and Cuervo Rosillo, 2021) and symptoms (Kumar et al., 2021).

In this work, we developed MicroBloggingNERo the first Romanian micro-blogging corpus enriched with health information.

3 Dataset

The MicroBloggingNERo dataset contains 7,800 messages manually annotated with the following named entity classes and expressions, totalling 11,099 annotations (see figure 1):

- Organization (ORG) entities representing companies, agencies, and political parties. Examples: *Facebook*, *Guvernul* ("the government"), *PSD* (Partidul Social Democrat "the Socialist Democratic Party"), *#ConsConRo*.
- Person (PER) entities are regularly limited to humans, but may include fictional characters and references to religious figures. Examples: *Adela*, *Moș Crăciun* ("Santa Claus"), *Niculina Stoican*.
- Location (LOC) entities represent addresses, geographical areas and landmasses, bodies of water, and geological formations, denoted by a proper name. Examples: *România* ("Romania"), *Parcul Tineretului* ("the Youth Park"), *Lacul Sfânta Ana* ("Lake Saint Ana").
- Time (TIME) expressions identify when something happened, how long something lasted, or how often something occurs. Sometimes the precise date cannot be determined, allowing for expressions indicating periods of time. Examples: *astăzi* ("today"), *15 septembrie* ("September 15"), *Crăciun* ("Christmas").
- Legal references (LEGAL) are expressions indicating a legal document. Examples: *legea 13/2021* ("law 13/2021"), *constituția* ("the Constitution").
- Anatomical parts (ANAT) class marks parts of the human body, organs, components of organs, tissues, cells, cellular components. Examples: *cap* ("head"), *mâini* ("hands"), *ficat* ("liver").
- Chemical and drugs (CHEM) class contains mentions of amino acids, peptides, proteins, antibiotics, active substances, drugs, enzymes, hormones, receptors. Examples: *sodiu* ("sodium"), *vaccin* ("vaccine").
- Disorders (DISO) class is intended primarily for diseases but includes also things such as anatomical abnormalities, congenital anomalies, syndromes, lesions, symptoms. Examples: *diabet* ("diabetes"), *COVID*.
- Medical devices (MED_DEVICE) class contains mentions of any device intended to be used for medical purposes. Example: *stetoscop* ("stethoscope").

The dataset was gathered by using a crawling process based on specific queries, aiming to gather messages containing the entities of interest. It was then cleaned by removing duplicates and very short messages (too short to be actual sentences) or messages containing only hashtags. Furthermore, all URLs were replaced with a special "<url>" tag.

The annotation process involved 7 annotators working under the guidance and supervision of 3 senior researchers. Parts of the corpus were common between several of the annotators, allowing us to compute inter-annotator agreement. Each pair of annotators had to annotate 300 common files. The Cohen Kappa coefficient, indicating the inter-annotator agreement was computed at the token level and led to an average Kappa of 0.80. According to (Cohen, 1960), a value between 0.61–0.80 indicates substantial agreement between the annotators. Moreover, Landis and Koch (1977) stated that a value of Cohen Kappa coefficient greater than 0.81 represents an almost perfect agreement and Fleiss et al. (2013) also characterizes kappas over 0.75 as excellent.

The remaining disagreements account for mistakes made by individual annotators when specific domain information was needed in order to identify and classify mainly entities from both medical and legal domains. For example, medical entities such as "virus" was annotated with labels "DISO" or "CHEM". In most situations the correct label is "DISO", however when the mechanism of the action of the virus is explained the correct label is "CHEM", thus creating confusion for the annotators. A similar situation can be found in the case of "vaccine" mentions when this entity was annotated with both "CHEM" and "MED_DEVICE" labels. In this case the correct label is "CHEM", but some annotators considered it "MEDICAL_DEVICE" because the term was used in a context of diagnosing Covid-19 together with tests, masks, gowns, gloves, sterilizers, and ventilators. Entity classes such as ORG, PER, LOC and TIME have an inter-annotator agreement greater than 0.85.

The annotation scheme, as well as the guidelines, were inspired by existing Romanian NE corpora (built without social media text), such as MoNERo (Mitrofan et al., 2019; Mitrofan, 2017), SiMoNERo (Mitrofan, 2019) and LegalNERo (Păiș et al., 2021; Păiş et al., 2021). The annotation guidelines[3] for the MicroBloggingNERo corpus had to be adjusted to take into account the particularities (Păiș et al., 2022b) of micro-blogging text, such as hashtags (*#Cluj, #LaculNoua*), text written with emojis instead of numbers, unusual abbreviation, elongated words (commonly used for emphasis in microblogging), code-mixed text, etc.

Since the MicroBloggingNERo corpus was created with multiple types of annotations, the health-related entities represent only 9% of the total entity classes, as depicted in Figure 1. This accounts for 958 annotations, half of them being in the DISO class (466 annotations).

After the annotation process was completed, the corpus was anonymized by replacing all person names, specific locations (such as addresses or street numbers), and organizations with randomly generated ones. Any user mentions were removed and replaced with a special tag "<user>", regardless of how it was included in the original message. Furthermore, any other identifiers (such as message IDs) were removed from the corpus.

The dataset was released with several usage scenarios in mind, reflected in the presence of multiple formats, including span-based annotations with all the entities, span-based annotations with general entities (PER, LOC, ORG, TIME), span-based annotations for the bio-medical domain, and tokenized text with token-based NE annotations attached (using tools available inside the RELATE platform (Păiș et al., 2019, 2020; Păiș, 2020)).

4 Experiments

We used the presented NER annotated micro-blogging dataset to train NER models, using different architectures and configurations. For this purpose, the corpus was split into train (70%), valid (15%) and test (15%). The splits were selected randomly while trying to preserve the general distribution of entities in each split. They are made available within the corpus release. Then, we experimented with classical word embedding representations. For this purpose, we used a recurrent neural network architecture, implemented with Long Short-Term Memory (LSTM) cells, as provided by the NeuroNER package (Dernoncourt et al., 2017). We used pre-trained word embeddings (Păiș and Tufis, 2018) trained on the Representative Corpus of the Contemporary Romanian Language (CoRoLa) (Tufiș et al., 2019). Then, we trained new word representations based on a larger corpus of raw micro-blogging text (comprising 853k messages), using the FastText tool (Bojanowski et al., 2017). Finally, based on observations such as those of (Păiș and Mitrofan, 2021), indicating that multiple representations can help the system achieve better results, we combine the two representations into a single, larger representation. We make the generated embedding models freely available[4].

Following the experiments with static word embeddings, we performed two additional experiments using contextualized embeddings provided by the XLM-RoBERTa model (Conneau et al., 2020). This model provides contextualized representations for multiple languages, including Romanian, and has proven to be a good choice for different NER experiments (Malmasi et al., 2022; Shaffer, 2021; Adelani et al., 2021; Suppa and Jariabka, 2021). Our experiments made use of a basic NER system, employing a linear layer followed by

[3]https://relate.racai.ro/resources/microblogging/Annotation_Guide_MicroBloggingNERo.pdf

[4]https://relate.racai.ro/resources/microblogging

System	ANAT	CHEM	DISO	LEG	LOC	MED DEV	ORG	PER	TIME	Total	Ep.
Neuroner CoRoLa	42.86	60.47	75.47	45.71	77.69	72.73	66.21	84.18	63.96	72.03	23
Neuroner Microblogging	22.54	58.82	71.43	47.37	81.27	61.54	65.95	80.95	63.64	71.26	28
Neuroner CoRoLa+ MicroBlogging	21.43	52.87	73.47	36.36	81.21	66.67	62.00	83.51	61.50	70.75	32
XLM-RoBERTa	**49.46**	**70.97**	**82.00**	68.29	86.88	62.50	**77.37**	88.56	68.32	**78.96**	35
XLM-RoBERTa with LI	45.24	67.42	81.00	**68.42**	**87.05**	**76.92**	75.39	**88.70**	**68.50**	78.62	61

Table 1: Results (% F1 scores) from different experiments using the MicroBloggingNERo corpus

a classification head, and the same NER system enhanced with a biologically inspired lateral inhibition layer, as introduced by (Păiș, 2022). This system was previously used for NER in the Romania bio-medical domain (Mitrofan and Păiș, 2022). Intuitively, similar to the biological process, we considered that the lateral inhibition layer would allow the system to better focus on difficult, and less represented, NE classes. Results are given in Table 1, in terms of F1 percent scores, while also indicating the training epoch associated with each model.

The results indicate that Romanian NER in a micro-blogging context is still far from achieving high-scores. Currently, even systems making use of contextualized embeddings do not achieve over 90% F1 score for any of the classes. On regular texts, previous works, such as (Păiş et al., 2021), indicate over 90% F1 scores for persons and legal references. Furthermore, Mitrofan and Păiș (2022) show an F1 score of 85% for health-related entities in regular Romanian texts.

An analysis of the errors associated with the best performing model, with regard to health-related entities, indicates that a large number of entities are not recognized (the predicted label is "O"). This suggests a need for a larger corpus with more diverse entities. Other types of errors account for predicting ORG as MED_DEVICE or CHEM, which is an error found also with the human annotators, where a company name is used also for a vaccine name.

5 Conclusion

Micro-blogging texts pose unique challenges in terms of natural language processing. This is particularly true with regard to bio-medical and other health-related entities. These are more difficult to detect when compared to more common NEs, such as person and organization names, due to their inherent complexity. Erhardt et al. (2006) consider this to be a general issue with regard to information extraction in the biomedical domain, given the complex entities and changing nomenclatures. This was also the case in the process of manual annotation, annotators having difficulties to identify and classify both health-related and legal NEs. Furthermore, as noted by Klein et al. (2020), imbalanced data remains a challenge for training deep neural network models. This is particularly true with the MicroBloggingNERo dataset, considering the distribution of classes depicted in Figure 1. Furthermore, this suggests that future work should focus on expanding the dataset, aiming for a more balanced class distribution.

When looking at micro-blogging word embeddings, these seem to be beneficial for recognizing certain entity classes, possibly accounting for spelling particularities in micro-blogging texts. General representations, trained on the large CoRoLa corpus, provide better results when detecting anatomical parts, chemicals and medical devices. This is due to the presence of bio-medical texts in the CoRoLa corpus, as well as to the reduced number of mentions in the micro-blogging corpus. Future research will explore more word representation models, particularly those trained on larger micro-blogging corpora. Currently, there is no contextualized word model available for the Romanian language built on micro-blogging texts or specifically on health-related texts, such as BioBERT (Lee et al., 2019).

References

David Ifeoluwa Adelani, Jade Abbott, Graham Neubig, Daniel D'souza, Julia Kreutzer, Constantine Lignos, Chester Palen-Michel, Happy Buzaaba, Shruti Rijhwani, Sebastian Ruder, Stephen Mayhew, Israel Abebe Azime, Shamsuddeen H. Muhammad, Chris Chinenye Emezue, Joyce Nakatumba-Nabende, Perez Ogayo, Aremu Anuoluwapo, Catherine Gitau, Derguene Mbaye, Jesujoba Alabi, Seid Muhie Yimam, Tajuddeen Rabiu Gwadabe, Ignatius Ezeani, Rubungo Andre Niyongabo, Jonathan Mukiibi, Verrah Otiende, Iroro Orife, Davis David, Samba Ngom, Tosin Adewumi, Paul Rayson, Mofetoluwa Adeyemi, Gerald Muriuki, Emmanuel Anebi, Chiamaka Chukwuneke, Nkiruka Odu, Eric Peter Wairagala, Samuel Oyerinde, Clemencia Siro, Tobius Saul Bateesa, Temilola Oloyede, Yvonne Wambui, Victor Akinode, Deborah Nabagereka, Maurice Katusiime, Ayodele Awokoya, Mouhamadane MBOUP, Dibora Gebreyohannes, Henok Tilaye, Kelechi Nwaike, Degaga Wolde, Abdoulaye Faye, Blessing Sibanda, Orevaoghene Ahia, Bonaventure F. P. Dossou, Kelechi Ogueji, Thierno Ibrahima DIOP, Abdoulaye Diallo, Adewale Akinfaderin, Tendai Marengereke, and Salomey Osei. 2021. MasakhaNER: Named entity recognition for African languages. *Transactions of the Association for Computational Linguistics*, 9:1116–1131.

Beatrice Alex, Claire Grover, Barry Haddow, Mijail Kabadjov, Ewan Klein, Michael Matthews, Stuart Roebuck, Richard Tobin, and Xinglong Wang. 2008. The iti txm corpora: Tissue expressions and protein-protein interactions.

Lama Altoaimy. 2018. Driving change on twitter: A corpus-assisted discourse analysis of the twitter debates on the saudi ban on women driving. *Social Sciences*, 7(5):81.

Michael Bada, Miriam Eckert, Donald Evans, Kristin Garcia, Krista Shipley, Dmitry Sitnikov, William A. Baumgartner, K. Bretonnel Cohen, Karin Verspoor, Judith A. Blake, and Lawrence E. Hunter. 2012. Concept annotation in the craft corpus. *BMC Bioinformatics*, 13.

Eckhard Bick. 2020. An annotated social media corpus for German. In *Proceedings of the 12th Language Resources and Evaluation Conference*, pages 6127–6135, Marseille, France. European Language Resources Association.

Piotr Bojanowski, Edouard Grave, Armand Joulin, and Tomas Mikolov. 2017. Enriching word vectors with subword information. *Transactions of the Association for Computational Linguistics*, 5:135–146.

Casimiro Pio Carrino, Jordi Armengol-Estapé, Ona de Gibert Bonet, Asier Gutiérrez-Fandiño, Aitor Gonzalez-Agirre, Martin Krallinger, and Marta Villegas. 2021. Spanish biomedical crawled corpus: A large, diverse dataset for spanish biomedical language models.

Mark Cieliebak, Jan Milan Deriu, Dominic Egger, and Fatih Uzdilli. 2017. A Twitter corpus and benchmark resources for German sentiment analysis. In *Proceedings of the Fifth International Workshop on Natural Language Processing for Social Media*, pages 45–51, Valencia, Spain. Association for Computational Linguistics.

Alexandra Ciobotaru, Mihai Vlad Constantinescu, Liviu P. Dinu, and Stefan Dumitrescu. 2022. Red v2: Enhancing red dataset for multi-label emotion detection. In *Proceedings of the Language Resources and Evaluation Conference*, pages 1392–1399, Marseille, France. European Language Resources Association.

Jacob Cohen. 1960. A coefficient of agreement for nominal scales. *Educational and psychological measurement*, 20(1):37–46.

Çağrı Çöltekin. 2020. A corpus of Turkish offensive language on social media. In *Proceedings of the 12th Language Resources and Evaluation Conference*, pages 6174–6184, Marseille, France. European Language Resources Association.

Alexis Conneau, Kartikay Khandelwal, Naman Goyal, Vishrav Chaudhary, Guillaume Wenzek, Francisco Guzmán, Édouard Grave, Myle Ott, Luke Zettlemoyer, and Veselin Stoyanov. 2020. Unsupervised cross-lingual representation learning at scale. In *Proceedings of the 58th Annual Meeting of the Association for Computational Linguistics*, pages 8440–8451.

Franck Dernoncourt, Ji Young Lee, and Peter Szolovits. 2017. NeuroNER: an easy-to-use program for named-entity recognition based on neural networks. *Conference on Empirical Methods on Natural Language Processing (EMNLP)*.

Wesley Ramos dos Santos, Amanda MM Funabashi, and Ivandré Paraboni. 2020. Searching brazilian twitter for signs of mental health issues. In *LREC*, pages 6111–6117.

Ramón A-A. Erhardt, Reinhard Schneider, and Christian Blaschke. 2006. Status of text-mining techniques applied to biomedical text. *Drug Discovery Today*, 11(7):315–325.

Joseph L Fleiss, Bruce Levin, and Myunghee Cho Paik. 2013. *Statistical methods for rates and proportions*. john wiley & sons.

Ari Klein, Ilseyar Alimova, Ivan Flores, Arjun Magge, Zulfat Miftahutdinov, Anne-Lyse Minard, Karen O'Connor, Abeed Sarker, Elena Tutubalina, Davy Weissenbacher, and Graciela Gonzalez-Hernandez. 2020. Overview of the fifth social media mining for health applications (#SMM4H) shared tasks at COLING 2020. In *Proceedings of the Fifth Social Media Mining for Health Applications Workshop & Shared Task*, pages 27–36, Barcelona, Spain (Online). Association for Computational Linguistics.

Deepak Kumar, Nalin Kumar, and Subhankar Mishra. 2021. NLP@NISER: Classification of COVID19 tweets containing symptoms. In *Proceedings of the Sixth Social Media Mining for Health (#SMM4H) Workshop and Shared Task*, pages 102–104, Mexico City, Mexico. Association for Computational Linguistics.

J Richard Landis and Gary G Koch. 1977. The measurement of observer agreement for categorical data. *biometrics*, pages 159–174.

Jinhyuk Lee, Wonjin Yoon, Sungdong Kim, Donghyeon Kim, Sunkyu Kim, Chan Ho So, and Jaewoo Kang. 2019. BioBERT: a pre-trained biomedical language representation model for biomedical text mining. *Bioinformatics*, 36(4):1234–1240.

Arjun Magge, Ari Klein, Antonio Miranda-Escalada, Mohammed Ali Al-Garadi, Ilseyar Alimova, Zulfat Miftahutdinov, Eulalia Farre, Salvador Lima López, Ivan Flores, Karen O'Connor, Davy Weissenbacher, Elena Tutubalina, Abeed Sarker, Juan Banda, Martin Krallinger, and Graciela Gonzalez-Hernandez. 2021. Overview of the sixth social media mining for health applications (#SMM4H) shared tasks at NAACL 2021. In *Proceedings of the Sixth Social Media Mining for Health (#SMM4H) Workshop and Shared Task*, pages 21–32, Mexico City, Mexico. Association for Computational Linguistics.

Shervin Malmasi, Anjie Fang, Besnik Fetahu, Sudipta Kar, and Oleg Rokhlenko. 2022. SemEval-2022 task 11: Multilingual complex named entity recognition (MultiCoNER). In *Proceedings of the 16th International Workshop on Semantic Evaluation (SemEval-2022)*, pages 1412–1437, Seattle, United States. Association for Computational Linguistics.

Mihai Manolescu and Çağrı Çöltekin. 2021. ROFF - a Romanian Twitter dataset for offensive language. In *Proceedings of the International Conference on Recent Advances in Natural Language Processing (RANLP 2021)*, pages 895–900, Held Online. INCOMA Ltd.

Béatrice Mazoyer, Julia Cagé, Nicolas Hervé, and Céline Hudelot. 2020. A French corpus for event detection on Twitter. In *Proceedings of the 12th Language Resources and Evaluation Conference*, pages 6220–6227, Marseille, France. European Language Resources Association.

Antonio Miranda-Escalada, Eulàlia Farré-Maduell, Salvador Lima-López, Luis Gascó, Vicent Briva-Iglesias, Marvin Agüero-Torales, and Martin Krallinger. 2021. The ProfNER shared task on automatic recognition of occupation mentions in social media: systems, evaluation, guidelines, embeddings and corpora. In *Proceedings of the Sixth Social Media Mining for Health (#SMM4H) Workshop and Shared Task*, pages 13–20, Mexico City, Mexico. Association for Computational Linguistics.

Maria Mitrofan. 2017. Bootstrapping a Romanian corpus for medical named entity recognition. In *RANLP*, pages 501–509.

Maria Mitrofan. 2019. *Extragere de cunoștințe din texte în limba română și date structurate cu aplicații în domeniul medical.* Ph.D. thesis, Romanian Academy.

Maria Mitrofan, Verginica Barbu Mititelu, and Grigorina Mitrofan. 2019. MoNERo: a biomedical gold standard corpus for the Romanian language. In *Proceedings of the 18th BioNLP Workshop and Shared Task*, pages 71–79.

Maria Mitrofan and Vasile Păiș. 2022. Improving Romanian BioNER using a biologically inspired system. In *Proceedings of the 21st Workshop on Biomedical Language Processing*, pages 316–322, Dublin, Ireland. Association for Computational Linguistics.

Sunil Mohan and Donghui Li. 2019. Medmentions: A large biomedical corpus annotated with umls concepts.

Tomoko Ohta, Yuka Tateisi, and Jin-Dong Kim. 2002. The genia corpus: An annotated research abstract corpus in molecular biology domain. In *Proceedings of the Second International Conference on Human Language Technology Research*, HLT '02, page 82–86, San Francisco, CA, USA. Morgan Kaufmann Publishers Inc.

Vasile Păiş, Maria Mitrofan, Carol Luca Gasan, Vlad Coneschi, and Alexandru Ianov. 2021. Named entity recognition in the Romanian legal domain. In *Proceedings of the Natural Legal Language Processing Workshop 2021*, pages 9–18, Punta Cana, Dominican Republic. Association for Computational Linguistics.

Vasile Păiș. 2020. Multiple annotation pipelines inside the relate platform. In *The 15th International Conference on Linguistic Resources and Tools for Natural Language Processing*, pages 65–75.

Vasile Păiș. 2022. Racai at semeval-2022 task 11: Complex named entity recognition using a lateral inhibition mechanism. In *Proceedings of the 16th International Workshop on Semantic Evaluation (SemEval-2022)*, pages 1562–1569, Seattle, United States. Association for Computational Linguistics.

Vasile Păiș, Radu Ion, and Dan Tufiș. 2020. A processing platform relating data and tools for Romanian language. In *Proceedings of the 1st International Workshop on Language Technology Platforms*, pages 81–88, Marseille, France. European Language Resources Association.

Vasile Păiș and Maria Mitrofan. 2021. Assessing multiple word embeddings for named entity recognition of professions and occupations in health-related social media. In *Proceedings of the Sixth Social Media Mining for Health (#SMM4H) Workshop and Shared Task*, pages 128–130, Mexico City, Mexico. Association for Computational Linguistics.

Vasile Păiș, Maria Mitrofan, Verginica Barbu-Mititelu, Elena Irimia, Carol Luca Gasan, Roxana Micu, Laura Marin, Maria Dicusar, Bianca Florea, and Ana Badila. 2022a. Romanian micro-blogging named entity recognition (MicroBloggingNERo).

Vasile Păiș, Maria Mitrofan, Verginica Barbu Mititelu, Elena Irimia, Roxana Micu, and Carol Luca Gasan. 2022b. Challenges in creating a representative corpus of Romanian micro-blogging text. In *Proceedings of the Workshop on Challenges in the Management of Large Corpora (CMLC-10)*, pages 1–7, Marseille, France. European Language Resources Association.

Vasile Păiș, Maria Mitrofan, Carol Luca Gasan, Alexandru Ianov, Corvin Ghiță, Vlad Silviu Coneschi, and Andrei Onuț. 2021. Romanian Named Entity Recognition in the Legal domain (LegalNERo).

Vasile Păiș and Dan Tufis. 2018. Computing distributed representations of words using the CoRoLa corpus. *Proceedings of the Romanian Academy Series A - Mathematics Physics Technical Sciences Information Science*, 19(2):185–191.

Vasile Păiș, Dan Tufiș, and Radu Ion. 2019. Integration of Romanian NLP tools into the RELATE platform. In *International Conference on Linguistic Resources and Tools for Natural Language Processing*, pages 181–192.

Kirk Roberts, Michael A. Roach, Joseph Johnson, Josh Guthrie, and Sanda M. Harabagiu. 2012. EmpaTweet: Annotating and detecting emotions on Twitter. In *Proceedings of the Eighth International Conference on Language Resources and Evaluation (LREC'12)*, pages 3806–3813, Istanbul, Turkey. European Language Resources Association (ELRA).

Manuela Sanguinetti, Fabio Poletto, Cristina Bosco, Viviana Patti, and Marco Stranisci. 2018. An Italian Twitter corpus of hate speech against immigrants. In *Proceedings of the Eleventh International Conference on Language Resources and Evaluation (LREC 2018)*, Miyazaki, Japan. European Language Resources Association (ELRA).

Sergio Santamaría Carrasco and Roberto Cuervo Rosillo. 2021. Word embeddings, cosine similarity and deep learning for identification of professions & occupations in health-related social media. In *Proceedings of the Sixth Social Media Mining for Health (#SMM4H) Workshop and Shared Task*, pages 74–76, Mexico City, Mexico. Association for Computational Linguistics.

Kyle Shaffer. 2021. Language clustering for multilingual named entity recognition. In *Findings of the Association for Computational Linguistics: EMNLP 2021*, pages 40–45, Punta Cana, Dominican Republic. Association for Computational Linguistics.

Rolly Sharma, Archana Verma, Reena Grover, Digvijay Pandey, Binay Pandey, and Lavanya A. 2020. Microbloging as a corpus for sentiment analysis structure and feeling mining. *Journal of Xi'an Shiyou University, Natural Science Edition*, pages 229–234.

Marek Suppa and Ondrej Jariabka. 2021. Benchmarking pre-trained language models for multilingual NER: TraSpaS at the BSNLP2021 shared task. In *Proceedings of the 8th Workshop on Balto-Slavic Natural Language Processing*, pages 105–114, Kiyv, Ukraine. Association for Computational Linguistics.

Dan Tufiș, Verginica Barbu Mititelu, Elena Irimia, Vasile Păiș, Radu Ion, Nils Diewald, Maria Mitrofan, and Onofrei Mihaela. 2019. Little strokes fell great oaks. creating CoRoLa, the reference corpus of contemporary Romanian. *Revue Roumaine de Linguistique*, 64(3):227–240.

Longyue Wang, Derek F. Wong, Lidia S. Chao, and Junwen Xing. 2012. CRFs-based Chinese word segmentation for micro-blog with small-scale data. In *Proceedings of the Second CIPS-SIGHAN Joint Conference on Chinese Language Processing*, pages 51–57, Tianjin, China. Association for Computational Linguistics.

N.M. White. 2022. The hmong medical corpus: a biomedical corpus for a minority language. *Language Resources and Evaluation*.

Ayman Al Zaatari, Rim El Ballouli, Shady ELbassouni, Wassim El-Hajj, Hazem Hajj, Khaled Shaban, Nizar Habash, and Emad Yahya. 2016. Arabic corpora for credibility analysis. In *Proceedings of the Tenth International Conference on Language Resources and Evaluation (LREC'16)*, pages 4396–4401, Portorož, Slovenia. European Language Resources Association (ELRA).

The Best of Both Worlds: Combining Engineered Features with Transformers for Improved Mental Health Prediction from Reddit Posts

Sourabh Zanwar
RWTH Aachen University
sourabh.zanwar@rwth-aachen.de

Daniel Wiechmann
University of Amsterdam
d.wiechmann@uva.nl

Yu Qiao
RWTH Aachen University
yu.qiao@rwth-aachen.de

Elma Kerz
RWTH Aachen University
elma.kerz@ifaar.rwth-aachen.de

Abstract

In recent years, there has been increasing interest in the application of natural language processing and machine learning techniques to the detection of mental health conditions (MHC) based on social media data. In this paper, we aim to improve the state-of-the-art (SotA) detection of six MHC in Reddit posts in two ways: First, we built models leveraging Bidirectional Long Short-Term Memory (BLSTM) networks trained on in-text distributions of a comprehensive set of psycholinguistic features for more explainable MHC detection as compared to black-box solutions. Second, we combine these BLSTM models with Transformers to improve the prediction accuracy over SotA models. In addition, we uncover nuanced patterns of linguistic markers characteristic of specific MHC.

1 Introduction

The last decade has seen a surge in digital mental health research aimed at using linguistic cues from social media data to predict mental health conditions (MHC). The ultimate goal of this research branch is to enable people to lead healthy lives and to support healthcare professionals in the diagnosis and treatment of mental disorders. Data from social media platforms are particularly compelling to the research community because of their scale and deep embedding in contemporary culture. The computational research on such data has yielded new insights into population mental health and has shown promise for incorporating data-driven analytics into the treatment of psychiatric disorders (see Guntuku et al. 2017; Thieme et al. 2020 for recent general overviews of this research; see Chancellor and De Choudhury 2020; Harrigian et al. 2021 for reviews focusing on methods and data used for mental health status on social media). The current state of the art (SotA) in MHC detection from text data in social media is achieved by approaches that employ deep learning techniques that use pre-trained word embeddings, in particular Transformer-based models. At the same time, there are increasing calls to move away from such models toward explainable models for critical industries (Rudin, 2019; Loyola-Gonzalez, 2019), and the black-box nature of such models is increasingly being recognized as a major hurdle in the use of AI models in clinical practice (Mullenbach et al., 2018). Effective supporting the diagnosis and treatment of mental disorders therefore requires both accurate and interpretable models.

In this paper, we make the following contributions to the dynamic area of research on detecting mental health conditions in social media texts: First, we present attention-based BLSTM models that leverage the within-text distributions ('textual contours') of a large array of engineered language features to predict the presence of six mental health conditions (MHC) from user posts on Reddit. We show that these models can perform competitively with SotA models (drop in F1 $\leq 3\%$) for three of the six MHC, while maintaining explainability. Second, we demonstrate that the accuracy of MHC detection can be increased considerably by integrating these models with Transformer-based models. In addition, we uncover nuanced patterns of linguistic markers characteristic of specific MHC.

2 Related work

In this section, we focus on the state-of-the-art approaches to MHC detection on the two Reddit datasets used in this paper, i.e SMHD (Cohan et al., 2018) and Dreaddit (Turcan and McKeown, 2019) (for details see Section 3.1). On the SMHD dataset the current state-of-the-art approach is based on a Hierarchical Attention Network (HAN) (Sekulic and Strube, 2019). The proposed model uses two layers of bidirectional GRU units with hidden size of 150, each of them followed by a 100 dimensional attention mechanism. The first layer was set up to encode posts, whereas the second one encodes

Proceedings of the 29th International Conference on Computational Linguistic, pages 197–202
October 12–17, 2022.

users in terms of a sequence of encoded posts. The input layer was initialized with 300 dimensional GloVe word embeddings (Pennington et al., 2014). The output layer is 50-dimensional fully connected network, with binary cross entropy as a loss function. Sekulic and Strube (2019) perform experiments with the number of posts per user available to the model (50, 100, 150, 200, 250) to examine the amount of data needed for reasonable performance. Their best-performing model achieved an average of 67.56% F1 across the five MHC. On the Dreaddit dataset the current state-of-the-art approach is based on a 'emotion-infused' BERT model (Turcan et al., 2021). In the best-performing model presented in the paper, the BERT representation was first augmented with emotion knowledge by fine-tuning it on the GoEmotions dataset (Demszky et al., 2020) before applying it to the stress detection task. The model reached 80.25% F1 in binary stress classification, performing slightly better than a BERT baseline model. It is interesting to note that both SotA papers highlight the importance of explainability in MHC detection and suggest ways to increase the interpretability of the predictions of their models: Sekulic and Strube (2019) perform attention weights analyses to identify posts, and words or phrases in those posts, that are relevant for classification, whereas Turcan et al. (2021) use LIME to identify the most important words used their models for stress classification and mapping these to LIWC categories (Pennebaker et al., 2015).

3 Experimental setup

3.1 Datasets

The data for this work comes from two recent corpora used for the detection of MHC: (1) the Self-Reported Mental Health Diagnoses (SMHD) dataset (Cohan et al., 2018) and (2) the Dreaddit dataset (Turcan and McKeown, 2019). Both SMHD and Dreaddit were constructed data from Reddit, a social media platform consisting of individual topic communities called subreddits, including those relevant to MHC detection. The length of Reddit posts makes them a particularly valuable resource, as it allows modeling of the distribution of linguistic features in the text (see the concept of 'text contours' in Section 3.2).

SMHD is a large dataset of social media posts from users with nine mental health conditions (MHC) corresponding to branches in the DSM-5 (APA, 2013), an authoritative taxonomy for psychiatric diagnoses. User-level MHC labels were obtained through carefully designed distantly supervised labeling processes based on diagnosis pattern matching. The pattern matching leveraged a seed list of diagnosis keywords collected from the corresponding DSM-5 headings and extended by synonym mappings. To prevent that target labels can be easily inferred from the presence of MHC indicating words/phrases in the posts, all posts made to mental health-related subreddits or containing keywords related to a mental health condition were removed from the diagnosed users' data.

Table 1: Datasets statistics (number of posts, means and standard deviations of post length (in words) across mental health conditions and control groups.

MHC	Dataset	N posts	M length	SD
Stress	Dreaddit	1857	91	35
Control	Dreaddit	1696	83.6	29.7
ADHD	SMHD	1849	91.4	57
Anxiety	SMHD	1846	91.7	56.3
Bipolar	SMHD	1848	93	57.7
Depression	SMHD	1846	92.4	58.7
PTSD	SMHD	1600	95.7	59.9
Control	SMHD	1805	78.8	48.6

Dreaddit is a dataset of lengthy social media posts from subreddits in five domains that include stressful and non-stressful text. For a subset of 3.5k users employed in this paper, binary labels (+/- stressful) were obtained from aggregated ratings of five crowdsourced human annotators.

Based on these two corpora, we constructed a dataset with the goal of obtaining sub-corpora of equal size for the six MHCs targeted in this paper. To this end, we downsampled SMHD to match the size of Dreaddit and to be balanced in terms of class distributions. The sampling procedure from the SMHD dataset was such that each post was produced by a distinct user. In doing so, we addressed a concerning trend described in recent review articles that points to the presence of a relatively small number of unique individuals, which may hinder the generalization of models to platforms that are already demographically skewed (Chancellor and De Choudhury, 2020; Harrigian et al., 2021). These constraints were met for five of the nine MHC in the SMHD dataset (ADHD, anxiety, bipolar, depression, PTSD). Statistics for these datasets are presented in Table 1.

3.2 Measurement of text contours of psycholinguistic features

A set of 435 psycholinguistic features used in our approach fall into five broad categories: (1) features of morpho-syntactic complexity (N=19), (2) features of lexical richness, diversity and sophistication (N=52), (3) register-based n-gram frequency features (N=25), (4) readability features (N=14), and (5) lexicon features designed to detect sentiment, emotion and/or affect (N=325). An overview of these features can be found in Table 3 in supplementary material. All measurements of these features were obtained using an automated text analysis system that employs a sliding window technique to compute sentence-level measurements. These measurements capture the within-text distributions of scores for a given psycholinguistic feature, referred to here as 'text contours' (for its recent applications, see e.g. Wiechmann et al. (2022) for predicting eye-moving patterns during reading and Kerz et al. (2022) for detection of Big Five personality traits and Myers–Briggs types). A visualization of these text contours is shown in Figure 1 in Supplementary material. Tokenization, sentence splitting, part-of-speech tagging, lemmatization and syntactic PCFG parsing were performed using Stanford CoreNLP (Manning et al., 2014).

4 Modeling approach

We conducted experiments with a total of five models: (1) a fine-tuned BERT model (Devlin et al., 2019), (2) a fine-tuned RoBERTa model (Zhuang et al., 2021), (3) a bidirectional long short-term memory (BLSTM) classifier trained on measurements of psycholinguistic features described in Section 3.2, and (4) and (5) two hybrid models integrating BERT and RoBERTa predictions with the psycholinguistic features. For (1) and (2) we used the pretrained 'bert-base-uncased' and 'roberta-base' models from the Huggingface Transformers library (Wolf et al., 2020) each with an intermediate BLSTM layer with 256 hidden units (Al-Omari et al., 2020). For (3) - the model based solely on psycholinguistic features, we constructed a 4-layer BLSTM with a hidden state dimension of 1024. The input to that model is a sequence $CM_1^N = (CM_1, CM_2 \ldots, CM_N)$, where CM_i, the output of CoCoGen for the ith sentence of a post, is a 435 dimensional vector and N is the sequence length. To predict the labels of a sequence, we concatenate the last hidden states of the last layer in forward ($\overrightarrow{h_n}$) and backward directions ($\overleftarrow{h_n}$). The result vector of concatenation $h_n = [\overrightarrow{h_n}|\overleftarrow{h_n}]$ is then transformed through a 2-layer feedforward neural network, whose activation function is Rectifier Linear Unit (Agarap, 2018). The output of this is then passed to a Fully Connected Layer FC with ReLu activation function and dropout of 0.2 and it is finally fed to a final FC layer. The output is finally passed through sigmoid function and finally a threshold is used to determine the labels. We trained these models for 100 epochs, with a batch size of 256 and a sequence length of 5. The architecture of the hybrid classification models - models (4) and (5) - consists of two parts: (i) a pre-trained Transformer-based model with a BLSTM layer and FC layer on top of it and (ii) the psycholinguistic features of the text fed into a BLSTM network and a subsequent FC layer. The FC layers of both parts take the concatenation of last hidden states of the last BLSTM layer in forward and backward direction. We concatenate the outputs of these layers before finally feeding them into a final FC layer with a sigmoid activation function. The model used to generate predictions for the test set was the RoBERTa-PsyLing hybrid model with the following configuration: 2-layer BLSTM, 256 hidden units and a dropout of 0.2; BLSTM-PsyLing: 3-layers, hidden size of 512 and dropout 0.2. We trained this model for 12 epochs, saving the model with the best performance (F1-Score) on the development set. The optimizer used is AdamW with a learning rate of 2e-5 and a weight decay of 1e-4. Structure diagrams of the model based solely on psycholinguistic features and the hybrid architectures are presented in Figures 2 and 3 in supplementary material. All models were trained using 5-fold CV of the training data as base classifiers and model stacking was performed using logistic regression as a meta-learner to adaptively combine the outputs of the base classifiers.

5 Results and Discussion

An overview of the results of our models in comparison to those reported in the previous studies reviewed above is presented in Table 2. The results show that our Psyling-BLSTM models outperform the Hierarchical Attention Networks that are based on 50 concatenated posts (HAN 50) across all five self-reported diagnosed MHC, with an average increase in performance of over 12% F1. This is significant considering that our Psyling-BLSTM models were trained in a much more challenging

Table 2: F1 scores averaged over five runs of the binary classification models across MHC.

	Stress	ADHD	Anxiety	Bipolar	Depression	PTSD
HAN (50) (Sekulic and Strube, 2019)	-	48	38	52	49	56
HAN (250) (Sekulic and Strube, 2019)	-	64.27	69.24	67.42	68.28	68.59
BERT (Turcan et al., 2021)	78.88	-	-	-	-	-
BERT-Emotion (Turcan et al., 2021)	80.25	-	-	-	-	-
BERT	76.24	60.82	64.52	65.11	62.91	69.40
RoBERTa	81.13	61.76	67.79	66.12	65.93	72.32
Psyling-BLSTM	70.20	58.2	61.19	61.19	59.54	65.18
Psyling-BLSTM-BERT	79.46	61.09	67.47	67.44	66.90	73.32
Psyling-BLSTM-RoBERTa	83.32	62.37	68.91	67.85	67.16	73.85

setting, namely the detection of the presence of an MHC based on a single post with an average length of less than 100 words. The Psyling-BLSTM models show a relative performance drop of -6.5% compared to a HAN that was trained on 250 times as much textual data. These results demonstrate that the Psyling-BLSTM approach is much more data-efficient than the HAN approach. Furthermore, our results show that the Psyling-BLSTM models are surpassed by the Transformer-based models, with BERT achieving on average a 3.9% higher F1 and RoBERTa achieving 6.6% higher F1. Importantly, the Psyling-BLSTM models perform competitively with BERT (drop in F1 $\approx 3\%$) for three of the six MHC (ADHD, anxiety, depression), indicating that strong detection models can be constructed for at least these MHC without compromising explainability. Another key finding is that the combination of Psyling-BLSTM models with Transformers consistently yielded the highest prediction accuracy for all six MHC, with the RoBERTa hybrid models outperforming the BERT hybrid models in all cases. The highest prediction accuracy was achieved in stress detection, with a +3.06% F1 improvement over the previous SotA, the emotion-loaded BERT model presented in Turcan et al. (2021). For diagnosed MHC, the RoBERTa hybrid showed a robust increase in F1 over a Transformer-only model by an average of 1.24%. For the BERT-based case, the improvement in F1 averaged +3.2% for stress and +2.7% for self-reported diagnosed MHC.

To further investigate the differences in the linguistic markers among MHC, we ran one-way ANOVAs comparing the features across the six MHC and the control group for each of the 435 linguistic features. Prior to being entered in the models, all feature scores were z-score normalized and Bonferroni correction was used to correct for family-wise type I errors. Follow up Tukey's HSD post-hoc test were run to measure the differences between all MHC pairs. The results revealed significant differences across groups for 337 of the linguistic features (see Table **??** in the appendix). We focused on the linguistic features with the greatest variance among the groups, as indicated by the highest F-statistics (N=68), and conducted hierarchical agglomerative cluster analyses over MHC and linguistic features to derive the linguistic profiles associated with the six MHC groups and the control group. The analyses show that the control group is distinctly separated from all MHC. Furthermore, while each of the six MHC were characterized by distinct patterns of language use, the linguistic markers of stress were clearly separated from those of the five self-reported diagnosed MHC. In light of space limitations, we present here only some selected general patterns (for details, see Figure 4 and Table 7 in supplementary material). For example, there were persistent differences between stressful and non-stressful Reddit posts, with stressful posts being generally characterized by much higher proportions of negative words and words related to sadness, fear, anxiety and anger. The language use of Reddit users diagnosed with depression was characterized by higher proportions of self-referencing words and lower lexical sophistication. Posts from users diagnosed with PTSD had significantly longer mean sentence lengths, greater syntactic complexity and lower readability, relative to controls, while the Reddit posts of users diagnosed with anxiety were characterized by high proportions of regular verbs and those of users diagnosed with bipolar disorder showed strong reliance on words signaling indifference and lower lexical diversity. Overall, our results show that mental health prediction from textual data benefits from the advancement of methods aimed at integrating Transformers with comprehensive sets of open- and closed-vocabulary features and general features of linguistic complexity and style.

References

Abien Fred Agarap. 2018. Deep learning using rectified linear units (relu). *arXiv preprint arXiv:1803.08375.*

Hani Al-Omari, Malak A. Abdullah, and Samira Shaikh. 2020. EmoDet2: Emotion detection in english textual dialogue using bert and bilstm models. In *2020 11th International Conference on Information and Communication Systems (ICICS)*, pages 226–232.

APA. 2013. Diagnostic and statistical manual of mental disorders. *American Psychiatric Association*, 21(21):591–643.

Stevie Chancellor and Munmun De Choudhury. 2020. Methods in predictive techniques for mental health status on social media: a critical review. *NPJ digital medicine*, 3(1):1–11.

Arman Cohan, Bart Desmet, Andrew Yates, Luca Soldaini, Sean MacAvaney, and Nazli Goharian. 2018. SMHD: a large-scale resource for exploring online language usage for multiple mental health conditions. In *Proceedings of the 27th International Conference on Computational Linguistics*, Santa Fe, New Mexico, USA. Association for Computational Linguistics.

Dorottya Demszky, Dana Movshovitz-Attias, Jeongwoo Ko, Alan Cowen, Gaurav Nemade, and Sujith Ravi. 2020. GoEmotions: A dataset of fine-grained emotions. In *Proceedings of the 58th Annual Meeting of the Association for Computational Linguistics*, Online. Association for Computational Linguistics.

Jacob Devlin, Ming-Wei Chang, Kenton Lee, and Kristina Toutanova. 2019. BERT: Pre-training of deep bidirectional transformers for language understanding. In *Proceedings of the 2019 Conference of the North American Chapter of the Association for Computational Linguistics: Human Language Technologies, Volume 1 (Long and Short Papers)*, Minneapolis, Minnesota. Association for Computational Linguistics.

Sharath Chandra Guntuku, David B Yaden, Margaret L Kern, Lyle H Ungar, and Johannes C Eichstaedt. 2017. Detecting depression and mental illness on social media: an integrative review. *Current Opinion in Behavioral Sciences*, 18:43–49.

Keith Harrigian, Carlos Aguirre, and Mark Dredze. 2021. On the state of social media data for mental health research. In *Proceedings of the Seventh Workshop on Computational Linguistics and Clinical Psychology: Improving Access*, Online. Association for Computational Linguistics.

Elma Kerz, Yu Qiao, Sourabh Zanwar, and Daniel Wiechmann. 2022. Pushing on personality detection from verbal behavior: A transformer meets text contours of psycholinguistic features. In *Proceedings of the 12th Workshop on Computational Approaches to Subjectivity, Sentiment & Social Media Analysis*, Dublin, Ireland. Association for Computational Linguistics.

Octavio Loyola-Gonzalez. 2019. Black-box vs. white-box: Understanding their advantages and weaknesses from a practical point of view. *IEEE Access*, 7:154096–154113.

Christopher Manning, Mihai Surdeanu, John Bauer, Jenny Finkel, Steven Bethard, and David McClosky. 2014. The stanford corenlp natural language processing toolkit. In *Proceedings of 52nd annual meeting of the association for computational linguistics: system demonstrations*, pages 55–60.

James Mullenbach, Sarah Wiegreffe, Jon Duke, Jimeng Sun, and Jacob Eisenstein. 2018. Explainable prediction of medical codes from clinical text. In *Proceedings of the 2018 Conference of the North American Chapter of the Association for Computational Linguistics: Human Language Technologies, Volume 1 (Long Papers)*, pages 1101–1111, New Orleans, Louisiana. Association for Computational Linguistics.

James W Pennebaker, Ryan L Boyd, Kayla Jordan, and Kate Blackburn. 2015. The development and psychometric properties of LIWC2015. Technical report.

Jeffrey Pennington, Richard Socher, and Christopher D Manning. 2014. Glove: Global vectors for word representation. In *Proceedings of the 2014 conference on empirical methods in natural language processing (EMNLP)*, pages 1532–1543.

Cynthia Rudin. 2019. Stop explaining black box machine learning models for high stakes decisions and use interpretable models instead. *Nature Machine Intelligence*, 1(5):206–215.

Ivan Sekulic and Michael Strube. 2019. Adapting deep learning methods for mental health prediction on social media. In *Proceedings of the 5th Workshop on Noisy User-generated Text (W-NUT 2019)*. Association for Computational Linguistics.

Anja Thieme, Danielle Belgrave, and Gavin Doherty. 2020. Machine learning in mental health: A systematic review of the hci literature to support the development of effective and implementable ml systems. *ACM Transactions on Computer-Human Interaction (TOCHI)*, 27(5):1–53.

Elsbeth Turcan and Kathy McKeown. 2019. Dreaddit: A Reddit dataset for stress analysis in social media. In *Proceedings of the Tenth International Workshop on Health Text Mining and Information Analysis (LOUHI 2019)*, pages 97–107. Association for Computational Linguistics.

Elsbeth Turcan, Smaranda Muresan, and Kathleen McKeown. 2021. Emotion-infused models for explainable psychological stress detection. In *Proceedings of the 2021 Conference of the North American Chapter of the Association for Computational Linguistics: Human Language Technologies*, pages 2895–2909.

Daniel Wiechmann, Yu Qiao, Elma Kerz, and Justus Mattern. 2022. Measuring the impact of (psycho-)linguistic and readability features and their spill over effects on the prediction of eye movement patterns. In *Proceedings of the 60th Annual Meeting of the Association for Computational Linguistics (Volume 1: Long Papers)*, Dublin, Ireland. Association for Computational Linguistics.

Thomas Wolf, Julien Chaumond, Lysandre Debut, Victor Sanh, Clement Delangue, Anthony Moi, Pierric Cistac, Morgan Funtowicz, Joe Davison, Sam Shleifer, et al. 2020. Transformers: State-of-the-art natural language processing. In *Proceedings of the 2020 Conference on Empirical Methods in Natural Language Processing: System Demonstrations*, pages 38–45.

Liu Zhuang, Lin Wayne, Shi Ya, and Zhao Jun. 2021. A robustly optimized BERT pre-training approach with post-training. In *Proceedings of the 20th Chinese National Conference on Computational Linguistics*, Huhhot, China. Chinese Information Processing Society of China.

A Supplementary material

Supplementary material can be found here https://bit.ly/3LgGYla.

Leveraging Social Media as a Source for Clinical Guidelines: A Demarcation of Experiential Knowledge

Jia-Zhen Chan j.m.chan@student.vu.nl

Florian Kunneman **Roser Morante** {f.a.kunneman,r.morantevallejo}@vu.nl

Lea Lösch and **Teun Zuiderent-Jerak**
{lea.loesch,teun.zuiderent-jerak}@vu.nl
Vrije Universiteit Amsterdam

Abstract

In this paper we present a procedure to extract posts that contain experiential knowledge from Facebook discussions in Dutch, using automated filtering, manual annotations and machine learning. We define guidelines to annotate experiential knowledge and test them on a subset of the data. After several rounds of (re-)annotations, we come to an inter-annotator agreement of $K = 0.69$, which reflects the difficulty of the task. We subsequently discuss inclusion and exclusion criteria to cope with the diversity of manifestations of experiential knowledge relevant to guideline development.

1 Introduction

Messages shared on social media platforms are widely acknowledged as a source to gain insights into current events and the related vox populi. Recently, this source of information has been extensively leveraged for identifying topics of interest (Kellert and Mahmud Uz Zaman, 2022), emotions (Aduragba et al., 2022), and stances (Hossain et al., 2020; Glandt et al., 2021) during the COVID-19 pandemic. A lesser scrutinized type of information is experiential knowledge in the medical domain: messages that describe practical contact with diseases or treatments (e.g.: having been contaminated, having received a vaccination), which may include contextual information that is relevant to the experience (e.g.: being a doctor, having certain allergies). Such information is valuable for clinical guideline development, where social media may provide insights into latent experiences that are relevant to sharpen guidelines on, for example, exemptions and communication to patients.

In this paper we discuss the task of automatically identifying experiences in Facebook messages in Dutch. As an input for clinical guidelines, experiences on social media are most valuable when they are provided with context and accompanied with value considerations. Discussions following news reports on dedicated Facebook pages are a relevant source in that respect, since these messages tend to be more extended than, for example, Twitter messages, and may express personal disclosures as well. There are, however, two prominent challenges to identifying experiences from these discussions. First, only a small part of these messages share an experience, making it challenging to locate examples to train and test a machine learning classifier on. Second, experiences take a diversity of manifestations, posing difficulties to merely reach a sufficient agreement among humans. This is also reported in several studies that aim to identify experiences in Twitter (Sarker et al., 2020) and patient fora (Liu et al., 2015). In this paper we present our approach to these challenges, contributing an iterative procedure to zoom in on experiences within Dutch Facebook discussions on the topic of COVID-19 vaccinations, as well as an annotation guideline catered for annotating experiences as input for vaccination guidelines.

2 Related work

The most common aim for targeting experiential information from social media messages is to inquire into unforeseen effects of medicine intake (Alvaro et al., 2015; Liu et al., 2015; Cavazos-Rehg et al., 2016; Klein et al., 2017; Oostdijk et al., 2019; Kim et al., 2020; Sarker et al., 2020). Other aims are to gain insight into the experiences of particular patient groups (Stemmer et al., 2021) and to inquire into topics of discussions related to HPV vaccinations, where the sharing of experiences is identified as one of these topics (Surian et al., 2016). Studies that aim at identifying experiences mostly focus on Twitter as a data source, while user fora are leveraged to a lesser extent (Liu et al., 2015, Oostdijk et al., 2019). In contrast to these studies, we set out to identify experiences as input for rapid medical guideline development, and make use of Facebook discussions as a data source. While ex-

Proceedings of the 29th International Conference on Computational Linguistic, pages 203–208
October 12–17, 2022.

periences shared on social media have not been used before as evidence for clinical guidelines, automated approaches have been applied to swiftly identify scientific papers on a particular medical context to facilitate rapid clinical guideline development (Whittington et al., 2019).

In alignment with most of the studies that target experiences, we deployed manual annotations to come to a ground truth and encountered difficulties to reach a sufficient inter-annotator agreement on this task. Klein et al. (2017) ascribe these difficulties to the wide range of linguistic patterns by which experiences are manifested. We follow their solution to develop an annotation guideline that includes the different variants. In line with Alvaro et al. (2015), we discuss the particular dilemma's that pose difficulties to decide on an annotation.

A hampering factor to our endeavour to gain insight into experiences related to COVID-19 vaccinations, is that the topic of vaccination is highly debated and therefore dominated by opinionated posts rather than experiences. Participants in the debate express strong attitudes in favor and against vaccines, discussing issues related to the safety of vaccinations, the side effects, and the moral aspects of enforcing mandatory vaccination (Wolfe and Sharp, 2002; Mollema et al., 2015). Nowadays, with collaborative media, anyone can join in a discussion and share information and opinions. This makes it difficult to attest the reliability of on-line content (Zummo, 2017). Doubts about the reliability of vaccines can easily grow among the population under the influence of negative information about vaccines or unbalanced reports of vaccine risk (Betsch and Sachse, 2012).

Knowing what people think about vaccines and why people decide to vaccinate or not has always been interesting for health practitioners who need to be adequately informed on public perception of the safety and necessity of vaccines. There have been efforts aimed at annotating data about the vaccination debate (Morante et al., 2020; Torsi and Morante, 2018) and at automatically finding opinions in favour or against vaccines (Kunneman et al., 2020; Chen and Crooks, 2022; Cascini et al., 2022).

3 Identifying experiences from Facebook discussions

Our study sets off from a set of Facebook posts discussing news articles related to COVID-19 vaccinations, without a clear notion of the frequency and characteristics of messages that share an experience in these discussions. Next we describe the iterative procedure that we conducted to identify experiential messages and collect enough examples to train a machine learning classifier on. An overview of the stages of this procedure and samples used is presented in Table 1.[1]

3.1 Data

In total we collected 230,863 Facebook comments on news articles about COVID-19 vaccination posted by the four Dutch news outlets (NOS, NU.nl, Telegraaf and NRC) on their public Facebook pages between 01.12.2020 and 08.06.2021. Relevant articles from the respective news outlets were located using the following query terms "(Corona* OR COVID* OR COVID-19) AND (vaccinatie* OR vaccin OR inenten [Dutch for 'vaccinate'] OR prik [Dutch for 'shot'])". Using the Facebook Pages API[2], the post ID, the comment text, the posting date and the number of likes and comments were retrieved.

3.2 Initial data annotation

We developed an initial three stage rudimentary filter to locate the parts of the data that are likely to contain experiences. First, considering that experiences tend to be lengthy to describe, only comments exceeding 250 characters were kept. Second and third, accounts of experiences tend to show a higher degree of subjectivity and sentiment polarity in their wording. In order to select comments with a higher degree of subjectivity and sentiment, a sentiment analysis was carried out with the python package Pattern (De Smedt and Daelemans, 2012).[3] All comments were thereby assigned a sentiment value between -1 (negative) and +1 (positive) and a value between 0 (objective) and 1 (subjective) indicating their degree of subjectivity. Comments were retained if they had an above-average subjectivity score ($\geq$ 0.4) and a sentiment score of $\leq$ -0.25 or $\geq$ 0.25.

The initial filtering step resulted in a set of 5,702 comments. A random sample of 500 comments was then coded independently by two researchers on the presence of experiential knowledge. The definition of experience-based knowledge related

[1] Ids of the messages and individual annotations will be made available.

[2] `https://developers.facebook.com/docs/pages/`

[3] `https://github.com/clips/pattern`

Data source	Size	Part of	Criteria	IAA
Complete	230,863			
Filter 1	5,702	Complete	Message size High subjectivity	
Sample 1	500	Filter 1		0.66
Filter 2	119	Filter 1	'My' + [family member]	0.60
Training	70,830	Complete		
Test	49,034	Complete	Disjoint from training	
Filter 3	1,258	Training	'My' + [family member] 'have...had' hypernym to 'feel'	
Sample 2	500	Test	> 0.90 classifier confidence (250 posts) > 0.50 < 0.90 classifier confidence (250 posts)	0.59
Sample 3	500	Test	> 0.90 classifier confidence (250 posts) > 0.50 < 0.90 classifier confidence (250 posts)	0.56[i] 0.69[ii]

Table 1: Overview of the samples that were drawn in the process of identifying experiential messages from Facebook discussions on COVID-19 vaccinations.Inter-annotator agreement (IAA) is measured in Cohen's Kappa. [i] First round. [ii]Second round.

to COVID-19 vaccination in comments was kept fairly broad, including both first-hand and second-hand experience. Our understanding is closest to the definition of "experience" from The Oxford Pocket Dictionary of Current English: "practical contact with and observation of facts or events". Whereas multiple experiences could be mentioned in a single post, the annotation target was only to assess whether at least one experience was mentioned.

The annotation effort indicated that 10% of these comments contained experiential knowledge, at a moderate inter-annotator agreement (0.66 Cohen's Kappa). We analysed the comments coded as "experience" to find specific features characteristic of these comments. The observation that these descriptions often contained references to a close contact was incorporated into the filter by additionally filtering for comments that included the combination of the words "my" and a close contact, such as a family member, e.g. "my grandma". This search resulted in 119 comments of which about 77% indicated experiential knowledge, which was confirmed by another round of annotation by the same two annotators who reached a moderate inter-annotator agreement (0.60 Cohen's Kappa).

To obtain a higher number of posts to train a machine learning classifier on, we expanded the latter filter rule with two additional filter rules and applied the filter to a sample of 70,830 Facebook posts (not complying with the initial filter based on length, subjectivity and sentiment). The additional filter rules were 1) including the pattern 'have ... had' (targeting expressions of having had COVID-19 or another disease) and 2) including a first- or second-level hypernym of the verb 'feel' (from the Open Dutch Wordnet (Postma et al., 2016)). This resulted in 1,258 (distantly supervised) examples of experiences as input for machine learning.

3.3 Experience modelling

We trained a machine learning classifier on the training set featuring 70,830 messages of which 1,258 were labeled as experience in a distantly supervised way. The performance of different algorithms and feature weightings was then compared by applying them to the 500 annotated posts in the previous phase, using Precision, Recall, F1 and Area under the ROC curve as evaluation metrics.

As preprocessing, the posts were cleaned of URLs, emojis, punctuation, numbers and symbols, which we did not consider predictive for identifying formulations of experiences, and the remaining tokens were lowercased. The cleaned and normalized texts were tokenized and lemmatized by means of the Dutch Stanza pipeline (Qi et al., 2020). We experimented with two different algorithms, Logistic Regression and XGBoost, which are common but distinct approaches to supervised machine learning, have yielded good results on several NLP tasks and lead to interpretable models. We also compared to feature weightings, tf*idf and binary. As a baseline we applied the three filters in a rule-based manner

System	P	R	F1	AUC
Baseline	0.32	0.14	0.20	
XGBoost (binary weighting)	0.39	0.59	0.47	0.74
XGBoost (tf*idf weighting)	0.43	0.53	0.47	0.73
Logistic Regression (binary weighting)	0.37	0.20	0.26	0.58
Logistic Regression (tf*idf weighting)	0.38	0.45	0.41	0.68

Table 2: Results on classifying Facebook posts sharing an experience in an annotated sample of 500 posts with a support of 49 posts labeled as experience (P=Precision, R=Recall, AUC=Area under the ROC Curve).

– if any of the given patterns matched the text in a post, the post was labeled as experience.

The outcomes of the machine learning experiment are presented in Table 2. The optimal machine learning set-up was an XGBoost classifier using a binary weighted lemma's representation, yielding an F1-score of 0.47 on predicting experiences in the annotated sample of 500 comments, considerably outperforming a rule-based filter (F1-score of 0.20). The significant difference in recall (0.59 vs. 0.14) is showing the generalisability of the machine learning approach. Note that the test set was considerably skewed with only 10% of the messages sharing an experience.

The best-performing classifier was subsequently used to identify experiences in a held-out set of 49,034 posts. From these classified posts, we extracted two samples to manually annotate for the number of experiences: a sample of 250 posts that were classified at a classifier confidence above 0.90, and a sample of 250 posts classified at a confidence between 0.50 and 0.90. This enabled us to inquire into the utility of the model in relation to classifier confidence when applied to unseen data. Based on the definition of experience used in the first annotation round the posts were coded by four annotators, who reached a slightly weak agreement (0.59 Cohen's Kappa).

To come to a higher agreement rate, the disagreements were discussed among the annotators and a set of inclusion and exclusion criteria were formulated in a more extensive annotation guideline. Three of the annotators then annotated an additional sample of again 500 posts (250 times at a classifier confidence above 0.90, 250 times between 0.50 and 0.90), surprisingly reaching an agreement of 0.56. They again discussed the disagreements, finding that some of the criteria in the annotation guidelines were leading to more confusion. Based on this discussion the guidelines were adapted, and the same set of posts were again annotated by the group of three. There were approximately 3 weeks in between the first and second round on these posts. A moderate agreement of 0.69 was reached on this set.

4 Annotation guideline and development

The annotation guideline including different manifestations of experiences and examples was written after the annotations of Sample 2 and refined after the annotations of Sample 3.[4] For reasons of space, we focus here on the cases that posed most difficulties to decide on an annotation:

- **Indirectness of experience**. Posts that describe a second-hand experience ('My mother had...') are arguably just as relevant as personal experiences. There is, however, a point where the described experience is too general to remain trustworthy ('I know of people who...'). We decided to exclude the most generic phrasing of experiences, but some cases in the annotation set still posed doubts ('A friend of my mother...').
- **Non-experiences**. Posts that state a non-experience ('I did not receive an invitation yet') were deemed relevant as well, since they may convey valuable information.
- **Personal disclosure**. A person may share information on group membership (working in a hospital) or having certain feelings ('I feel pressure to make a decision'), which may be of relevance as guideline input. Such personal disclosures are not strictly an experience, but we chose to include it.
- **Routines**. Some posts share a person's routine ('visiting my mother every 1,5 weeks), which are not experiences from a bounded period in the past, but do convey information about events that take place and may be of relevance.
- **Sources of information**. In a discussion on the topic of vaccinations, many posts discuss sources of information and their reliability. In some cases, obtaining information from a doctor or medical specialist is mentioned, which could be seen as an experience. We

[4]The final version of the guideline can be downloaded from the appendix: `https://surfdrive.surf.nl/files/index.php/s/r1UKVCWhAwVUo2P`

chose to exclude this as it relates too much to sources of information.

5 Conclusion

We set out to identify experiential knowledge from Facebook discussions in Dutch in relation to COVID-19 vaccinations, using filtering criteria, manual annotations and machine learning in order to broaden the filter and draw more samples for manual annotation. After several annotation rounds, we defined annotation guidelines for experiential knowledge. Annotators reached an agreement of 0.69 Cohen's Kappa, which indicates that the task is rather subjective with diverse manifestations of the targeted type of posts.

Messages on social media provide information that is difficult to obtain in other ways for rapid guideline development, with access to a diversity of backgrounds and experiences. Before being able to integrate this type of information, insight is needed into what separates utterances baring experiential information from other types of messages in Facebook discussions. The current endeavour to filter for such messages and discuss their characteristics has been a crucial step towards this point, but needs to be further refined to come to a machine learning classifier that yields a sufficient performance to be integrated in a hybrid workflow where a clinical guideline expert can further inspect the collected messages. Future work to come to this point includes experimenting with different algorithms, feature representations and distant supervision filters, and further refining the annotation guideline.

References

Olanrewaju Tahir Aduragba, Jialin Yu, Alexandra I. Cristea, and Lei Shi. 2022. Detecting fine-grained emotions on social media during major disease outbreaks: Health and well-being before and during the covid-19 pandemic. In *AMIA Annu Symp Proc. 2021*, pages 187–0–196. AMIA.

Nestor Alvaro, Mike Conway, Son Doan, Christoph Lofi, John Overington, and Nigel Collier. 2015. Crowdsourcing twitter annotations to identify first-hand experiences of prescription drug use. *Journal of biomedical informatics*, 58:280–287.

C. Betsch and K. Sachse. 2012. Dr. jekyll or mr. hyde? how the internet influences vaccination decisions: recent evidence and tentative guidelines for online vaccine communication. *Primary Care: Clinics in Office Practice*, 30(25):3723–3726.

Fidelia Cascini, Ana Pantovic, Yazan A. Al-Ajlouni, Giovanna Failla, Valeria Puleo, Andriy Melnyk, Alberto Lontano, and Walter Ricciardi. 2022. Social media and attitudes towards a covid-19 vaccination: A systematic review of the literature. *eClinical Medicine*, 48(101454):3723–3726.

Patricia A Cavazos-Rehg, Shaina J Sowles, Melissa J Krauss, Vivian Agbonavbare, Richard Grucza, and Laura Bierut. 2016. A content analysis of tweets about high-potency marijuana. *Drug and alcohol dependence*, 166:100–108.

Qingqing Chen and Andrew Crooks. 2022. Analyzing the vaccination debate in social media data pre- and post-covid-19 pandemic. *International Journal of Applied Earth Observation and Geoinformation*, 110:102783.

Tom De Smedt and Walter Daelemans. 2012. " vreselijk mooi!"(terribly beautiful): A subjectivity lexicon for dutch adjectives. In *LREC*, pages 3568–3572.

Kyle Glandt, Sarthak Khanal, Yingjie Li, Doina Caragea, and Cornelia Caragea. 2021. Stance detection in COVID-19 tweets. In *Proceedings of the 59th Annual Meeting of the Association for Computational Linguistics and the 11th International Joint Conference on Natural Language Processing (Volume 1: Long Papers)*, pages 1596–1611, Online. Association for Computational Linguistics.

Tamanna Hossain, Robert L. Logan IV, Arjuna Ugarte, Yoshitomo Matsubara, Sean Young, and Sameer Singh. 2020. COVIDLies: Detecting COVID-19 misinformation on social media. In *Proceedings of the 1st Workshop on NLP for COVID-19 (Part 2) at EMNLP 2020*, Online. Association for Computational Linguistics.

Olga Kellert and Md Mahmud Uz Zaman. 2022. Using neural topic models to track context shifts of words: a case study of COVID-related terms before and after the lockdown in April 2020. In *Proceedings of the 3rd Workshop on Computational Approaches to Historical Language Change*, pages 131–139, Dublin, Ireland. Association for Computational Linguistics.

Myeong Gyu Kim, Jungu Kim, Su Cheol Kim, and Jaegwon Jeong. 2020. Twitter analysis of the nonmedical use and side effects of methylphenidate: machine learning study. *Journal of medical Internet research*, 22(2):e16466.

Ari Klein, Abeed Sarker, Masoud Rouhizadeh, Karen O'Connor, and Graciela Gonzalez. 2017. Detecting personal medication intake in twitter: an annotated corpus and baseline classification system. In *BioNLP 2017*, pages 136–142.

Florian Kunneman, Mattijs Lambooij, Albert Wong, Antal van den Bosch, and Liesbeth Mollema. 2020. Monitoring stance towards vaccination in twitter messages. *BMC medical informatics and decision making*, 20(1):1–14.

Yunzhong Liu, Yi Chen, Jiliang Tang, and Huan Liu. 2015. Context-aware experience extraction from online health forums. In *2015 International Conference on Healthcare Informatics*, pages 42–47. IEEE.

Liesbeth Mollema, Irene Anhai Harmsen, Emma Broekhuizen, Rutger Clijnk, Hester De Melker, Theo Paulussen, Gerjo Kok, Robert Ruiter, and Enny Das. 2015. Disease detection or public opinion reflection? Content analysis of tweets, other social media, and online newspapers during the measles outbreak in The Netherlands in 2013. *Journal of medical Internet research*, 17(5).

Roser Morante, Chantal van Son, Isa Maks, and Piek Vossen. 2020. Annotating perspectives on vaccination. In *Proceedings of the 12th Language Resources and Evaluation Conference*, pages 4964–4973, Marseille, France. European Language Resources Association.

Nelleke HJ Oostdijk, Mattijs S Lambooij, Peter Beinema, Albert Wong, Florian A Kunneman, and Peter HJ Keizers. 2019. Fora fuelling the discovery of fortified dietary supplements–an exploratory study directed at monitoring the internet for contaminated food supplements based on the reported effects of their users. *Plos one*, 14(5):e0215858.

Marten Postma, Emiel van Miltenburg, Roxane Segers, Anneleen Schoen, and Piek Vossen. 2016. Open Dutch WordNet. In *Proceedings of the 8th Global WordNet Conference (GWC)*, pages 302–310, Bucharest, Romania. Global Wordnet Association.

Peng Qi, Yuhao Zhang, Yuhui Zhang, Jason Bolton, and Christopher D. Manning. 2020. Stanza: A Python natural language processing toolkit for many human languages. In *Proceedings of the 58th Annual Meeting of the Association for Computational Linguistics: System Demonstrations*.

Abeed Sarker, Sahithi Lakamana, Whitney Hogg-Bremer, Angel Xie, Mohammed Ali Al-Garadi, and Yuan-Chi Yang. 2020. Self-reported COVID-19 symptoms on Twitter: an analysis and a research resource. *Journal of the American Medical Informatics Association*, 27(8):1310–1315.

Maya Stemmer, Yisrael Parmet, and Gilad Ravid. 2021. What are ibd patients talking about on twitter? In *International Conference on ICT for Health, Accessibility and Wellbeing*, pages 206–220. Springer.

Didi Surian, Dat Quoc Nguyen, Georgina Kennedy, Mark Johnson, Enrico Coiera, Adam G Dunn, et al. 2016. Characterizing twitter discussions about hpv vaccines using topic modeling and community detection. *Journal of medical Internet research*, 18(8):e6045.

Benedetta Torsi and Roser Morante. 2018. Annotating claims in the vaccination debate. In *Proceedings of the 5th Workshop on Argument Mining*, pages 47–56, Brussels, Belgium. Association for Computational Linguistics.

Craig Whittington, Todd Feinman, Sandra Zelman Lewis, Greg Lieberman, and Michael del Aguila. 2019. Clinical practice guidelines: Machine learning and natural language processing for automating the rapid identification and annotation of new evidence.

Robert M Wolfe and Lisa K Sharp. 2002. Anti-vaccinationists past and present. *BMJ: British Medical Journal*, 325(7361):430.

Marianna Lya Zummo. 2017. A linguistic analysis of the online debate on vaccines and use of fora as information stations and confirmation niche. *International Journal of Society, Culture & Language*, 5(1):44.

COVID-19-related Nepali Tweets Classification in a Low Resource Setting

Rabin Adhikari[1,2], **Safal Thapaliya**[1,2], **Nirajan Basnet**[1,2], **Samip Poudel**[1,2]
Aman Shakya[2], **Bishesh Khanal**[1]

[1]NepAl Applied Mathematics and Informatics Institute for research (NAAMII)
[2]Institute of Engineering, Pulchowk Campus, Tribhuvan University

Abstract

Billions of people across the globe have been using social media platforms in their local languages to voice their opinions about the various topics related to the COVID-19 pandemic. Several organizations, including the World Health Organization, have developed automated social media analysis tools that classify COVID-19-related tweets to various topics. However, these tools that help combat the pandemic are limited to very few languages, making several countries unable to take their benefit. While multi-lingual or low-resource language-specific tools are being developed, there is still a need to expand their coverage, such as for the Nepali language. In this paper, we identify the eight most common COVID-19 discussion topics among the Twitter community using the Nepali language, set up an online platform to automatically gather Nepali tweets containing the COVID-19-related keywords, classify the tweets into the eight topics, and visualize the results across the period in a web-based dashboard. We compare the performance of two state-of-the-art multi-lingual language models for Nepali tweet classification, one generic (mBERT) and the other Nepali language family-specific model (MuRIL). Our results show that the models' relative performance depends on the data size, with MuRIL doing better for a larger dataset. The annotated data, models, and the web-based dashboard are open-sourced at `https://github.com/naamiinepal/covid-tweet-classification`.

1 Introduction

The COVID-19 pandemic has caused a global rise in social media users who express their opinions and share information on various topics related to the pandemic. Public health organizations and relevant agencies could analyze the social media data for early warning on potentially new virus variants based on symptoms discussion, for understanding the impact of various intervention measures, the efficacy of vaccination programs, etc. Social media data analysis can help develop strategies for combating the pandemic (Yigitcanlar et al., 2020), and improve the efficiency of the health industry (Scanfeld et al., 2010; Signorini et al., 2011; Harris et al., 2013; Paul and Dredze, 2014; Eichstaedt et al., 2015).

Several studies performed sentiment analysis of tweets to understand people's views towards the pandemic (Dubey, 2020; Jelodar et al., 2020; Samuel et al., 2020; Alamoodi et al., 2021). Since sentiment analysis provides limited coarse level information, recently there is an interest in building tools for early warning and topic-level discourse analysis. Most notably, the World Health Organization (WHO) tracks internet discourse by examining global pandemic-related Twitter data and news using tools like COVID-19 News Map[1] and EARS[2]. Although a large fraction of the global population uses local languages in social media, most of these tools are limited to English or Anglo-European languages. For instance, the WHO EARS works in only nine languages being piloted in 30 countries.

In recent years, there has been a growing interest in building multi-lingual language models, building low-resource language datasets, and exploring NLP methods with smaller language models and smaller data (Conneau et al., 2019; Wang et al., 2020; Ogueji et al., 2021). Nepali is a low-resource language where there is still a large gap in terms of advances, data availability, and the development of NLP tools. While there has been some work on low-resource languages for sentiment analysis in low-resource languages (Addawood et al., 2020; Hosseini et al., 2020) including Nepali (Sitaula et al., 2021; Shahi et al., 2022), to our knowledge there is no work on COVID-19 tweet topics classification for discourse analysis in the Nepali language.

[1]`https://portal.who.int/eios-coronavirus-newsmap/`

[2]`https://www.who-ears.com/`

Proceedings of the 29th International Conference on Computational Linguistic, pages 209–215
October 12–17, 2022.

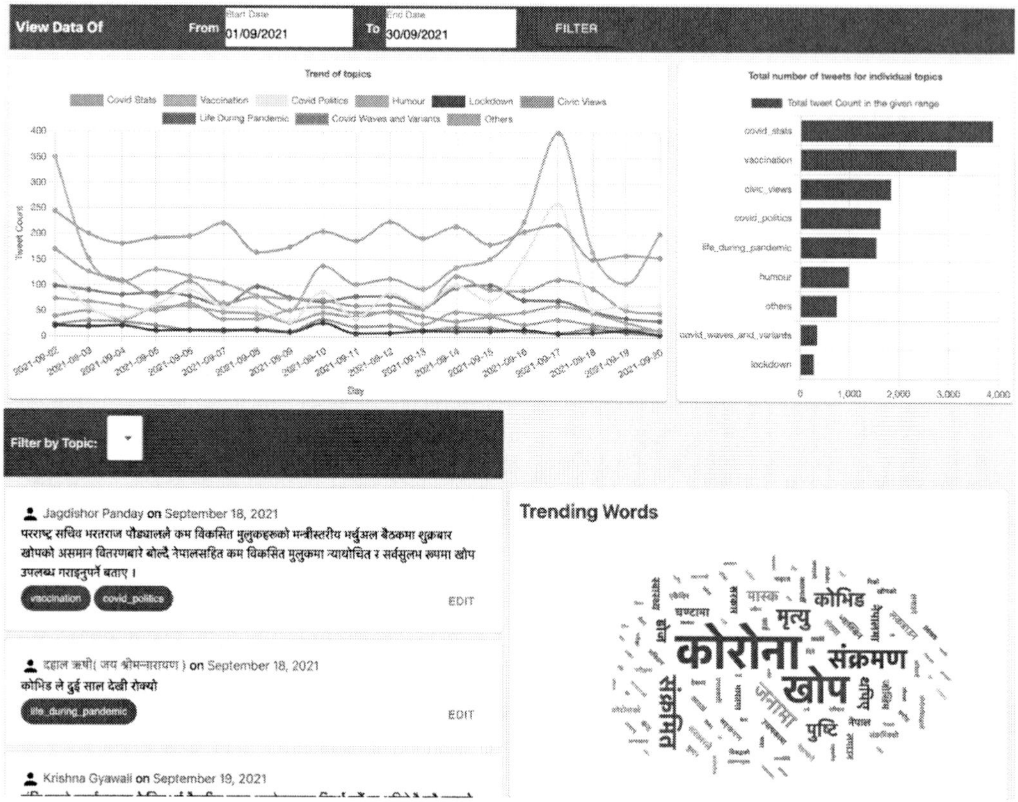

Figure 1: The web app dashboard shown above uses infographics viz. bar graphs and line charts to track the trend of various topics. We can filter the tweets based on time and topics to make analysis easier. Additionally, we have incorporated humans in the loop by developing an administrator interface to validate the predicted tweet labels from the model and proofread the validated ones.

In this work, we propose a new dataset, deep learning classification models based on multi-lingual language models, and an interactive dashboard for incremental learning and visualization of COVID-19 tweets topic classification in the Nepali language. Figure 1 shows a snapshot of our dashboard. In addition to visualizing topics classification in real-time, the dashboard can be used to manually verify the ML model's prediction, correct the predictions to annotate more data, and re-train the model via GUI for improvement as more data becomes available.

The followings are our contributions to the scientific community.

- We release a multi-annotator multi-label Nepali Annotated Tweets with COVID-19 Topics Classification (NAT-CTC) dataset that contains 12,241 tweets in Devanagari script, manually tagged with eight simplified topics. We also provide inter-annotator agreement results on this dataset using four annotators labeling 400 common tweets.

- We release our open-source web-based platform with GUI for automatic keywords-based tweets collection; tweet pre-processing, topic classification, and visualization. This platform can be used for AI-assisted annotation and incremental learning, where human annotators can correct the labels predicted by ML models and then retrain ML models. We use this approach during the dataset preparation as well.

- We show that the benefit of using a Nepali language family-specific model compared to using generic multi-lingual language models may come only if there is a certain minimum number of annotated data for the downstream task.

- We analyze 98,849 tweets using our topic classification model and the dashboard and show how the frequency of discussion in eight topics varied over time during the pandemic.

2 NAT-CTC Dataset

2.1 Keyword-based Filter for Tweet Collection

We identified 48 keywords in the Devanagari script that cover the majority of COVID-19-related tweets and used *twarc*[3] to extract tweets that contain the keywords and is tagged as Nepali by Twitter. New keywords were iteratively added into an initial set using *twarc* and manual review until finally settling to 48 keywords.

2.2 Eight COVID-19-related Topics

While the WHO EARS has 30 topics, to reduce the complexity and due to very limited tweets for Nepali language in some topics, we contextualized and developed these eight topics suitable to describe the specific narratives in Nepal: *COVID Stats*, *Vaccination*, *COVID Politics*, *Humor*, *Lockdown*, *Civic Views*, *Life during Pandemic*, and *Waves and Variants*.

2.3 Multi-annotator Manual Annotation with Incremental Learning

Seven annotators initially used *Label Studio*[4] to annotate the tweets (one tweet annotated by only one person), after which we trained a machine learning model (see subsection 3.2) to predict labels that were then corrected by the annotators using our dashboard. This increased annotation speed substantially and enabled improving the ML model as more data came in. Each tweet could be tagged with multiple topics. Moreover, for studying inter-rater agreement, we randomly selected 400 tweets each of which was annotated by four annotators. The tweets chosen for the agreement are sampled uniformly at random from the labeled dataset. Their exact number and the proportion to the larger labeled dataset can be seen from Figure 2.

3 Methods

3.1 Tweets Pre-processing

The tweets were pre-processed in the given order using *Pandas*[5]:

1. Remove user mentions and links, and change Latin characters to lowercase.

[3]`https://github.com/DocNow/twarc`

[4]`https://labelstud.io/`

[5]`https://pandas.pydata.org/`

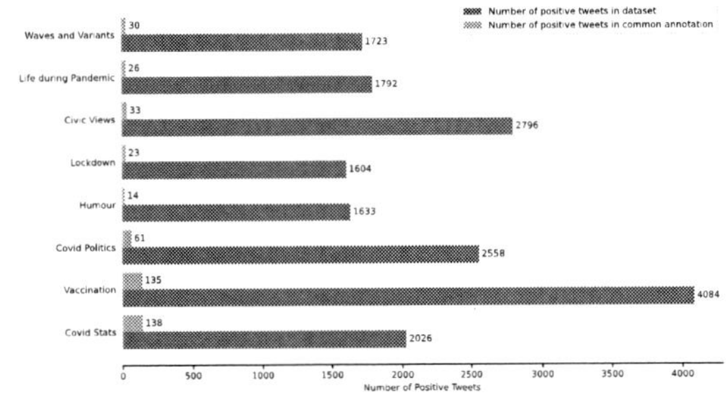

Figure 2: Number of positive tweets for each label. The imbalanced nature of the multi-label classification can be clearly seen. *Vaccination* has only 4084 positive samples out of 12,241 tweets; other topics are even more imbalanced.

2. Reduce spaces to a single space between words.

3. To eliminate bias from source information, remove text followed by the word *via*.

4. Remove leading and trailing spaces and tweets with three or fewer words.

5. Normalize the Unicode strings to NFKC standards[6].

3.2 Tweet's Topic Classification

We utilized Indic Language multi-lingual model MuRIL's (Khanuja et al., 2021) preprocessing and encoder models with a batch normalization (Ioffe and Szegedy, 2015) layer, and a linear classifier with a dropout rate of 0.5 to categorize the preprocessed tweets into the eight topics. For the training approach, we followed the blog, *A Recipe for Training Neural Networks*[7]. We adjusted the initial bias of the output layer to $\log\left(\frac{pos}{neg}\right)$ to reflect the imbalance in the dataset and facilitate the initial convergence of the model. We utilized the AdamW (Loshchilov and Hutter, 2017) with 0.01 weight decay, a learning rate of 5×10^{-5} and used the Polynomial Decay Scheduler with 10% of the total training steps as Warmup[8]. Since Precision and Recall are unaffected by class imbalance (Saito and Rehmsmeier, 2015; Branco et al., 2016), the preferred metrics to evaluate the model's performance were the F1 score with weighted averaging and the Area under the PR Curve (AUPR).

[6]`https://unicode.org/reports/tr15/`

[7]`https://karpathy.github.io/2019/04/25/recipe/`

[8]`https://github.com/tensorflow/models/blob/v2.7.2/official/nlp/optimization.py`

Labels	F1 Score	Area under PR Curve	Fleiss' Kappa Score
COVID Stats	0.913 ± 0.008	0.964 ± 0.003	0.87
Vaccination	0.974 ± 0.002	0.984 ± 0.002	0.88
COVID Politics	0.711 ± 0.012	0.763 ± 0.013	0.42
Humor	0.737 ± 0.014	0.766 ± 0.016	0.65
Lockdown	0.967 ± 0.005	0.988 ± 0.004	0.79
Civic Views	0.729 ± 0.007	0.757 ± 0.01	0.61
Life During Pandemic	0.61 ± 0.03	0.616 ± 0.043	0.34
Waves and Variants	0.851 ± 0.006	0.915 ± 0.005	0.53

Table 1: Area under PR-curve and F1-Score for each label, along with the corresponding Fleiss' Kappa score. In addition to depicting the mean value for each metric and its standard deviation for 5-fold cross-validation, it helps to find the correlation between the metrics and the corresponding Kappa score.

4 Experiments and Results

4.1 Topics Inter-rater Agreement

With 400 tweets annotated by four annotators each, we calculated Fleiss' Kappa (Fleiss, 1971; Fleiss et al., 2013) score for each of the eight categories. Since our dataset contains multi-label classification, we have reported the Kappa score for the individual categories (shown in Table 1). We averaged the Fleiss' Kappa scores of the individual categories to obtain a mean of 0.64, which shows substantial agreement between the four annotators (Artstein and Poesio, 2008; McHugh, 2012).

4.2 Label-wise Model Performance

Table 1 shows mean and standard deviation of model performance for 5-fold cross-validation where each fold consisted of 2,448 tweets. From the inter-rater agreement Kappa score for the eight labels shown aside, we can infer that the performance seems to drop as the agreement among the annotators for the target class reduces (reduced Kappa score), with a few exceptions such as *Waves and Variants*.

Avg. Type	F1 Score	AUPR
Micro	0.817 ± 0.005	-
Macro	0.811 ± 0.004	0.841 ± 0.007
Weighted	0.823 ± 0.004	0.854 ± 0.006

Table 2: Micro, Macro, and Weighted scores to capture the overall performance across all the target topics.

Table 2 presents the overall performance across all the topics using various averaging of AUPR and F1 Score[9]. The model's prediction for some tweets are shown in Table 3.

[9]https://scikit-learn.org/stable/modules/generated/sklearn.metrics.f1_score.html

4.3 Performance of mBERT and MuRIL Changes Differently When Training Dataset Size Changes

Figure 3: MuRIL performed better than mBERT when dataset size was increased from 6,952 to 12,241. For the smaller dataset, MuRIL has 0.64 mean AUPR whereas mBERT has 0.65. MuRIL has 0.84 mean AUPR for the larger dataset whereas mBERT has 0.81 only.

We compared mBERT (Devlin et al., 2019) and MuRIL (Khanuja et al., 2021) for the same normalization and dropout for various training data sizes. As shown in Figure 3, MuRIL had slightly lower performance on a smaller dataset of 6,952 tweets but outperformed mBERT by a greater margin when the dataset size increased to 12,241 tweets. The language family-specific models seem to provide a greater benefit than generic models only when finetuning training dataset size is of a certain minimum number, albeit this has to be further explored with additional languages and language models.

Example	English Translation	Label Prediction
कोरोनाको परीक्षण किन घटाउँदै छ नालायक सरकार?	Why is the worthless government reducing the testing of Corona?	*COVID Politics, Civic views*
पछिल्लो २४ घण्टामा थप ९४४ जनामा कोरोना संक्रमण पुष्टि, संक्रिय संक्रमितको संख्या ६ हजार नाघ्यो	In the last 24 hours, 944 more Corona infections have been confirmed, and the number of active infected has exceeded 6 thousand	*COVID Stats*
कोरोनाको ग्राफ उकालो लागि सक्यो, बेलैमा झार्नु पर्‍यो, भिडभाडमा नजाने र मास्क लगाउने नै प्रमुख उपाय।	The graph of Corona has gone up, it has to be realized early, the best solution is not to go to crowded places and wear a mask.	*Civic Views, Life during Pandemic, Waves and Variants*
फेरि लकडाउनको हल्लाले उद्योगी चिन्तित, कोरोना खोप लगाएर व्यवसाय सन्चालन गर्न पाउनुपर्ने माग	Again, the rumors of the lockdown are worrying the industrialists, they demand to be able to operate the business with Corona vaccination.	*Vaccination, Lockdown, Life during Pandemic*
तेत्रो लकडाउन त एक्लै कटाइयो जाबो February 14 एक दिन त कसो कटाउन नसकिएला र ?	That lockdown was spent alone, I can spend February 14 alone, can't I?	*Humour, Lockdown*

Table 3: Examples of some model predictions

4.4 Trend Analysis

We present here a couple of examples of the insights we can see from our ML model's classification results on 98,849 tweets from June 2021 to February 2022. Discussions related to *Vaccination* surged in mid-September 2021 which was just after Nepal's Government decided to provide vaccines to students and youths[10], as can be seen in Figure 1. Similarly, tweets related to *COVID Stats* increased during late January and early February of 2022 when Nepal was hit by the third wave of the virus[11].

5 Discussion and Conclusion

We have developed an online platform to gather, analyze, and categorize tweets about COVID-19 in Nepali, written in the Devanagari script. We selected eight topics pertinent to online discussions in Nepal based on the various categories of online discourse established by the WHO. Seven different people have annotated 12,241 tweets from our dataset. We arrived at our best model architecture using MuRIL (Khanuja et al., 2021) as the encoder and a batch normalization (Ioffe and Szegedy, 2015) layer immediately before the final output layer after fine-tuning several hyperparameters and utilizing two encoder backbones, i.e., mBERT (Devlin et al., 2019) and MuRIL (Khanuja et al., 2021). The web app dashboard uses infographics like bar graphs and area charts to track the development of online discussions. We can filter the tweets based on time and topics to make analysis easier. Additionally, we developed an administrator interface to validate the predicted tweet labels from the model and proofread the validated ones.

Although our dataset is not large, it may be valuable for transfer learning and semi-supervised learning (Lwowski and Najafirad, 2020). Our dataset can help to make multi-lingual datasets more inclusive and the models trained on them more robust. Translating and transliterating to and from our dataset can help in augmentation in various settings.

[10] https://kathmandupost.com/health/2021/09/20/all-students-above-18-to-be-jabbed-with-covid-19-vaccine

[11] https://english.onlinekhabar.com/nepal-covid-19-third-wave-signs.html

References

Aseel Addawood, Alhanouf Alsuwailem, Ali Alohali, Dalal Alajaji, Mashail Alturki, Jaida Alsuhaibani, and Fawziah Aljabli. 2020. Tracking and understanding public reaction during COVID-19: Saudi Arabia as a use case. In *Proceedings of the 1st Workshop on NLP for COVID-19 (Part 2) at EMNLP 2020*, Online. Association for Computational Linguistics.

Abdullah Hussein Alamoodi, Bilal Bahaa Zaidan, Aws Alaa Zaidan, Osamah Shihab Albahri, KI Mohammed, Rami Qays Malik, Esam Motashar Almahdi, Mohammed A Chyad, Ziadoon Tareq, Ahmed Shihab Albahri, et al. 2021. Sentiment analysis and its applications in fighting covid-19 and infectious diseases: A systematic review. *Expert systems with applications*, 167:114155.

Ron Artstein and Massimo Poesio. 2008. Inter-coder agreement for computational linguistics. *Computational linguistics*, 34(4):555–596.

Paula Branco, Luís Torgo, and Rita P Ribeiro. 2016. A survey of predictive modeling on imbalanced domains. *ACM Computing Surveys (CSUR)*, 49(2):1–50.

Alexis Conneau, Kartikay Khandelwal, Naman Goyal, Vishrav Chaudhary, Guillaume Wenzek, Francisco Guzmán, Edouard Grave, Myle Ott, Luke Zettlemoyer, and Veselin Stoyanov. 2019. Unsupervised cross-lingual representation learning at scale. *arXiv preprint arXiv:1911.02116*.

Jacob Devlin, Ming-Wei Chang, Kenton Lee, and Kristina Toutanova. 2019. BERT: Pre-training of deep bidirectional transformers for language understanding. In *Proceedings of the 2019 Conference of the North American Chapter of the Association for Computational Linguistics: Human Language Technologies, Volume 1 (Long and Short Papers)*, pages 4171–4186, Minneapolis, Minnesota. Association for Computational Linguistics.

Akash Dutt Dubey. 2020. Twitter sentiment analysis during covid-19 outbreak. *Available at SSRN 3572023*.

Johannes C Eichstaedt, Hansen Andrew Schwartz, Margaret L Kern, Gregory Park, Darwin R Labarthe, Raina M Merchant, Sneha Jha, Megha Agrawal, Lukasz A Dziurzynski, Maarten Sap, et al. 2015. Psychological language on twitter predicts county-level heart disease mortality. *Psychological science*, 26(2):159–169.

Joseph L Fleiss. 1971. Measuring nominal scale agreement among many raters. *Psychological bulletin*, 76(5):378.

Joseph L Fleiss, Bruce Levin, and Myunghee Cho Paik. 2013. *Statistical methods for rates and proportions*. john wiley & sons.

Jenine K Harris, Nancy L Mueller, and Doneisha Snider. 2013. Social media adoption in local health departments nationwide. *American journal of public health*, 103(9):1700–1707.

Pedram Hosseini, Poorya Hosseini, and David Broniatowski. 2020. Content analysis of Persian/Farsi tweets during COVID-19 pandemic in Iran using NLP. In *Proceedings of the 1st Workshop on NLP for COVID-19 (Part 2) at EMNLP 2020*, Online. Association for Computational Linguistics.

Sergey Ioffe and Christian Szegedy. 2015. Batch normalization: Accelerating deep network training by reducing internal covariate shift. In *International conference on machine learning*, pages 448–456. PMLR.

Hamed Jelodar, Yongli Wang, Rita Orji, and Shucheng Huang. 2020. Deep sentiment classification and topic discovery on novel coronavirus or covid-19 online discussions: Nlp using lstm recurrent neural network approach. *IEEE Journal of Biomedical and Health Informatics*, 24(10):2733–2742.

Simran Khanuja, Diksha Bansal, Sarvesh Mehtani, Savya Khosla, Atreyee Dey, Balaji Gopalan, Dilip Kumar Margam, Pooja Aggarwal, Rajiv Teja Nagipogu, Shachi Dave, et al. 2021. Muril: Multilingual representations for indian languages. *arXiv preprint arXiv:2103.10730*.

Ilya Loshchilov and Frank Hutter. 2017. Decoupled weight decay regularization. *arXiv preprint arXiv:1711.05101*.

Brandon Lwowski and Peyman Najafirad. 2020. COVID-19 surveillance through Twitter using self-supervised and few shot learning. In *Proceedings of the 1st Workshop on NLP for COVID-19 (Part 2) at EMNLP 2020*, Online. Association for Computational Linguistics.

Mary L McHugh. 2012. Interrater reliability: the kappa statistic. *Biochemia medica*, 22(3):276–282.

Kelechi Ogueji, Yuxin Zhu, and Jimmy Lin. 2021. Small data? no problem! exploring the viability of pretrained multilingual language models for low-resourced languages. In *Proceedings of the 1st Workshop on Multilingual Representation Learning*, pages 116–126.

Michael J Paul and Mark Dredze. 2014. Discovering health topics in social media using topic models. *PloS one*, 9(8):e103408.

Takaya Saito and Marc Rehmsmeier. 2015. The precision-recall plot is more informative than the roc plot when evaluating binary classifiers on imbalanced datasets. *PloS one*, 10(3):e0118432.

Jim Samuel, GG Ali, Md Rahman, Ek Esawi, Yana Samuel, et al. 2020. Covid-19 public sentiment insights and machine learning for tweets classification. *Information*, 11(6):314.

Daniel Scanfeld, Vanessa Scanfeld, and Elaine L Larson. 2010. Dissemination of health information through social networks: Twitter and antibiotics. *American journal of infection control*, 38(3):182–188.

TB Shahi, C Sitaula, and N Paudel. 2022. A hybrid feature extraction method for nepali covid-19-related tweets classification. *Computational Intelligence and Neuroscience*, 2022.

Alessio Signorini, Alberto Maria Segre, and Philip M Polgreen. 2011. The use of twitter to track levels of disease activity and public concern in the us during the influenza a h1n1 pandemic. *PloS one*, 6(5):e19467.

Chiranjibi Sitaula, Anish Basnet, A Mainali, and Tej Bahadur Shahi. 2021. Deep learning-based methods for sentiment analysis on nepali covid-19-related tweets. *Computational Intelligence and Neuroscience*, 2021.

Zihan Wang, Stephen Mayhew, Dan Roth, et al. 2020. Extending multilingual bert to low-resource languages. *arXiv preprint arXiv:2004.13640*.

Tan Yigitcanlar, Nayomi Kankanamge, Alexander Preston, Palvinderjit Singh Gill, Maqsood Rezayee, Mahsan Ostadnia, Bo Xia, and Giuseppe Ioppolo. 2020. How can social media analytics assist authorities in pandemic-related policy decisions? insights from australian states and territories. *Health Information Science and Systems*, 8(1):1–21.

SMM4H 2022 Task 2: Dataset for stance and premise detection in tweets about health mandates related to COVID-19

Vera Davydova
Sber AI
Moscow, Russia
veranchos@gmail.com

Elena Tutubalina
Kazan Federal University, Kazan, Russia
Sber AI, Moscow, Russia
AIRI, Moscow, Russia
tutubalinaev@gmail.com

Abstract

This paper is an organizers' report of the competition on argument mining systems dealing with English tweets about COVID-19 health mandates. This competition was held within the framework of the SMM4H 2022 Workshop. During the competition, the participants were offered two subtasks: stance detection and premise classification. We present a manually annotated corpus containing 6,156 short posts from Twitter on three topics related to the COVID-19 pandemic: school closures, stay-at-home orders, and wearing masks. We hope the prepared dataset will support further research on argument mining in the health field.

1 Introduction

Nowadays, people are actively sharing their views on various issues related to the COVID-19 pandemic on social media. For example, users express their attitude towards a quarantine and wearing masks in public places. Some statements are reasoned by arguments, and other statements are just emotional claims. Automated approaches for detecting people's stances and their premises towards health orders related to COVID-19 can help to estimate the level of cooperation within the health mandates announced by the government.

Thereby, since the beginning of the pandemic, new datasets for argument mining in the health field have been created. The first and the largest dataset of Twitter users' stances in the context of the COVID-19 pandemic is COVID-CQ (Mutlu et al., 2020). It contains controversial tweets about the efficacy of hydroxychloroquine as a treatment. Similarly, Wührl and Klinger (2021) presented a dataset for biomedical claim detection in Twitter posts. Miao et al. (2020) created the Lockdown-Tweets – mostly unlabelled tweet dataset, which is related to the lockdown policy in New York State during the pandemic. People's opinions towards health mandates in Germany are discovered in Beck et al. (2021): first, relevant tweets were identified and then the expressed stances were detected.

While most researchers concentrate on the stance detection task, there are far fewer datasets for premise classification (Kotelnikov et al., 2022).

In this work, we aim to fill in this gap and focus not only on stance detection, but also on premise classification. Therefore, we organised a competition on detecting both of these argument structures from English tweets related to COVID-19 health mandates. This competition was carried out as one of the shared tasks during the Social Media Mining for Health Applications (#SMM4H) 2022 Workshop. The SMM4H shared tasks aim to take a community-driven approach to address NLP challenges of utilising social media data for health informatics, including informal, colloquial expressions of clinical concepts, noise, data sparsity, ambiguity, and multilingual posts (Klein et al., 2020), (Magge et al., 2021). In 2022, the seventh iteration of the SMM4H shared tasks, including Task 2 on automatic classification of stance and premise in tweets about health mandates related to COVID-19 (in English), was held (Weissenbacher et al., 2022).

The dataset for this task contains tweets that express views towards three claims/topics: (i) keeping schools closed, (ii) stay-at-home orders, and (iii) wearing a face mask. The main task consists of two sub-tasks: (i) **Task 2a** on stance detection, and (ii) **Task 2b** on premise classification.

Task 2a: stance detection The first task aims to determine the point of view (stance) of the text's author concerning the given claim (e.g., wearing a face mask). The tweets are manually annotated for stance according to three categories: in-favor, against, and neither.

Task 2b: premise classification The second subtask is to predict whether at least one premise/argument is mentioned. A given tweet

Proceedings of the 29th International Conference on Computational Linguistic, pages 216–220
October 12–17, 2022.

is considered as having a premise if it contains a statement that can be used as an argument in a discussion. For instance, the annotator could use it to convince an opponent about the given claim. The tweets are manually annotated for binary classification: participants of this task are required to submit whether each tweet has a premise (1) or not (0).

The main contributions of this work are the following. Firstly, we complemented existing dataset of COVID-19 Stance detection (Glandt et al., 2021) with premise classification labels. Furthermore, guidelines for annotating texts that contain premises are prepared. Secondly, we released the baselines for both subtasks that use RoBERTa architecture. Finally, we compared and analysed the results of the participants of the shared task and proposed steps for further improvement. All the materials can be found on SMM4H Workshop page[1] and on Codalab[2] competition page.

Claim/Topic	Stance			Premise	
	favor	against	neither	1	0
train set					
face masks	652	324	343	508	811
close school	526	217	307	535	515
home orders	168	333	686	288	899
validation set					
face masks	121	51	36	82	126
close school	91	35	51	80	97
home orders	32	72	111	58	157
test set					
face masks	209	208	260	253	424
close school	215	192	263	294	376
home orders	102	170	381	169	484

Table 1: Summary of statistics of stance and premise classification datasets. The topic on *school closures* and *stay at home orders* has been shortened to *close school* an *home orders*, respectively.

2 Datasets

We provided the participants of the SMM4H 2022 competition the training and validation sets. During the evaluation phase, all participants had five days to submit their predictions for test sets on Codalab for evaluation. The training set included 3,556 tweets, a good mix of three topics: 37%, 33% and 30% of tweets are about face masks, school closures and stay-at-home orders, respectively. The validation and test sets contained 600 and 2,000 tweets, respectively.

As a source of tweets for training and validation sets, we leveraged a COVID-19 stance detection dataset (Glandt et al., 2021) along with stance annotation guidelines and stance labels. Tweets for test sets were collected using 33 keywords such as #OpenSchools, #LockdownNow, #NoMasks. After this, we removed the hashtags to exclude annotation bias toward specific classes. The test set for Task 2a and all three sets for Task 2b were manually annotated. Each tweet was labelled by three *Yandex.Toloka*[3] annotators. We followed annotation guidelines of an argument mining shared task RuArg-2022 (Kotelnikov et al., 2022). Below we highlight some of the key features of our guidelines:

- A statement is evaluated as an argument if it contains a statement that can be used in a dispute to persuade an opponent. For example, *masks help prevent the spread of the disease.* (1)
- It is also necessary to distinguish sentiment (positive and/or negative) from argumentation. For example, *and the fact that Trump did not introduce a suffocating quarantine is well done!* (0)
- The argument should not be a fragment that needs to be thought out. For example, *It is effective if you declare a quarantine.* (0)
- An example of an argument could be such a common sense statement. For example, *in all countries of the world, everyone is wearing masks, but ours... this is not a joke.* (1)
- The position of the author "favor" or "against" should be clear – only under this condition it is possible to detect an argument. The annotator should not think for the author. For example, the author's position on quarantine is unclear in the text *here are the words of my classmate from Annecy, France, from today's Facebook correspondence - "France introduced quarantine, and immediately everyone poured out to barbecue in nature."* (0).

To measure annotation quality on this platform, we mixed raw tweets with control tasks (tweets

[1] https://healthlanguageprocessing.org/smm4h-2022/
[2] https://codalab.lisn.upsaclay.fr/competitions/5067
[3] https://toloka.yandex.ru/

Tweet	Claim/Topic	Stance	Premise
The fact that anti-masking is a thing is a completely terrifying insight into the nature of some beings who look, walk and breathe just like us.	face masks	favor	0
Masks help prevent the spread of the disease. Please, #WEARAMASK	face masks	favor	1
We are now experiencing a surge in the number of infected health care workers, with two deaths already. Prior to Covid19, we were experiencing a shortage and this is worsening with them in quarantine. You can help us by staying safe and staying home.	stay at home orders	favor	1
0.02% chance of dying of #Covid and @GovInslee keeps our state in an "indefinite" lockdown. I'll take those odd, thanks.	stay at home orders	against	1
I see that @BBCOne are still showing the people on their tandem bikes before programmes. Don't you lot not know that there is a lockdown and no one can go out right now? #coronavirus	stay at home orders	neither	0
Close the damn schools until there is a vaccine.	school closures	favor	1

Table 2: Examples of tweets annotated for stance and premise classification.

from the validation set with correct responses annotated by both authors). These tasks were used for calculating the Toloker's percentage of correct responses. Depending on the annotator's result, the system blocked access to tasks. The internal quality of the control tasks was 0.68. The obtained crowdsourced labels were aggregated into a single label (Dawid and Skene, 1979). Samples of annotated tweets are shown in Table 2.

Table 1 shows statistics of experimental datasets across topics. The training set includes 38% and 25% in-favor and against tweets, respectively. Relative class balance is also present for argumentation: 63% of train tweets contain a premise (1). 34% of tweets in the test set contain a premise; 26% of tweets in the test set are annotated as in-favor. As shown in Table 1, the distribution of classes by topic is different. In particular, the topic about staying-at-home orders contains more "against" tweets than tweets "in-favor".

3 Experiments

We used F_1 as the main evaluation metric in each of the two subtasks, which is calculated according to the following formula: $F_1 = \frac{1}{n}\sum_{c \in C} F_{1_{rel_c}}$, where $C =$ {"face masks", "stay at home orders", "school closures"}, n is the size of C, $F_{1_{rel_c}}$ is macro F_1-score averaged over two classes for each task (in-favor & against classes for stance; 0 & 1 classes for premise).

We used two models as baselines: Random and RoBERTa. Random baseline is rather simplistic: we randomly assigned labels for each tweet in both tasks according to the label distribution. The second baseline leverages RoBERTa architecture (Liu et al., 2019) for multiclass (Task 2a) and binary classification (Task 2b). We fine-tuned pretrained RoBERTa-base model from HuggingFace[4] on our data. The validation set was utilised to select appropriate hyperparameters for the models. For each model, AdamW optimizer (Loshchilov and Hutter, 2019) was used with a learning rate of 4e-5, and gradient clipping with a max norm of 1.0. Each model was trained for 5 epochs, with a mini-batch size of 8 in each iteration and a maximum sequence length of 128.

The results of both baseline models in terms of F_1 scores are described in Table 3. As we can see, RoBERTa-based baseline showed relatively strong results in both set-ups: F_1-score is 0.566 (validation) and 0.45 (test) on the more difficult Stance detection task; and 0.75 (validation) and 0.722 (test) on Premise classification.

[4] https://huggingface.co/roberta-base

Claim/Topic	Stance		Premise	
	Rnd.	RoBERTa	Rnd.	RoBERTa
validation set				
face masks	0.309	0.599	0.423	0.770
close school	0.325	0.513	0.413	0.731
home orders	0.334	0.589	0.406	0.755
All tweets	0.323	0.566	0.414	0.750
test set				
face masks	0.342	0.439	0.506	0.708
close school	0.332	0.345	0.526	0.719
home orders	0.301	0.566	0.521	0.738
All tweets	0.325	0.450	0.518	0.722

Table 3: Macro F_1 scores for both Random (Rnd.) and RoBERTa baselines. The topic on *school closures* and *stay at home orders* has been shortened to *close school* an *home orders*, respectively.

3.1 Official SMM4H 2022 Task 2 Results

We observed a strong interest in Task 2, with 47 participants registered in Codalab. Among these participants, 14 teams submitted their prediction to Codalab for both tasks (15 teams for Task 2b). We summarized their performance in (Weissenbacher et al., 2022). Further, we highlight the key observations. The median F_1 scores of all team's best-performing submissions are 0.55 for Task 2a and 0.65 for Task 2b. The mean F_1 scores of all team's best-performing submissions are 0.49 for Task 2a and 0.57 for Task 2b. The best performance achieved in task 2a is 0.64 F_1 which is 0.19 higher than the RoBERTa baseline model. The three top-performing systems achieved 0.70 F_1 in Task 2b, which is 0.02 less than the baseline model. The majority of teams used COVID-related BERT models. Two teams tried to combine data from both tasks into one unified model. Leading teams on both tasks tried to aggregate claim and tweet texts: the leading team in Task 2a appended the claim text to the tweet, while the second-best team in Task 2a with the highest F_1 in Task 2b proposed a new pre-training task constructed by the tweets and claims similarly to next sentence prediction.

4 Conclusion

In this paper, we have presented the dataset for stance and premise detection in tweets written in English about health mandates related to COVID-19. We hosted a shared task on SMM4H 2022 workshop and released this dataset to the research community. The 15 teams that took part in the task proposed a variety of classification architectures, ranging from just fine-tuning BERT models to multi-task learning on both subtasks and combining tweets with claims. We plan to extend our dataset for future work with a broader set of health-related claims.

Acknowledgements

The work was supported by the Russian Science Foundation [grant number 18-11-00284]. The authors would also like to thank Natalia Loukachevitch for her suggestions on task definition, Dmitry Ustalov and other members of the *Yandex.Toloka* team for providing credits for the crowd-sourced annotation of tweets.

References

Tilman Beck, Ji-Ung Lee, Christina Viehmann, Marcus Maurer, Oliver Quiring, and Iryna Gurevych. 2021. Investigating label suggestions for opinion mining in German covid-19 social media. In *Proceedings of the 59th Annual Meeting of the Association for Computational Linguistics and the 11th International Joint Conference on Natural Language Processing (Volume 1: Long Papers)*, pages 1–13, Online. Association for Computational Linguistics.

Alexander Philip Dawid and Allan M. Skene. 1979. Maximum likelihood estimation of observer error-rates using the EM algorithm. *Journal of the Royal Statistical Society: Series C (Applied Statistics)*, 28(1):20–28.

Kyle Glandt, Sarthak Khanal, Yingjie Li, Doina Caragea, and Cornelia Caragea. 2021. Stance detection in covid-19 tweets. In *Proceedings of the 59th Annual Meeting of the Association for Computational Linguistics and the 11th International Joint Conference on Natural Language Processing (Volume 1: Long Papers)*, pages 1596–1611.

Ari Klein, Ilseyar Alimova, Ivan Flores, Arjun Magge, Zulfat Miftahutdinov, Anne-Lyse Minard, Karen O'Connor, Abeed Sarker, Elena Tutubalina, Davy Weissenbacher, and Graciela Gonzalez-Hernandez. 2020. Overview of the fifth social media mining for health applications (#SMM4H) shared tasks at COLING 2020. In *Proceedings of the Fifth Social Media Mining for Health Applications Workshop & Shared Task*, pages 27–36, Barcelona, Spain (Online). Association for Computational Linguistics.

Evgeny Kotelnikov, Natalia Loukachevitch, Irina Nikishina, and Alexander Panchenko. 2022. Ruarg-2022: Argument mining evaluation. *arXiv preprint arXiv:2206.09249*.

Yinhan Liu, Myle Ott, Naman Goyal, Jingfei Du, Mandar Joshi, Danqi Chen, Omer Levy, Mike Lewis,

Luke Zettlemoyer, and Veselin Stoyanov. 2019. Roberta: A robustly optimized bert pretraining approach.

Ilya Loshchilov and Frank Hutter. 2019. Decoupled weight decay regularization. In *ICLR*.

Arjun Magge, Ari Klein, Antonio Miranda-Escalada, Mohammed Ali Al-garadi, Ilseyar Alimova, Zulfat Miftahutdinov, Eulalia Farre-Maduell, Salvador Lima Lopez, Ivan Flores, Karen O'Connor, Davy Weissenbacher, Elena Tutubalina, Abeed Sarker, Juan M Banda, Martin Krallinger, and Graciela Gonzalez-Hernandez, editors. 2021. *Proceedings of the Sixth Social Media Mining for Health (#SMM4H) Workshop and Shared Task*. Association for Computational Linguistics, Mexico City, Mexico.

Lin Miao, Mark Last, and Marina Litvak. 2020. Twitter data augmentation for monitoring public opinion on COVID-19 intervention measures. In *Proceedings of the 1st Workshop on NLP for COVID-19 (Part 2) at EMNLP 2020*, Online. Association for Computational Linguistics.

Ece Mutlu, Toktam Oghaz, Jasser Jasser, Ege Tütüncüler, Amirarsalan Rajabi, Aida Tayebi, Ozlem Ozmen, and Ivan Garibay. 2020. A stance data set on polarized conversations on twitter about the efficacy of hydroxychloroquine as a treatment for covid-19.

Davy Weissenbacher, Ari Z. Klein, Luis Gascó, Darryl Estrada-Zavala, Martin Krallinger, Yuting Guo, Yao Ge, Abeed Sarker, Ana Lucia Schmidt, Raul Rodriguez-Esteban, Mathias Leddin, Arjun Magge, Juan M. Banda, Vera Davydova, Elena Tutubalina, and Graciela Gonzalez-Hernandez. 2022. Overview of the seventh social media mining for health applications smm4h shared tasks at coling 2022. In *Proceedings of the Seventh Social Media Mining for Health Applications (#SMM4H) Workshop Shared Task*, pages –.

Amelie Wührl and Roman Klinger. 2021. Claim detection in biomedical Twitter posts. In *Proceedings of the 20th Workshop on Biomedical Language Processing*, pages 131–142, Online. Association for Computational Linguistics.

Overview of the Seventh Social Media Mining for Health Applications (#SMM4H) Shared Tasks at COLING 2022

Davy Weissenbacher,[1]† Juan M. Banda,[2] Vera Davydova,[3] Darryl Estrada-Zavala,[4] Luis Gascó,[4] Yao Ge,[5] Yuting Guo,[5] Ari Z. Klein,[6] Martin Krallinger,[4] Mathias Leddin,[7] Arjun Magge,[6] Raul Rodriguez-Esteban,[7] Abeed Sarker,[5] Ana Lucía Schmidt,[7] Elena Tutubalina,[8,9] Graciela Gonzalez-Hernandez[1]

[1]Cedars-Sinai Medical Center, Los Angeles, CA, USA
[2]Georgia State University, Atlanta, GA, USA
[3]Sber AI, Moscow, Russia
[4]Barcelona Supercomputing Center (BSC), Barcelona, Spain
[5]Emory University, Atlanta, GA, USA
[6]University of Pennsylvania, Philadelphia, PA, USA
[7]Roche Innovation Center, Basel, Switzerland
[8]Kazan Federal University, Kazan, Russia
[9]AIRI, Moscow, Russia
†Corresponding author: `davy.weissenbacher@cshs.org`

Abstract

For the past seven years, the Social Media Mining for Health Applications (#SMM4H) shared tasks have promoted the community-driven development and evaluation of advanced natural language processing systems to detect, extract, and normalize health-related information in public, user-generated content. This seventh iteration consists of ten tasks that include English and Spanish posts on Twitter, Reddit, and WebMD. Interest in the #SMM4H shared tasks continues to grow, with 117 teams that registered and 54 teams that participated in at least one task—a 17.5% and 35% increase in registration and participation, respectively, over the last iteration. This paper provides an overview of the tasks and participants' systems. The data sets remain available upon request, and new systems can be evaluated through the post-evaluation phase on CodaLab.

1 Introduction

For the past seven years, the Social Media Mining for Health (#SMM4H) Workshop has been a competitive platform to promote the development and evaluation of advanced natural language processing systems to detect, extract, and normalize health-related information in user generated content publicly available online. For the seventh iteration of the Workshop shared tasks, researchers from international institutions challenged the community with ten tasks. The tasks run on various media (tweets, reviews and Reddit posts) and languages (English and Spanish): classification, detection and normalization of adverse events mentions in tweets (Task 1), classification of stance and premise in tweets about health mandates related to COVID-19 (Task 2), classification of changes in medication treatments in tweets and WebMD reviews (Task 3), classification of tweets self-reporting exact age (Task 4), classification of tweets containing self-reported COVID-19 symptoms - in Spanish (Task 5), classification of tweets which indicate self-reported COVID-19 vaccination status (Task 6), classification of self-reported intimate partner violence on Twitter (Task 7), classification of self-reported chronic stress on Twitter (Task 8), classification of Reddit posts self-reporting exact age (Task 9), detection of disease mentions in tweets – in Spanish (Task 10).

This iteration shows that the interest of the community for the shared tasks continues to grow with 117 teams registered, representing a 17.5% growth compare to the last iteration in 2021. The growth of interest translated in an increase of participation with 54 teams having submitted their predictions for at least one task, a growth of 35% compared to last year participation. While all task organizers were free to change the modality for their task, we recommended generic guidelines to organize the tasks. We allowed all participants to register in one or more tasks. Upon acceptance, we provided

Proceedings of the 29th International Conference on Computational Linguistic, pages 221–241
October 12–17, 2022.

the participants with a training and a validation set for each task they registered in during the practice period. We released unlabeled test sets at the beginning of the evaluation period of the competition, a period which was 4 days long. During this period, each team could upload up to 3 predictions sets for the labels of the test sets to the web-based platform, Codalab. Codalab automatically evaluated their performance. We report in this overview the results achieved with the best set of three predictions sets submitted. We left the post-evaluation period opened indefinitely on Codalab for all tasks run during this iteration. Interested readers can request the datasets and upload their own predictions to the Codalab server to compare their performance with the winners of each task.

In Section 2, we briefly describe and motivate the ten tasks of the competition. In Section 3, we present the performance and a short interpretation of the results for each task. In Appendix 4, we provide the list of the publications describing the systems of the competing teams along with the team IDs referring them in Section 3.

2 Tasks

2.1 Task 1: Classification, detection and normalization of Adverse Events (AE) mentions in tweets (in English)

For Task 1, participants were asked to develop methods to extract adverse drug effects (ADE) from tweets containing drug mentions. Such automated methods are considered necessary for social media pharmacovigilance efforts for monitoring emergence of ADEs in post-market surveillance of drugs, especially in vulnerable populations such as children, pregnant women and the elderly who are often excluded from clinical trials. Reliable signals generated from such social media pharmacovigilance pipelines can be complementary to traditional pharmacovigilance efforts such as voluntary consumer and health provider reporting. ADE mining in tweets has been the longest running tasks at the SMM4H shared task. Task 1 presents three subtasks in increasing order of complexity: (Task 1a) Identify tweets that contain one or more ADEs, (Task 1b) in addition to Task 1a, extract the text span of reported ADEs in tweets, and (Task 1c) in addition to Tasks 1a and 1b, normalize extracted spans to MedDRA ontology's preferred terms `https://meddra.org/`. For this task, we used the same dataset as the previous year. During the training state, the dataset contains a total of 18,300 tweets with 17,385 tweets for training and 915 tweets for validation (Magge et al., 2020). During the evaluation stage, we performed a blind evaluation on 10,984 tweets. The tweets were manually annotated by experts with: (a) binary labels of *hasADE* and *noADE* indicating presence of one or more ADEs, (b) starting and ending indices of the ADE mention(s) in the text, and (c) the normalized MedDRA lower-level term (LLT) that were evaluated at the higher preferred term (PT) level. MedDRA contains about 79,000 LLT terms and about 23,000 preferred terms, making it important for the developed systems to be capable of identifying and normalizing ADEs that were not occurring in the training set. Task 1 presents multiple technical challenges such as class imbalance and normalizing into a large potential label space. Submissions were evaluated and ranked based on the F_1-score for the *ADE* class for subtask 1a, F_1-score for overlapping ADE mentions for subtask 1b, and F_1-score for overlapping ADE mentions with matching preferred term IDs for subtask 1c. Task 1 is hosted on Codalab at `https://codalab.lisn.upsaclay.fr/competitions/2073`.

2.2 Task 2: Classification of stance and premise in tweets about health mandates related to COVID-19 (in English)

For Task 2, we focus on argument mining (or argumentation mining) for extracting arguments from COVID-related tweets. Tweets express views towards three claims/topics associated with governments' restrictions: (i) keeping schools closed, (ii) stay-at-home orders, and (iii) wearing a face mask. Systems for detecting people's stances and their premises about governments' health mandates related to COVID-19 can help to gauge the level of cooperation in society which is essential for stopping the spread of the virus. The task consists of two sub-tasks: (i) **Task 2a** on stance detection, and (ii) **Task 2b** on premise classification.

Task 2a: stance detection The first sub-task aims to determine the point of view (stance) of the text's author in relation to the given claim. The tweets are manually annotated for stance according to three categories: in-favor, against, and neither.

Task 2b: premise classification The second sub-task is to predict whether at least one premise/argument is mentioned. A given tweet is considered as having a premise if it contains

Set	Stance classes			Premise classes	
	in-favor	against	neither	1	0
Train	1346	874	1336	1331	2225
Valid	244	158	198	221	379
Test	526	570	904	716	1284

Table 1: Statistics of the Tasks 2a (stance) & 2b (premise) datasets

a statement that can be used as an argument in a discussion. For instance, the annotator could use it to convince an opponent about the given claim. For example, both tweets "Petition to stop calling people who don't wear masks "anti-maskers". Instead, let's just call them terrorists. #coronavirus" and "Masks help prevent the spread of the disease. Please, #WEARAMASK" are in favor of the claim, yet only the second tweet contains the argument. The tweets were manually annotated for binary classification, and participants of this sub-task were required to submit whether each tweet has a premise (1) or not (0).

We split an existing corpus of 4,269 COVID-related tweets (Glandt et al., 2021) into a training set of 3,669 tweets and a validation set of 600 tweets. This corpus includes annotations for Task 2a. We added a new annotation layer for Task 2b to this corpus. In order to create the test set, we collected new tweets using 33 keywords such as #OpenSchools, #LockdownNow, #NoMasks. We removed the hashtags from the tweets to exclude obvious signals (e.g., #SayNoToMasks can be seen as the "against" hashtag). The test set for Task 2a and all three sets for Task 2b were manually annotated. Three *Yandex.Toloka*[1] annotators' crowd-sourced labels were aggregated into a single label (Dawid and Skene, 1979). Table 1 shows statistics of the experimental datasets. More details on the datasets are presented in (Davydova and Tutubalina, 2022). We used the F_1-score as the main evaluation metric in both sub-tasks: $F_1 = \frac{1}{n}\sum_{c \in C} F_{1_{rel_c}}$, where C ={"face masks", "stay at home orders", "school closures"}, n is the size of C, $F_{1_{rel_c}}$ is macro F_1-score averaged over two classes for each task (in-favor & against classes for Task 2a; 0 & 1 classes for Task 2b). The Task 2 is hosted on Codalab at `https://codalab.lisn.upsaclay.fr/competitions/5067`.

[1]`https://toloka.yandex.ru/`

2.3 Task 3: Classification of changes in medication treatments in tweets and WebMD reviews (in English)

In Task 3, we challenged the participants to design binary classifiers to detect posts where social media users self-declare changing their medication treatments, regardless of being advised by a health care professional to do so. Such changes are, for example, not filling a prescription, stopping a treatment, changing a dosage, forgetting to take the drugs, etc. This is an important task since it is the first step toward detecting patients who may be non-adherent to their treatments and it would allow us to further understand their reasons when they are expressed on social media (Weissenbacher et al., 2021). We released two corpora for the task, 9,830 tweets from Twitter and 12,972 drug reviews from WebMD, a website which provides an opportunity for users to comment anonymously on their personal experience with a drug in a free text form. Two annotators labeled the posts with "1" when the posts state a change in medication regimen, the positive examples, "0" otherwise, the negative examples. The Inter-annotator agreements were moderate on both corpora with 0.65 and 0.74 Cohen Kappa scores on the corpus of tweets and reviews, respectively. With 7,222 positive examples and 5,750 negative examples, the corpus of WebMD is naturally balanced with an Imbalance Ratio of 0.80. The corpus of tweets is more challenging for training classifiers with supervision, which is the default approach to solve the task. The corpus of tweets has a strong imbalance with only 864 positive examples for 8,966 negative examples, that is a 10.38 Imbalance Ratio. We split randomly each corpus into a training, a validation, and a test set with 90%/10%/10% of the total reviews available and 60%/16%/24% of the total tweets available in each set. During the practice period of the competition, we provided the participants the training and validation sets. During the evaluation period, all participants had 5 days to compute their predictions on the test set and submit them on Codalab for evaluation. We added 11,835 reviews and 13,000 tweets as decoys in the test sets to prevent manual correction of the predictions. We evaluated participants' systems with the Precision, Recall and F_1-score for the positive class, that is tweets or reviews mentioning a change in medication treatments. The Task 3 is hosted on Codalab at `https://codalab.lisn.upsaclay.fr/competitions/2138`.

2.4 Task 4: Classification of tweets self-reporting exact age (in English)

A limitation of using Twitter data for research applications is that users may not be representative of the general population. Therefore, advancing the utility of Twitter data for research applications requires methods for automatically detecting demographic information, including users' age. Automatically identifying the exact age of Twitter users, rather than their age groups, would enable the large-scale use of Twitter data for applications that do not align with the predefined age groupings of extant models, including health applications such as identifying specific age-related risk factors for observational studies (Golder et al., 2019), or selecting age-based study populations (Davies et al., 2022). As a first step, this binary classification task involves automatically distinguishing tweets that self-report the user's exact age ("age" tweets) from those that do not ("no age" tweets). The training set contains 8800 annotated tweets: 2834 (32%) "age" tweets and 5966 (68%) "no age" tweets. The validation set contains 2200 annotated tweets: 709 (32%) "age" tweets and 1491 (68%) "no age" tweets. The test set also contains 2200 annotated tweets: 768 (35%) "age" tweets and 1432 (65%) "no age" tweets. Inter-annotator agreement (Fleiss' kappa), based on 1000 tweets annotated by five annotators, was 0.80. The Task 4 is hosted on Codalab at `https://codalab.lisn.upsaclay.fr/competitions/3566`.

2.5 Task 5: Classification of tweets containing self-reported COVID-19 symptoms (in Spanish)

The purpose of this task is to bridge the gap in NLP and social media for COVID-19 research performed in languages other than English. While there has been an increased amount of non-English datasets and tasks using social media proposed in the last couple of years, there is still a need for different applications on pressing topics. This shared task is similar to the 2021 #SMM4H shared task 6 (Magge et al., 2021), which involved identifying personal mentions of COVID-19 symptoms tweets. The annotated set of tweets for this task is a set of manually curated Spanish-native language tweets. The task is a three-way classification problem, requiring participants to distinguish personal symptom mentions (self-reports) from other mentions such as symptoms reported by others (non-personal reports) and references to external sources (literature/news mentions). We provided a training dataset consisting of 1,654 tweets labeled as self-reports, 2,413 tweets labeled as non-personal reports, and 5,985 labeled as literature/news mentions. The test set consisted of 6,851 tweets, 1,096 self-reports, 1,644 non-personal reports, and 4,111 literature/news mentions. The complete annotated dataset (train, validation, testing) has a Cohen Kappa score inter annotator agreement of 0.85. The systems submitted for this task where evaluated using micro-average F1-score. The Task 5 is hosted on Codalab at `https://codalab.lisn.upsaclay.fr/competitions/3535`.

2.6 Task 6: Classification of tweets which indicate self-reported COVID-19 vaccination status (in English)

With the widespread roll-out of COVID-19 vaccines, vaccine surveillance became a very pressing research issue. While some vaccinated people report adverse events via their healthcare providers to systems like Vaccine Adverse Event Reporting System (VAERS), or are found documented in their electronic health record (EHR), a more robust and convenient method could be devised using self-reports from social media. In this task we provided an annotated dataset of Tweets with users personally reporting vaccination status or discussing vaccination status but not revealing their own, extracted from Banda et al. (Banda et al., 2021). This task is challenging since users often discuss vaccination status of others or from news reports in similar ways and at a higher rate than they discuss their own (1 to 8 on average). The dataset presents as the positive class, unambiguous tweets of users clearly stating that they have been vaccinated (vaccination confirmation). All other tweets are of users discussing vaccination status. This task involved the identification of self-reported COVID-19 vaccination status in English tweets (vaccine related chatter). This task is posed as a two-way classification task, where the systems submitted were evaluated on precision, recall and F1-score. The class imbalance in this dataset is roughly 1 to 8, meaning that for training we provided 1,496 tweets of vaccination confirmation, and 12,197 of vaccine chatter tweets. The test set is comprised of 652 tweets labeled as self-reports and 5,271 tweets of vaccine chatter. The complete annotated dataset (train, validation, testing) has a Cohen Kappa score inter annotator agreement of 0.82. The Task 6 is hosted on Codalab at `https://codalab.lisn.upsaclay.fr/competitions/3536`.

2.7 Task 7: Classification of self-reported intimate partner violence on Twitter (in English)

Intimate partner violence (IPV), which refers to abuse or aggression that occurs in a romantic relationship, is a serious health problem that can have a lifelong impact on health and well-being (Smith et al., 2018). Social media platforms are often used by IPV victims to share experiences and seek for help (Westbrook, 2015; Cravens et al., 2015; McCauley et al., 2018; Chu et al., 2021; Al-Garadi et al., 2022). To potentially provide timely intervention and support to victims, we have the need for an effective automatic classifier to detect self-reports of IPV on social media platforms. Task 7 is a binary classification task that involves identifying the IPV self-report posts on Twitter. This task presents two specific challenges. First, the annotated data is significantly imbalanced where only around 11% of the tweets are identified as self-reports of IPV. Second, the negative tweets include non-IPV domestic violence and non-self-reported IPV, which can be very difficult to distinguish from self-reported IPV for an automatic system. The data (Al-Garadi et al., 2022) include a total of 6,348 annotated posts from Twitter, of which 4,523 were provided for training, 534 provided for validation, and 1,291 provided for testing. IAA was found to be 0.86 (Cohen's kappa) among 1,834 double-annotated tweets. Systems were evaluated and ranked based on the F_1-score of the positive class (self-reported IPV). The Task 7 is hosted on Codalab at `https://codalab.lisn.upsaclay.fr/competitions/1535`.

2.8 Task 8: Classification of self-reported chronic stress on Twitter (in English)

Chronic stress is defined as the *physiological or psychological response to a prolonged internal or external stressful event* (VandenBos, 2007), which can lead to poor mental health, including depression and anxiety, and can also take a toll on the body, resulting in the dysfunctions of cardiovascular, metabolic, endocrine, and immuno-inflammatory systems. Traditional methods for assessing stress, including interviews, questionnaires/surveys, etc., have limitations associated with accurately measuring stress at the population level (Epel et al., 2018). Therefore, there is a need to identify new sources of data and new methods for assessing chronic stress. One potential source of information for analyzing chronic stress related information is social media such as Twitter. The first step to do so is to accurately detect the tweets that report personal experiences of chronic stress. Task 8 is a binary classification that involves automatically identifying tweets that are self-disclosures of chronic stress from those that are not.

For Task 8, we released a corpus of tweets where about 37% of the tweets are positive (self-disclosures; P) and 63% are negative (non-self-disclosures; N). We split the corpus in three set, the training set which contains 2,936 tweets, the validation set, 420 tweets, and the test set, 839 tweets. The pairwise inter-annotator agreement among 695 double-annotated tweets was κ=0.83 (Cohen's kappa (Cohen, 1968)), which can be interpreted as a substantial agreement. Further details about the data and the annotation process are provided in Yang et al. (2022b). Classifiers were evaluated based on the F_1-scores for the "positive" class (i.e., tweets that are self-disclosure of chronic stress). The Task 8 is hosted on Codalab at `https://codalab.lisn.upsaclay.fr/competitions/1542`.

2.9 Task 9: Classification of Reddit posts self-reporting exact age (in English)

Pharmaceutical companies, with the encouragement of regulatory agencies (Donegan et al., 2019; U.S. Food and Drug Administration, 2020), have started using social media listening (SML) to integrate the patient perspective in the clinical development process to ensure relevant treatments and outcomes. Traditionally, SML in the pharmaceutical setting has been done through manual, qualitative methods. However, it has been shown that quantitative SML (QSML) can enhance the value of social media data and enable a patient-centric approach to understanding disease burden and influence drug discovery decisions at all stages (Schmidt et al., 2022).

The detection of self-reported demographic information on social media can help in assessing the demographic characteristics (e.g. age, gender, ethnicity, medical history) of patients on social media in comparison to patients in target clinical populations. Task 9 is a binary classification that aims to distinguish automatically posts in Reddit forum where users that self-report their exact age at the time of posting from those where they do not. The dataset is disease-specific and consists of posts collected via a series of keywords associated with dry eye disease. The training set contains 9,000 annotated posts mentioning numbers that were randomly selected: 2,921 (32.5%) with self-reported ages (annotated as "1") and 6,079 (67.5%) with no self-reported ages (annotated as "0"). The test set contains 2,000 annotated posts: 629 (31.5%) with self-reported ages (annotated as "1") and 1,371 (68.5%) with no self-reported ages (annotated as "0"). Inter-annotator agreement (Cohen's kappa) was 0.939. Systems were evaluated based on the F1-score for the target class (self-reported age mentioned). The Task 9 is hosted on Codalab at `https://codalab.lisn.upsaclay.fr/competitions/3646`.

2.10 Task 10: SocialDisNER - Detection of disease mentions in tweets (in Spanish)

Since diseases is an important category of named entities to help recognizing health-related content in social media, the community has invested time and effort to collect and annotate large corpora to train Named-Entity Recognition systems to perform the task automatically. However, most corpora are written in English, for example (Scepanovic et al., 2020), CADED (Karimi et al., 2015), Micromed (Jimeno-Yepes et al., 2015), or TwiMed (Alvaro et al., 2017). Fewer corpora written in Spanish, like SpanishADR (Segura-Bedmar et al., 2014) or ProfNER (Miranda-Escalada et al., 2021), are available. Task 10, hereafter called SocialDisNER, is a first attempt to fill the gap. The goal of SocialDisNER is the automatic recognition of disease mentions in tweets. We used the LINKAGE methodology (Gasco et al., 2022) to select a set of disease-related tweets, written in Spanish and with first-hand experience from patients and their relatives. The corpus also contains relevant health information tweets written by medical professionals. The corpus consists of 9,500 tweets, with 5,000 tweets used for training, 2,500 for development and 2,000 for evaluation. The corpus was annotated by a medical professional following an adaptation of the DisTEMIST annotation guidelines, which were tested with several rounds of inter-annotation agreement (IAA) to achieved a final human-IAA of 0.823 (Gasco et al., 2022). The primary evaluation metric for the task was the micro-averaged F1-score. Participants were also compared to a baseline system that was computed using a Levenshtein lexical lookup approach with a sliding window of variable length. In addition to the Gold Standard of 9,500 tweets, we also provided the participants with a Silver Standard. It is a large-scale corpus of 80,000 tweets where we automatically annotated the mentions of diseases, symptoms, procedures, drugs, species and professions, among others. Annotation was carried out through NER systems trained on clinical data in Spanish and post-processing to eliminate false negatives such as URLs or twitter mentions. In order to encourage the use of the large-scale corpus and foster the use of mined content, the co-mention matrices of the Large scale corpus were calculated and shared. These matrices represent the co-occurrences of disease mentions to each other, as well as the co-occurrences of diseases with other terms such as symptoms or professions. An example of these matrices can be seen in the co-mention network in Figure 1. The Task 10 is hosted on Codalab at `https://codalab.lisn.upsaclay.fr/competitions/3531`.

Figure 1: Simplified SocialDisNER co-mention network between disease and symptoms. Network filtered for fibromyalgia symptoms.

3 Results

3.1 Task 1: Classification, detection and normalization of Adverse Events (AE) mentions in tweets (in English)

A total of 34 teams participated in at least one of the subtasks of Task 1 making 80 valid submissions. 19 out of the 34 teams submitted system descriptions. Tables 2, 3, 4 presents F_1 scores for the participating teams' best submissions along with brief descriptions of the architecture choices made by the teams. We encourage readers to refer to the original papers for detailed descriptions of the systems. Similar to last years' submissions, all teams for Task 1a used transformer models, while for Tasks 1b and 1c, about 70% of the submissions used transformer models. RoBERTa and BERTweet (based on RoBERTa) were the most popular choice of models and model ensembles were used in many of the top submissions. Variations of off-the-shelf models were used in the tasks for addressing the class imbalance problem and the large label space problem presented in the normalization task. Some of these variations included data augmentation techniques, multi-corpus training, optimizer adaptations and adversarial training.

3.2 Task 2: Classification of stance and premise in tweets about health mandates related to COVID-19 (in English)

Table 5 presents F_1 scores for each of the 15 team's best-performing system. Most teams used COVID-related BERT models with additional techniques, such as regularized dropout (R-drop), to alleviate the unbalanced label distribution and overfitting. Two teams (# 22, 28) tried to combine the data to train a model that perform simultaneously both tasks. Leading teams on both tasks tried to aggregate claim and tweet texts: the leading team #7 with 0.64 F_1 in Task 2a appended the claim text to the end of the tweet, while the second-best team #12 in Task 2a with highest F_1 (0.7) in Task 2b proposed a new pre-training task constructed by the tweets and claims similarly to next sentence prediction.

3.3 Task 3: Classification of changes in medication treatments in tweets and WebMD reviews (in English)

We observed a good interest in Task 3 with 43 teams registered. Among these teams, 7 teams submitted their predictions to Codalab. In Table 6, we summarized their performance and compare them with our baseline systems described in (Weissenbacher et al., 2021). The main focus of the participants during the competition was the generalizability of their classifiers. These past years, the community released large pre-trained embeddings, such as glove, fasttext, or transformers, along with convenient interfaces to integrate them into neural networks. These interfaces propose default architectures for these neural networks that can be programmatically customized by advanced users. Such networks provide new opportunities since they can be easily trained with supervision to solve any classification tasks at hand by fine-tuning their models on small - or larger - annotated corpora and still achieving competitive results. All but one participants took advantages of these neural networks and evaluate their abilities to solve multiple tasks of the #SMM4H 2022 competition, among which Task 3. We note that few teams adapted their training process to improve performances on imbalance corpora, which was the key of success for sub-task 3a. This corpus of tweets remains challenging with the best scores plateauing around 0.65 F_1-score compare to the balanced corpus of WebMD reviews were transformer-based approaches achieved high F_1 scores, scores above 0.80.

3.4 Task 4: Classification of tweets self-reporting exact age (in English)

Table 7 presents the precision, recall, and F_1-score for each team's best-performing system for Task 4. The four top-performing systems achieved F_1-scores within less than 0.01 points of one another using BERTweet or RoBERTa pretrained transformer models. Among these four systems, using a model pretrained on tweets did not seem to improve performance for this task. These four teams marginally outperformed a RoBERTa-Large baseline classifier (Klein et al., 2022) by using techniques such as pseudo-labeling, ensemble learning, multi-corpus training, adversarial attacks, and child-tuning. All seven systems that used BERTweet or RoBERTa models, including the baseline classifier, outperformed systems that used BERT or BioBERT models.

3.5 Task 5: Classification of tweets containing self-reported COVID-19 symptoms (in Spanish)

With 26 participants registered in CodaLab and seven final team submissions highlighted on Table 8, this task presented an interesting challenge to the teams. Most teams, with distinct approaches ranging from ensemble models to tuned BERT-based model, performed well and only 3 percentage points separated the best from the last ranking teams. The baseline model, which featured a hefty augmented training set using weak supervision (Tekumalla and Banda, 2021), still man-

Team	F_1	P	R	System Summary
25	0.698	0.839	0.598	BERTweet-large with data augmentation
23	0.693	0.772	0.629	Majority ensemble of 10 RoBERTa-large models
45	0.689	0.790	0.611	DeBERTa-v3-large model with adversarial training
-	0.671	0.799	0.578	-
-	0.669	0.791	0.579	-
14	0.662	0.785	0.573	RoBERTa and BERTweet with exponential moving average
-	0.662	0.790	0.570	-
46	0.662	0.765	0.584	Ensemble of BERTweet, DeBERTa and BioBERT with data augmentation
-	0.660	0.787	0.569	-
37	0.655	0.688	0.625	Ensemble modeling from T5 gpt2-large templates with over/under sampling
41	0.652	0.737	0.585	RoBERTa model pretrained on in-domain tweets
11	0.642	0.554	0.765	Glove embeddings and DeepADEMiner (RoBERTa) classifier
-	0.638	0.807	0.528	-
10	0.637	0.787	0.536	Fine-tuned RoBERTa-base with Adversarial Training
27	0.610	0.606	0.614	Ensemble of finetuned BERTweet-large, RoBERTa-large, CT-BERT
40	0.601	0.705	0.524	RoBERTa with Fast Gradient Method and Project Gradient Descent
38	0.567	0.674	0.489	RoBERTa with adaptive learning and mixut
-	0.537	0.724	0.427	-
18	0.491	0.384	0.681	RoBERTa with data augmentation and downsampling techniques
28	0.472	0.607	0.386	BERT fine-tuned on medically relevant data
-	0.470	0.659	0.365	-
8	0.433	0.614	0.334	Template-augmented training using BERTweet model
17	0.413	0.677	0.297	BERT, RoBERTa and ERNIE 2.0 with voting ensembler
-	0.396	0.593	0.297	-
-	0.333	0.398	0.287	-
-	0.326	0.603	0.223	-
1	0.299	0.235	0.409	Ensemble of BERT, BioBERT, XLNet, RoBERTa
-	0.224	0.485	0.145	-
36	0.077	0.041	0.547	RoBERTa and BERTweet with label-distribution aware margin loss

Table 2: Evaluation results for Task 1a: Classification of ADE mentions in English tweets. Metrics show F_1-scores (F_1), precision (P), and recall (R) for the *ADE* class.

Team	F_1	P	R	System Summary
45	0.651	0.684	0.621	W2NER model with character and location features
-	0.642	0.688	0.601	-
-	0.640	0.686	0.600	-
-	0.639	0.683	0.599	-
-	0.637	0.681	0.599	-
11	0.624	0.569	0.691	DeepADEMiner(RoBERTa-large) and Flair embeddings
23	0.568	0.671	0.492	Majority ensemble of 5 BERT-large models
37	0.519	0.644	0.434	Ensemble modeling from T5 gpt2-large templates with over/under sampling
25	0.484	0.828	0.341	BERTweet-large with positive examples from WEBRADR dataset
38	0.435	0.562	0.354	Question Answering methodology using RoBERTa-base fine-tuned on SQuAD2.0 dataset
28	0.404	0.489	0.344	Fine-tuned BERT developed for SMM4H 2019
1	0.220	0.178	0.288	Fine-tuned RoBERTa+CRF model
-	0.164	0.096	0.576	-
-	0.132	0.074	0.651	-

Table 3: Evaluation results for Task 1b: Extraction of ADE mentions in English tweets. Metrics show overlapping F_1-scores (F_1), precision (P), and recall (R) for the *ADE* spans.

Team	F_1	P	R	System Summary
11	0.387	0.350	0.432	DeepADEMiner(bert-based-uncased) and mcn-en-smm4h model (BioBERT)
45	0.367	0.405	0.336	Ranked average pooling of DeBERTa model word vectors
28	0.243	0.294	0.207	GPT-2 model trained to predict MedDRA term from ADE span
38	0.172	0.232	0.137	Fuzzy matching with Levenshtein distance
1	0.116	0.094	0.152	Ranked similarity metrics of extracted spans and MedDRA dicitionary
23	0.070	0.087	0.058	Stop-word removal and simple string matching in MedDRA
-	0.013	0.019	0.009	-
-	0.011	0.017	0.008	-
-	0.010	0.015	0.007	-
-	0.008	0.013	0.006	-
-	0.007	0.011	0.005	

Table 4: Evaluation results for Task 1c: Extraction and normalization of ADE mentions in English tweets. Metrics show overlapping F_1-scores (F_1), precision (P), and recall (R) for the *ADE* spans with exact matches for MedDRA preferred term IDs.

Team	F1 stance (Task 2a)	F1 premise (Task 2b)	System Summary
7	**0.64**	0.66	CovidTwitterBERT (for 2a) / BART-base (for 2b) + cross-entropy and contrastive losses + features
12	<u>0.63</u>	**0.70**	COVID-Twitter-BERT-v2 / RoBERTa-large + Tweet Claim Matching (TCM) pre-training
22	0.62	0.62	CT-BERT V2 + related data, sentiment classification along with multi-task learning
42	0.62	0.63	Dual-view attention neural network + COVID-Twitter-BERT-v2
14	0.59	**0.70**	BERTweet + regularized dropout (R-drop), exponential moving average (EMA), and focal loss
23	0.58	**0.70**	RoBERTa-large + majority ensemble (2a), BERT-large + weighted average ensemble (2b)
46	0.55	0.64	BioBERT + R-drop, poly loss and focal loss, pseudo labels
28	0.53	0.65	6-way joint classification + ensemble of RoBERTa and BERT
17	0.50	0.62	A voting-ensemble model that comprises fine-tuned BERT, RoBERTa, and ERNIE 2.0 / fine-tuned single models
31	0.48	0.64	A voting classifier to ensemble the predictions of BERT, RoBERTa, PubMedBERT, SciBERT and SPECTER
19	0.43	0.63	bioBERT-base
-	0.32	-	-
9	0.23	0.23	An ensemble of fine-tuned GAN-BERT models
6	0.08	0.00	RoBERTa-large/BERTweet-large
29	-	<u>0.67</u>	SqueezeBERT

Table 5: Evaluation results for Task 2: Classification of stance and premise in tweets about health mandates related to COVID-19. The best scores for each task are in bold and second best scores are underlined.

aged to rank the highest with a micro F_1-score of 0.90. Team 14 and their pre-trained BERT-base model with R-drop, exponential moving average, and pseudo-labeling was the highest ranking team with a micro F_1 score of 0.86.

3.6 Task 6: Classification of tweets which indicate self-reported COVID-19 vaccination status (in English)

The task of classifying self-reported COVID-19 vaccination status, in English, attracted 44 participants in the competition, with eight teams submitting their predictions on the unseen test set on Codalab. Table 9 shows that team 28 managed to match the baseline system as the best performing systems, with an improvement in precision and a

Team	F_1	P	R	System Summary
				Task 3a
14	**0.66**	0.68	**0.63**	Ensemble of BERTweet classifiers trained with 5-fold cross validation + Cost sensitive learning and data augmentation
46	0.64	**0.68**	0.60	BERTweet + data augmentation
33	0.61	0.66	0.57	RoBERTa embeddings input for a neural network with a stacked LSTM and linear layer unified branches
17	0.59	0.62	0.56	Bio-RoBERTa
baseline	0.50	0.47	0.53	CNN trained with transfer and active learning
19	0.45	0.53	0.39	BioBERT
11	0.19	0.55	0.12	Glove embeddings pretrained on tweets input for Bi-LSTM + Cost sensitive learning
41	0.06	0.03	0.34	RoBERTa embeddings pretrained on tweets
				Task 3b
baseline	**0.87**	**0.87**	0.88	BERT-based
33	0.86	0.85	0.88	RoBERTa embeddings input for a neural network with a stacked LSTM and linear layer unified branches
46	0.86	0.86	0.85	BERTweet + data augmentation
19	0.83	0.84	0.82	BioBERT
41	0.72	0.57	**1.0**	RoBERTa embeddings pretrained on tweets

Table 6: System summaries and scores (F_1), precision (P), and recall (R)) for Task 3: Classification of changes in medication treatments in tweets and WebMD reviews - in English

Team	F_1	P	R	System Summary
46	**0.92**	**0.93**	**0.91**	BERTweet, pseduo-labeling, R-Drop, PolyLoss
10	0.92	0.93	0.90	RoBERTa-Base, sub-corpus ensemble, adversarial training, child-tuning
14	0.91	0.92	0.91	Ensemble of BERTweet classifiers trained with 5-fold cross validation, pseudo-labeling, R-Drop, EMA
6	0.91	0.92	0.90	RoBERTa-Large, additional training data from Task 9
Baseline	0.91	0.93	0.88	RoBERTa-Large
18	0.89	0.90	0.89	RoBERTa
35	0.85	0.80	0.89	RoBERTa-Large
15	0.82	0.77	0.87	BERT-Base-Uncased
19	0.81	0.82	0.80	BioBERT-Base-Cased
-	0.73	0.67	0.79	-

Table 7: System summaries and F_1-scores (F_1), precision (P), and recall (R) for Task 4: Classification of tweets self-reporting exact age (in English).

Team	F_1	P	R	System Summary
baseline	**0.90**	**0.90**	**0.90**	CT-BERT + data augmentation (labeled using weak supervision)
14	0.86	0.86	0.86	Custom pre-training BERT-based model + R-drop, exponential moving average, and pseudo-labeling
26	0.85	0.85	0.85	XLM-RoBERTa-base + post-processing step
46	0.85	0.85	0.85	BioBERT + R-drop + focal loss
13	0.85	0.85	0.85	Majority ensemble of: BERT BASE-multilingual, BETO, XLM-RoBERTa
5	0.84	0.84	0.84	Fine-tuned RoBERTuito + data pre-processing
28	0.84	0.84	0.84	Ensemble of two differently configured BERT + one RoBERTa models
17	0.83	0.83	0.83	XLM-R

Table 8: Evaluation results for Task 5: Classification of tweets containing self-reported COVID-19 symptoms (in Spanish). Metrics shows F_1-scores (F_1), precision (P), and recall (R) for the self-reports class. The table shows the best submission of each team and their system.

drop in recall using an ensemble model. Similar to other classification shared tasks, most systems used BERT-based models with variants like CT-BERT and BioBERT being popular. The two approaches, baseline and team 27 with augmented data performed better (on average) than the systems that did not use additional data, so this is an interesting result from the presented approaches.

3.7 Task 7: Classification of self-reported intimate partner violence on Twitter (in English)

Table 10 presents F_1-scores, precision, and recall for the Intimate Partner Violence (IPV) self-report class for the participating teams. The median F_1-score, precision, and recall of all the submissions are 0.763, 0.79, and 0.716, respectively. All of the participants used pre-trained transformer-based models, 4 out of 7 teams used multiple BERT variants, 4 teams used RoBERTa, and 3 teams used ensemble techniques. The leading team achieved F_1-score of 0.851 which is 0.8 higher than the baseline system. The leading team also achieved the best recall, which is 0.4 higher than the second highest recall score. The leading team used RoBERTa-Large to encode tweet text and make a binary prediction according to the corresponding pooling vector. The leading team trained on 5 RoBERTa models and applied a weighted ensemble strategy. The leading team achieved lower precision but higher recall than the second-placed team, which also used RoBERTa and domain-adaptive pre-training. The leading team achieved higher precision and recall than those of the third-placed team, which applied an averaging ensemble on different transformer-based models. Team 3 and 10 achieved significantly higher F_1-scores compared to other teams, which might be attributed to the efficient weighted ensemble strategy and domain adaptive pre-training.

3.8 Task 8: Classification of self-reported chronic stress on Twitter (in English)

Table 11 presents the F_1-scores, precision, and recall for the positive class, for each of the 11 team's best-performing system for Task 8. The best performance achieved in task 8 was an F_1-score of 0.792, which is comparable to the benchmark reported in the literature on the same corpus (Yang et al., 2022b). The median F_1-score, precision and recall of all submissions are 0.75, 0.72 and 0.76, respectively. Only 7 of 11 teams submitted their system descriptions, among which 6 of the 7 teams used RoBERTa. 3 teams used ensemble systems built from multiple pre-trained transformer-based models, among them, 2 teams included BERT and 2 teams included BERTweet. The leading team preprocessed the texts by lowercasing, deleting URLs, replacing emojis with their text strings; and used 5-fold CV during the training phase. The leading team used three pre-trained BERT-based models including BERT, RoBERTa and BERTweet then fine-tuned them using task 8's datasets. To further improve the models' performance, the leading team also adopted pseudo-labeling in the post-processing phase. The leading team's results showed that BERTweet outperformed BERT and RoBERTa for this task. Both top teams included BERTweet in their systems, indicating the benefit of pre-training the embeddings on social media based corpora for this task.

Team	F_1	P	R	System Summary
baseline	**0.83**	**0.9**	**0.77**	CT-BERT + data augmentation (labeled using weak supervision)
28	0.83	0.93	0.75	Ensemble of three differently configured BERT models.
27	0.82	0.86	0.78	CT-BERT fine-tuned from scratch + Data augmentation (manual and automatic)
46	0.81	0.90	0.74	BioBERT + R-drop
14	0.80	0.90	0.71	Custom pre-training BERT-based model + R-drop, exponential moving average, and pseudo-labeling
10	0.78	0.91	0.68	Continued pre-trained RoBERTa$_b asemodel$
11	0.69	0.87	0.87	LSTM + GLOVE Twitter Embeddings
17	0.68	0.76	0.87	BERT
47	0.66	0.77	0.93	Voting ensemble of: fine-tuned BERT, RoBERTa, and XLNet models

Table 9: Evaluation results for Task 6: Classification of tweets which indicate self-reported COVID-19 vaccination status (in English). Metrics shows F_1-scores (F_1), precision (P), and recall (R) for the self-reports class. The table shows the best submission of each team and their system.

Team	F_1	P	R	System Summary
3	0.851	0.86	0.841	RoBERTa-Large with weighted ensemble
10	0.833	0.875	0.795	RoBERTa with domain adaptive pre-training
14	0.795	0.795	0.795	BERT, RoBERTa, and BERTweet with averaging ensemble
29	0.791	0.903	0.705	Domain specific BERT variants with different loss functions
Baseline	0.756	0.823	0.699	RoBERTa
46	0.734	0.784	0.689	Six BERT variants including RoBERTa with voting ensemble
19	0.707	0.763	0.659	BioBERT
36	0.625	0.549	0.727	RoBERTa and BERTweet with different loss functions

Table 10: Evaluation results for Task 7: Classification of self-reported intimate partner violence on Twitter (in English). Metrics show F_1-scores (F_1), precision (P), and recall (R) for the self-reports class.

Team	F_1	P	R	System Summary
14	**0.792**	0.734	0.859	BERT, RoBERTa, and BERTweet with averaging ensemble
46	0.783	**0.797**	0.769	BioBERT, pubmedBERT, DeBERTa and BERTweet with different loss functions
10	0.781	0.739	0.827	RoBERTa with domain adaptive pre-training
-	0.773	0.731	0.821	-
-	0.764	0.793	0.737	-
19	0.764	0.677	0.878	RoBERTa
44	0.750	0.718	0.785	RoBERTa with BiLSTM
21	0.730	0.703	0.760	BERT, RoBERTa, ALBERT, XLNet and ELECTRA transformers
-	0.719	0.743	0.696	-
-	0.643	0.599	0.692	-
41	0.542	0.372	**1.000**	RoBERTa

Table 11: Evaluation results for Task 8: Classification of self-reported chronic stress on Twitter. Metrics show F_1-scores (F_1), precision (P), and recall (R) for the self-disclosure class.

3.9 Task 9: Classification of Reddit posts self-reporting exact age (in English)

Table 12 presents the precision, recall, and F_1-score for each team's best-performing system for Task 9. The median F1-score, precision, and recall of all the submissions were 0.891, 0.896 and 0.919, respectively. All participating teams had a common approach which was to create a generic framework to fine-tune pre-trained transformer-base models, with which they participated in multiple classifications tasks. Two teams (6 and 35), nonetheless, focused their attention only on tasks 4 and 9, which were highly similar. These 2 teams obtained contrasting results, with one being the winner of the competition and the other being second to last. The best submissions of each team ended up using RoBERTa (3 teams), BERTweet (2 teams), BioBERT (1 team) and BERT-Large (1 team). Furthermore, 2 teams used ensemble models and 4 teams used a combination of additional strategies to improve performance (R-drop, FocalLoss, pseudo-labeling, oversampling, among others).

The best performance was obtained by team 6 with an F_1-score of 0.956, including the best recall, 0.963. The winner team used RoBERTa-Large with an augmented training corpus that included the training data from Task 4. This team was the only team that augmented the training dataset.

Finally, 2 teams additionally performed error analysis on the validation set. One of them found sufficient mislabeled posts, both as false positives and false negatives, that suggest that the dataset could benefit from a revision. The other team found that, unsurprisingly, their model puts special focus on numbers which might lead to incorrect predictions as it fails to understand complex semantics where the age is indirectly reported.

3.10 Task 10: SocialDisNER - Detection of disease mentions in tweets (in Spanish)

SocialDisNER has achieved good participation results with a total of 47 registered teams and 17 participating teams. Table 13 shows the ranking of the teams that uploaded predictions to the task according to their micro-average F1-score. There were 11 teams that achieved better performance than the baseline system. The top-performing team (Fu et al., 2022) obtained an F_1-score of 0.891 by developing a Unified Named Entity Recognition system through domain-adaptive pre-training using a domain-general Spanish BERT model (Canete et al., 2020) fine-tuned by adversarial training, child adjustment and model fusion.

Most participants used systems built from pre-trained transformer-based models and obtained the final predictions using token classification layers, CRFs or ensembles. A significant number of the participants (6 out of 17) have used or tested some of the language models trained with biomedical texts in Spanish published in the last few months (Carrino et al., 2022; Chizhikova et al.; Lange et al., 2021). Two teams opted to use general domain models in Spanish (Canete et al., 2020; Gutiérrez Fandiño et al., 2022) tuned using domain adaptation or weak training. A total of six teams chose to use multilingual models such as XLM-RoBERTa or multilingualBERT (Devlin et al., 2018; Conneau et al., 2019), one of them finishing in the top 3, showing how a good fine-tuning and processing strategy can deliver reliable results. It is worth highlighting that several teams used models that were pretrained on Spanish tweets (Huertas-Tato et al., 2022; Barbieri et al., 2022), although, they all achieved moderate performances. A possible explanation for these results may be that, to efficiently extract biomedical entities in Tweets, the models needs to transfer the linguistic knowledge learned during their pre-training on biomedical content, a knowledge which seems missing in general domain content like general tweets.

Six teams used the SocialDisNER large-scale corpus to solve the task. The top-performing team (Fu et al., 2022) used it to apply continual pre-training on the Spanish generalist language model to achieve better adaptation to the task domain. SINAI team (Chizhikova et al., 2022) used it to carry out a weak-supervision approach when training their system.

Most teams used the DisteMIST disease gazetteer for preprocessing and postprocessing disease mentions contained within special tweet tokens such as hashtags. In addition, they also used other resources such as CodiESP data (Miranda-Escalada et al., 2020) or biomedical resources such as Snomed-CT, UMLS or ICD-10 terminologies.

4 Conclusion

This paper presented an overview of the #SMM4H 2022 shared tasks. With ten tasks proposed this year, the interest and the participation in the #SMM4H shared tasks continues to grow. All best

Team	F_1	P	R	System Summary
6	**0.956**	0.948	**0.963**	RoBERTa-Large augemented with Task 4 training data
46	0.938	**0.957**	0.919	BERTweet and DeBERTa with R-drop, FocalLoss, PolyLoss and pseudo-labeling
17	0.918	0.896	0.941	Fine-tuned RoBERTa
14	0.891	0.893	0.889	BERTweet with averaging ensemble wirth R-drop, Exponential Moving Average, FocalLoss and pseudo-labeling
10	0.885	0.909	0.862	Continue RoBERTa-Base with averaging sub-corpus ensemble, FGM adversarial training, child-tuning and oversampling
35	0.865	0.797	0.946	BERT-Large with Synthetic Minority Oversampling Technique
19	0.856	0.862	0.851	BioBERT

Table 12: Evaluation results for Task 9: Classification of Reddit posts self-reporting exact age. Metrics show F_1-scores (F_1), precision (P), and recall (R) for the self-reported age class. The table shows the best submission of each team and their system.

Team	P	R	F1	System summary
10	**0.906**	<u>0.876</u>	**0.891**	Spanish BERT with domain adaptive pre-training + adversarial training + child tuning + model fusion
43	0.868	0.875	<u>0.871</u>	RoBERTa-base trained on biomedical corpus in Spanish + post-processing
39	0.851	**0.888**	0.869	Pre-processing and gazetteer lookup + XLM RoBERTa-Large
30	<u>0.882</u>	0.843	0.862	RoBERTa-base trained on biomedical corpus in Spanish with contextualized embeddings at document level.
34	0.842	0.860	0.851	Multilingual BERT + post-processing
26	0.828	0.845	0.836	XLM-RoBERTa-base + rule-based system
2	0.868	0.779	0.821	XLM-RoBERTa-base + lateral inhibitory layer
20	0.809	0.798	0.803	RoBERTa-base trained on biomedical corpus in Spanish + post-processing
5	0.756	0.795	0.775	Pre-processing + RoBERTa-base trained on Spanish general-domain corpus + Weak supervision + post-processing
24	0.779	0.769	0.774	Pre-processing and gazetteer lookup + Transformer-based ensemble
4	0.680	0.805	0.738	Data Augmentation + Joint Learning + post-processing
-	0.759	0.644	0.697	Terminology-based system
32	0.640	0.655	0.647	static and contextual embeddings with Flair NER framework.
16	0.836	0.494	0.621	Data Augmentation + BioBERT
-	0.505	0.625	0.559	-
28	0.504	0.461	0.481	Multilingual BERT
19	0.004*	0.004*	0.004*	Multilingual BERT
baseline	0.776	0.701	0.737	Terminology-based system

*Team 19 had problems in the evaluation phase due to the format used in the submission. The best system of team 51 was a post-workshop evaluation.

Table 13: Evaluation results for Task 10, Socialdisner: ranking with the best submission per team. Best result bolded, second best underlined. Metrics show micro-averaged F_1-score (F_1), precision (P), and recall (R)s.

performing teams opted for an architecture based on transformer models, often trained with heuristics chosen according to the characteristics of the task at hand, e.g. imbalance or texts not written in English. We note that among the 47 teams that submitted at least one predictions and wrote a paper description, 0.43% (20 teams) participated in multiple tasks, often with the same systems that were fine tuned to perform the different tasks. We believe this marks an important effort for the community to develop high-performing classifiers/label sequencers that are generalizable and reusable. The system description papers that are cited in Appendix 4 were each peer-reviewed by two reviewers and provide further details about the 47 teams' systems.

Acknowledgements

The work for #SMM4H 2022 at Cedars-Sinai Medical Center was supported by the National Institutes of Health (NIH) National Library of Medicine (NLM) [grant number R01LM011176]. The work on the set of tweets for Task 2 was done at Kazan Federal University and supported by the Russian Science Foundation [grant number 18-11-00284]. The work on Task 10-SocialDisNER has been supported by: the Encargo of PlanTL (SEDIA) to BSC; BIOMATDB, a European Union Horizon Europe coordination and support action under grant agreement no. 101058779; and the AI4ProfHealth project (PID2020-119266RA-I00)

The authors would also like to thank all those who reviewed the system description papers.

References

Omar Adjali, Fréjus A. A. Laleye, and Umang Aggarwal. 2022. Ofu@smm4h'22: Mining advent drug events using pretrained language models. In *Proceedings of the Seventh Social Media Mining for Health Applications (#SMM4H) Workshop & Shared Task*, pages 171–175.

Mohammed Ali Al-Garadi, Sangmi Kim, Yuting Guo, Elise Warren, Yuan-Chi Yang, Sahithi Lakamana, and Abeed Sarker. 2022. Natural language model for automatic identification of intimate partner violence reports from twitter. *Array*, 15:100217.

Nestor Alvaro, Yusuke Miyao, Nigel Collier, et al. 2017. Twimed: Twitter and pubmed comparable corpus of drugs, diseases, symptoms, and their relations. *JMIR public health and surveillance*, 3(2):e6396.

Andrei-Marius Avram, Vasile Pais, and Maria Mitrofan. 2022. Racai@smm4h'22: Tweets disease mention detection using a neural lateral inhibitory mechanism. In *Proceedings of the Seventh Social Media Mining for Health Applications (#SMM4H) Workshop & Shared Task*, pages 1–3.

Juan M. Banda, Ramya Tekumalla, Guanyu Wang, Jingyuan Yu, Tuo Liu, Yuning Ding, Ekaterina Artemova, Elena Tutubalina, and Gerardo Chowell. 2021. A large-scale covid-19 twitter chatter dataset for open scientific research—an international collaboration. *Epidemiologia*, 2(3):315–324.

Francesco Barbieri, Luis Espinosa-Anke, and Jose Camacho-Collados. 2022. Xlm-t: Multilingual language models in twitter for sentiment analysis and beyond. *Proceedings of the LREC, Marseille, France*, pages 20–25.

Alec Louis Clemente Candidato, Akshat Gupta, Xiaomo Liu, and Sameena Shah. 2022. Airjpmc@smm4h'22: Classifying self-reported intimate partner violence in tweets with multiple bert-based models. In *Proceedings of the Seventh Social Media Mining for Health Applications (#SMM4H) Workshop & Shared Task*, pages 135–137.

José Canete, Gabriel Chaperon, Rodrigo Fuentes, Jou-Hui Ho, Hojin Kang, and Jorge Pérez. 2020. Spanish pre-trained bert model and evaluation data. *Pml4dc at iclr*, 2020:1–10.

Casimiro Pio Carrino, Joan Llop, Marc Pàmies, Asier Gutiérrez-Fandiño, Jordi Armengol-Estapé, Joaquín Silveira-Ocampo, Alfonso Valencia, Aitor Gonzalez-Agirre, and Marta Villegas. 2022. Pretrained biomedical language models for clinical nlp in spanish. In *Proceedings of the 21st Workshop on Biomedical Language Processing*, pages 193–199.

Kendrick Cetina and Nuria García-Santa. 2022. Fre at socialdisner: Joint learning of language models for named entity recognition. In *Proceedings of the Seventh Social Media Mining for Health Applications (#SMM4H) Workshop & Shared Task*, pages 68–70.

Mariia Chizhikova, Jaime Collado-Montañéz, Pilar López-Úbeda, Manuel C. Díaz-Galiano, L. Alfonso Ureña-López, and M. Teresa Martín-Valdivia. Sinai at clef 2022: Leveraging biomedical transformers to detect and normalize disease mentions. pages 265–273.

Mariia Chizhikova, Pilar López-Úbeda, Manuel C. Díaz-Galiano, L. Alfonso Ureña-López, and M. Teresa Martín-Valdivia. 2022. Sinai@smm4h'22: Transformers for biomedical social media text mining in spanish. In *Proceedings of the Seventh Social Media Mining for Health Applications (#SMM4H) Workshop & Shared Task*, pages 27–30.

Tsz Hang Chu, Youzhen Su, Hanxiao Kong, Jingyuan Shi, and Xiaohui Wang. 2021. Online Social Support for Intimate Partner Violence Victims in China: Quantitative and Automatic Content Analysis. *Violence Against Women*, 27(3-4):339–358.

Daniel Claeser and Samantha Kent. 2022. Fraunhofer fkie @ smm4h 2022: System description for shared tasks 2, 4 and 9. In *Proceedings of the Seventh Social Media Mining for Health Applications (#SMM4H) Workshop & Shared Task*, pages 103–107.

Jacob Cohen. 1968. Weighted kappa: nominal scale agreement provision for scaled disagreement or partial credit. *Psychological bulletin*, 70(4):213.

Alexis Conneau, Kartikay Khandelwal, Naman Goyal, Vishrav Chaudhary, Guillaume Wenzek, Francisco Guzmán, Edouard Grave, Myle Ott, Luke Zettlemoyer, and Veselin Stoyanov. 2019. Unsupervised cross-lingual representation learning at scale. *arXiv preprint arXiv:1911.02116*.

Jaclyn D Cravens, Jason B Whiting, and Rola O Aamar. 2015. Why I Stayed/Left: An Analysis of Voices of Intimate Partner Violence on Social Media. *Contemporary Family Therapy*, 37(4):372–385.

Millon Das, Archit Mangrulkar, Ishan Manchanda, Manav Kapadnis, and Sohan Patnaik. 2022. Enolp musk@smm4h'22 : Leveraging pre-trained language models for stance and premise classification. In *Proceedings of the Seventh Social Media Mining for Health Applications (#SMM4H) Workshop & Shared Task*, pages 156–159.

Shelby H. Davies, Miriam D. Langer, Ari Klein, Graciela Gonzalez-Hernandez, and Nadia Dowshen. 2022. Adolescent perceptions of menstruation on Twitter: Opportunities for advocacy and education. *Journal of Adolescent Health*, 71(1):94–104.

Vera Davydova and Elena Tutubalina. 2022. Smm4h 2022 task 2: Dataset for stance and premise detection in tweets about health mandates related to covid-19. In *Proceedings of the Seventh Social Media Mining for Health Applications (#SMM4H) Workshop & Shared Task*, pages 216–220.

Alexander Philip Dawid and Allan M. Skene. 1979. Maximum likelihood estimation of observer error-rates using the EM algorithm. *Journal of the Royal Statistical Society: Series C (Applied Statistics)*, 28(1):20–28.

Jacob Devlin, Ming-Wei Chang, Kenton Lee, and Kristina Toutanova. 2018. Bert: Pre-training of deep bidirectional transformers for language understanding. *arXiv preprint arXiv:1810.04805*.

Katherine Donegan, Hans Ovelgonne, Gavril Flores, Per Fuglerud, and Ada Georgescu. 2019. Social media and m-health data, subgroup report. *European Medicines Agency*.

Elissa S Epel, Alexandra D Crosswell, Stefanie E Mayer, Aric A Prather, George M Slavich, Eli Puterman, and Wendy Berry Mendes. 2018. More than a feeling: A unified view of stress measurement for population science. *Frontiers in neuroendocrinology*, 49:146–169.

Sumam Francis and Marie-Francine Moens. 2022. Kul@smm4h'22: Template augmented adaptive pre-training for tweet classification. In *Proceedings of the Seventh Social Media Mining for Health Applications (#SMM4H) Workshop & Shared Task*, pages 153–155.

Raphael Antonius Frick and Martin Steinebach. 2022. Fraunhofer sit@smm4h'22: Learning to predict stances and premises in tweets related to covid-19 health orders using generative models. In *Proceedings of the Seventh Social Media Mining for Health Applications (#SMM4H) Workshop & Shared Task*, pages 111–113.

Jia Fu, Sirui Li, Hui Ming Yuan, Zhucong Li, Zhen Gan, Yubo Chen, Kang Liu, Jun Zhao, and Shengping Liu. 2022. Casia@smm4h'22: A uniform health information mining system for multilingual social media texts. In *Proceedings of the Seventh Social Media Mining for Health Applications (# SMM4H) Workshop & Shared Task*, pages 143–147.

Luis Gasco, Darryl Estrada-Zavala, Eulàlia Farré-Maduell, Salvador Lima-López, Antonio Miranda-Escalada, and Martin Krallinger. 2022. Overview of the SocialDisNER shared task on detection of diseases mentions from healthcare related and patient generated social media content: methods, evaluation and corpora. In *Proceedings of the Seventh Social Media Mining for Health Applications (#SMM4H) Workshop & Shared Task*.

Kyle Glandt, Sarthak Khanal, Yingjie Li, Doina Caragea, and Cornelia Caragea. 2021. Stance detection in covid-19 tweets. In *Proceedings of the 59th Annual Meeting of the Association for Computational Linguistics and the 11th International Joint Conference on Natural Language Processing (Volume 1: Long Papers)*, pages 1596–1611.

Su Golder, Stephanie Chiuve, Davy Weissenbacher, Ari Klein, Karen O'Connor, Martin Bland, Murray Malin, Mondira Bhattacharya, Linda J. Scarazzini, and Graciela Gonzalez-Hernandez. 2019. Pharmacoepidemiologic evaluation of birth defects from health-related postings in social media during pregnancy. *Drug Safety*, 42(3):389–400.

Imane Guellil, Jinge Wu, Honghan Wu, Tony Sun, and Beatrice Alex. 2022. Edinburgh_ucl_health@smm4h'22: From glove to flair for handling imbalanced healthcare corpora related to adverse drug events, change in medication and self-reporting vaccination. In *Proceedings of the Seventh Social Media Mining for Health Applications (#SMM4H) Workshop & Shared Task*, pages 148–152.

Asier Gutiérrez Fandiño, Jordi Armengol Estapé, Marc Pàmies, Joan Llop Palao, Joaquin Silveira Ocampo, Casimiro Pio Carrino, Carme Armentano Oller, Carlos Rodriguez Penagos, Aitor Gonzalez Agirre, and Marta Villegas. 2022. Maria: Spanish language models. *Procesamiento del Lenguaje Natural*, 68.

Pan He, Chen YuZe, and Yanru Zhang. 2022. Zhegu@smm4h-2022: The pre-training tweet claim matching makes your prediction better. In *Proceedings of the Seventh Social Media Mining for Health Applications (#SMM4H) Workshop & Shared Task*, pages 38–41.

Adrian Garcia Hernandez, Leung Wai Liu, Akshat Gupta, Vineeth Ravi, Saheed O. Obitayo, Xiaomo Liu, and Sameena Shah. 2022. Air-jpmc@smm4h'22: Identifying self-reported spanish covid-19 symptom tweets through multiple-model ensembling. In *Proceedings of the Seventh Social Media Mining for Health Applications (#SMM4H) Workshop & Shared Task*, pages 160–162.

Chenghao Huang, Xiaolu Chen, Yuxi Chen, Yutong Wu, Weimin Yuan, Yan Wang, and Yanru Zhang. 2022. zydhjh4593@smm4h'22: A generic pre-trained bert-based framework for social media health text classification. In *Proceedings of the Seventh Social Media Mining for Health Applications (#SMM4H) Workshop & Shared Task*, pages 11–15.

Javier Huertas-Tato, Alejandro Martin, and David Camacho. 2022. Bertuit: Understanding spanish language in twitter through a native transformer. *arXiv preprint arXiv:2204.03465*.

Antonio Jimeno-Yepes, Andrew MacKinlay, Bo Han, and Qiang Chen. 2015. Identifying diseases, drugs, and symptoms in twitter. In *MEDINFO 2015: eHealth-enabled Health*, pages 643–647. IOS Press.

Keshav Kapur, Rajitha Harikrishnan, and Sanjay Singh. 2022. Manlp@smm4h'22: Bert for classification of twitter posts. In *Proceedings of the Seventh Social Media Mining for Health Applications (#SMM4H) Workshop & Shared Task*, pages 42–43.

Akbar Karimi and Lucie Flek. 2022. Caisa@smm4h'22: Robust cross-lingual detection of disease mentions on social media with adversarial methods. In *Proceedings of the Seventh Social Media Mining for Health Applications (#SMM4H) Workshop & Shared Task*, pages 168–170.

Sarvnaz Karimi, Alejandro Metke-Jimenez, Madonna Kemp, and Chen Wang. 2015. Cadec: A corpus of adverse drug event annotations. *Journal of biomedical informatics*, 55:73–81.

Prabsimran Kaur, Guneet Singh Kohli, and Jatin Bedi. 2022. Arguably@smm4h'22: Classification of health related tweets using ensemble, zero-shot and fine-tuned language model. In *Proceedings of the Seventh Social Media Mining for Health Applications (#SMM4H) Workshop & Shared Task*, pages 138–142.

Roshan Vivek Khatri, Sougata Saha, Souvik Das, and Rohini K Srihari. 2022. Ub health miners@smm4h'22: Exploring pre-processing techniques to classify tweets using transformer based pipelines. In *Proceedings of the Seventh Social Media Mining for Health Applications (#SMM4H) Workshop & Shared Task*, pages 114–117.

Ari Z. Klein, Arjun Magge, and Graciela Gonzalez-Hernandez. 2022. Reportage: Automatically extracting the exact age of twitter users based on self-reports in tweets. *PLoS One*, 17(1):e0262087.

Veysel Kocaman, Cabir Celik, Damla Gurbaz, Gursev Pirge, Bunyamin Polat, Halil Saglamlar, Meryem Vildan Sarikaya, Gokhan Turer, and David Talby. 2022. John_snow_labs@smm4h'22: Social media mining for health (#smm4h) with spark nlp. In *Proceedings of the Seventh Social Media Mining for Health Applications (#SMM4H) Workshop & Shared Task*, pages 44–47.

Antoine Lain, Wonjin Yoon, Hyunjae Kim, Jaewoo Kang, and Ian Simpson. 2022. Ku_ed at socialdisner: Extracting disease mentions in tweets written in spanish. In *Proceedings of the Seventh Social Media Mining for Health Applications (#SMM4H) Workshop & Shared Task*, pages 78–80.

Lukas Lange, Heike Adel, Jannik Strötgen, and Dietrich Klakow. 2021. Clin-x: pre-trained language models and a study on cross-task transfer for concept extraction in the clinical domain. *arXiv preprint arXiv:2112.08754*.

Tzi-Mi Lin, Chao-Yi Chen, Yu-Wen Tzeng, and Lung-Hao Lee. 2022. Ncuee-nlp@smm4h'22: Classification of self-reported chronic stress on twitter using ensemble pre-trained transformer models. In *Proceedings of the Seventh Social Media Mining for Health Applications (#SMM4H) Workshop & Shared Task*, pages 62–64.

Oscar William Lithgow-Serrano, Joseph Cornelius, Fabio Rinaldi, and Ljiljana Dolamic. 2022. mattica@smm4h'22: Leveraging sentiment for stance premise joint learning. In *Proceedings of the Seventh Social Media Mining for Health Applications (#SMM4H) Workshop & Shared Task*, pages 75–77.

Leung Wai Liu, Akshat Gupta, Saheed Obitayo, Xiaomo Liu, and Sameena Shah. 2022a. Air-jpmc@smm4h'22: Bert + ensembling = too cool: Using multiple bert models together for various covid-19 tweet identification tasks. In *Proceedings of the Seventh Social Media Mining for Health Applications (#SMM4H) Workshop & Shared Task*, pages 163–167.

Xi Liu, Han Zhou, and Chang Su. 2022b. Pingantech at smm4h task1: Multiple pre-trained model approaches for adverse drug reactions. In *Proceedings of the Seventh Social Media Mining for Health Applications (#SMM4H) Workshop & Shared Task*, pages 4–6.

Arjun Magge, Ari Klein, Antonio Miranda-Escalada, Mohammed Ali Al-Garadi, Ilseyar Alimova, Zulfat Miftahutdinov, Eulalia Farre, Salvador Lima López, Ivan Flores, Karen O'Connor, Davy Weissenbacher,

Elena Tutubalina, Abeed Sarker, Juan Banda, Martin Krallinger, and Graciela Gonzalez-Hernandez. 2021. Overview of the sixth social media mining for health applications (#SMM4H) shared tasks at NAACL 2021. In *Proceedings of the Sixth Social Media Mining for Health (#SMM4H) Workshop and Shared Task*, pages 21–32, Mexico City, Mexico. Association for Computational Linguistics.

Arjun Magge, Elena Tutubalina, Zulfat Miftahutdinov, Ilseyar Alimova, Anne Dirkson, Suzan Verberne, Davy Weissenbacher, and Graciela Gonzalez-Hernandez. 2020. Deepademiner: A deep learning pharmacovigilance pipeline for extraction and normalization of adverse drug effect mentions on twitter. *medRxiv*.

Heather L McCauley, Amy E Bonomi, Megan K Maas, Katherine W Bogen, and Teagen L O'Malley. 2018. # MaybeHeDoesntHitYou: Social Media Underscore the Realities of Intimate Partner Violence. *Journal of Women's Health*, 27(7):885–891.

Antonio Miranda-Escalada, Eulàlia Farré-Maduell, Salvador Lima-López, Luis Gascó, Vicent Briva-Iglesias, Marvin Agüero-Torales, and Martin Krallinger. 2021. The profner shared task on automatic recognition of occupation mentions in social media: systems, evaluation, guidelines, embeddings and corpora. In *Proceedings of the Sixth Social Media Mining for Health (# SMM4H) Workshop and Shared Task*, pages 13–20.

Antonio Miranda-Escalada, Aitor Gonzalez-Agirre, Jordi Armengol-Estapé, and Martin Krallinger. 2020. Overview of automatic clinical coding: Annotations, guidelines, and solutions for non-english clinical cases at codiesp track of clef ehealth 2020. *CLEF (Working Notes)*, 2020.

Rosa M. Montañés-Salas, Irene López-Bosque, Luis García-Garcés, and Rafael del Hoyo-Alonso. 2022. Itainnova at socialdisner: A transformers cocktail for disease identification in social media in spanish. In *Proceedings of the Seventh Social Media Mining for Health Applications (#SMM4H) Workshop & Shared Task*, pages 71–74.

Edgar Morais, José Luis Oliveira, Alina Trifan, and Olga Fajarda. 2022. Bioinfo@uavr@smm4h'22: Classification and extraction of adverse event mentions in tweets using transformer models. In *Proceedings of the Seventh Social Media Mining for Health Applications (#SMM4H) Workshop & Shared Task*, pages 65–67.

Miguel Ortega-Martín, Alfonso Ardoiz, Jorge Álvarez, Oscar García-Sierra, and Adrián Alonso. 2022. dezzai@smm4h'22: Tasks 5 10 - hybrid models everywhere. In *Proceedings of the Seventh Social Media Mining for Health Applications (#SMM4H) Workshop & Shared Task*, pages 7–10.

Christopher Palmer, Sedigheh Khademi, and Muhammad Javed. 2022. Chaai@smm4h'22: Roberta, gpt-2 and sampling - an interesting concoction. In *Proceedings of the Seventh Social Media Mining for Health Applications (#SMM4H) Workshop & Shared Task*, pages 81–84.

Beatrice Portelli, Simone Scaboro, Emmanuele Chersoni, Enrico Santus, and Giuseppe Serra. 2022. Ailab-udine@smm4h'22: Limits of transformers and bert ensembles. In *Proceedings of the Seventh Social Media Mining for Health Applications (#SMM4H) Workshop & Shared Task*, pages 130–134.

Vadim A. Porvatov and Natalia Semenova. 2022. Transformer-based classification of premise in tweets related to covid-19. In *Proceedings of the Seventh Social Media Mining for Health Applications (#SMM4H) Workshop & Shared Task*, pages 108–110.

Matias Rojas, Jose Barros, Kinan R. Martin, Mauricio Araneda-Hernandez, and Jocelyn Dunstan. 2022. Pln cmm at socialdisner: Improving detection of disease mentions in tweets by using document-level features. In *Proceedings of the Seventh Social Media Mining for Health Applications (#SMM4H) Workshop & Shared Task*, pages 52–54.

Vatsal Savaliya, Aakash Bhatnagar, Nidhir Bhavsar, and Muskaan Singh. 2022. Innovators@smm4h'22: An ensembles approach for stance and premise classification of covid-19 health mandates tweets. In *Proceedings of the Seventh Social Media Mining for Health Applications (#SMM4H) Workshop & Shared Task*, pages 126–129.

Sanja Scepanovic, Enrique Martin-Lopez, Daniele Quercia, and Khan Baykaner. 2020. Extracting medical entities from social media. In *Proceedings of the ACM Conference on Health, Inference, and Learning*, pages 170–181.

Ana Lucía Schmidt, Raul Rodriguez-Esteban, Juergen Gottowik, and Mathias Leddin. 2022. Applications of quantitative social media listening to patient-centric drug development. *Drug Discovery Today*.

Isabel Segura-Bedmar, Ricardo Revert, and Paloma Martínez. 2014. Detecting drugs and adverse events from spanish social media streams. In *Proceedings of the 5th international workshop on health text mining and information analysis (LOUHI)*, pages 106–115.

Aman Sinha, Cristina Garcia Holgado, Marianne Clausel, and Matthieu Constant. 2022. Iai @ socialdisner : Catch me if you can! capturing complex disease mentions in tweets. In *Proceedings of the Seventh Social Media Mining for Health Applications (#SMM4H) Workshop & Shared Task*, pages 85–89.

Sharon Smith, Xinjian Zhang, Kathleen Basile, Melissa Merrick, Jing Wang, Marcie-jo Kresnow, and Jieru Chen. 2018. The National Intimate Partner and Sexual Violence Survey: 2015 Data Brief — Updated Release. Technical report, National Center for Injury Prevention and Control, Centers for Disease Control and Prevention, Atlanta, Georgia.

Afrin Sultana, Nihad Karim Chowdhury, and Abu Nowshed Chy. 2022. Csecu-dsg@smm4h'22: Transformer based unified approach for classification of changes in medication treatments in tweets and webmd reviews. In *Proceedings of the Seventh Social Media Mining for Health Applications (#SMM4H) Workshop & Shared Task*, pages 118–122.

Antonio Tamayo, Diego Burgos, and Alexander Gelbukh. 2022. Nlp-cic-wfu at socialdisner: Disease mention extraction in spanish tweets using transfer learning and search by propagation. In *Proceedings of the Seventh Social Media Mining for Health Applications (#SMM4H) Workshop & Shared Task*, pages 19–22.

Ramya Tekumalla and Juan M Banda. 2021. Using weak supervision to generate training datasets from social media data: a proof of concept to identify drug mentions. *Neural Comput. Appl.*, pages 1–9.

Atnafu Lambebo Tonja, Olumide Ebenezer Ojo, Mohammed Arif, Abdul Gafar Manuel Meque, Olga Kolesnikova, Grigori Sidorov, and Alexander Gelbukh. 2022. Cic nlp at smm4h 2022: a bert-based approach for classification of social media forum posts. In *Proceedings of the Seventh Social Media Mining for Health Applications (#SMM4H) Workshop & Shared Task*, pages 58–61.

Paul Trust, Rosane Minghim, Ahmed Zahran, Provia Kadusabe, and Kizito Omala. 2022. Ucc-nlp@smm4h'22:label distribution aware long-tailed learning with post-hoc posterior calibration applied to text classification. In *Proceedings of the Seventh Social Media Mining for Health Applications (#SMM4H) Workshop & Shared Task*, pages 90–94.

Gökçe Uludoğan and Zeynep Yirmibeşoğlu. 2022. Boun-tabi@smm4h'22: Text-to-text adverse drug event extraction with data balancing and prompting. In *Proceedings of the Seventh Social Media Mining for Health Applications (#SMM4H) Workshop & Shared Task*, pages 31–34.

Reshma Unnikrishnan, Kamath S. Sowmya, and V. S. Ananthanarayana. 2022. Halelab_nitk@smm4h'22: Adaptive learning model for effective detection, extraction and normalization of adverse drug events from social media data. In *Proceedings of the Seventh Social Media Mining for Health Applications (#SMM4H) Workshop & Shared Task*, pages 95–97.

U.S. Food and Drug Administration. 2020. *Patient-Focused Drug Development: Collecting Comprehensive and Representative Input Guidance for Industry, Food and Drug Administration Staff,and Other Stakeholders.*

Gary R VandenBos. 2007. *APA dictionary of psychology.* American Psychological Association.

Harsh Verma, Parsa Bagherzadeh, and Sabine Bergler. 2022. Claclab at socialdisner: Using medical gazetteers for named-entity recognition of disease mentions in spanish tweets. In *Proceedings of the Seventh Social Media Mining for Health Applications (#SMM4H) Workshop & Shared Task*, pages 55–57.

Chunchen Wei, Ran Bi, and Yanru Zhang. 2022. uestcc@smm4h'22: Roberta based adverse drug events classification on tweets. In *Proceedings of the Seventh Social Media Mining for Health Applications (#SMM4H) Workshop & Shared Task*, pages 35–37.

Davy Weissenbacher, Suyu Ge, Ari Klein, Karen O'Connor, Robert Gross, Sean Hennessy, and Graciela Gonzalez-Hernandez. 2021. Active neural networks to detect mentions of changes to medication treatment in social media. *Journal of the American Medical Informatics Association*, 28(12):2551–2561.

Lynn Westbrook. 2015. Intimate Partner Violence Online: Expectations and Agency in Question and Answer Websites. *Journal of the Association for Information Science and Technology*, 66.

Orest Xherija and Hojoon Choi. 2022. Complx@smm4h'22: In-domain pretrained language models for detection of adverse drug reaction mentions in english tweets. In *Proceedings of the Seventh Social Media Mining for Health Applications (#SMM4H) Workshop & Shared Task*, pages 176–181.

Huabin Yang, Zhongjian Zhang, and Yanru Zhang. 2022a. yiriyou@smm4h'22: Stance and premise classification in domain specific tweets with dual-view attention neural networks. In *Proceedings of the Seventh Social Media Mining for Health Applications (#SMM4H) Workshop & Shared Task*, pages 23–26.

Yuan-Chi Yang, Angel Xie, Sangmi Kim, Jessica Hair, Mohammed Ali Al-Garadi, and Abeed Sarker. 2022b. Automatic detection of Twitter users who express chronic stress experiences via supervised machine learning and natural language processing. *Computers, Informatics, Nursing*, Article in press.

Antonio Jimeno Yepes and Karin Verspoor. 2022. Readbiomed@socialdisner: Adaptation of an annotation system to spanish tweets. In *Proceedings of the Seventh Social Media Mining for Health Applications (#SMM4H) Workshop & Shared Task*, pages 48–51.

Sourabh Satish Zanwar, Daniel Wiechmann, Yu Qiao, and Elma Kerz. 2022. Mantis at smm4h'2022: Pretrained language models meet a suite of psycholinguistic features for the detection of self-reported chronic stress. In *Proceedings of the Seventh Social Media Mining for Health Applications (#SMM4H) Workshop & Shared Task*, pages 16–18.

Yan Zhuang and Yanru Zhang. 2022. Yet@smm4h'22: Improved bert-based classification models with rdrop and polyloss. In *Proceedings of the Seventh Social Media Mining for Health Applications (#SMM4H) Workshop & Shared Task*, pages 98–102.

Mohammad Zohair, Nidhir Bhavsar, Aakash Bhatnagar, and Muskaan Singh. 2022. Innovators @ smm4h'22: An ensembles approach for self-reporting of covid-19 vaccination status tweets. In *Proceedings of the Seventh Social Media Mining for Health Applications (#SMM4H) Workshop & Shared Task*, pages 123–125.

Appendix A. Team numbers and System Description Papers

#Team	System description paper
1	(Adjali et al., 2022)
2	(Avram et al., 2022)
3	(Candidato et al., 2022)
4	(Cetina and García-Santa, 2022)
5	(Chizhikova et al., 2022)
6	(Claeser and Kent, 2022)
7	(Das et al., 2022)
8	(Francis and Moens, 2022)
9	(Frick and Steinebach, 2022)
10	(Fu et al., 2022)
11	(Guellil et al., 2022)
12	(He et al., 2022)
13	(Hernandez et al., 2022)
14	(Huang et al., 2022)
15	(Kapur et al., 2022)
16	(Karimi and Flek, 2022)
17	(Kaur et al., 2022)
18	(Khatri et al., 2022)
19	(Kocaman et al., 2022)
20	(Lain et al., 2022)
21	(Lin et al., 2022)
22	(Lithgow-Serrano et al., 2022)
23	(Liu et al., 2022a)
24	(Montañés-Salas et al., 2022)
25	(Morais et al., 2022)
26	(Ortega-Martín et al., 2022)
27	(Palmer et al., 2022)
28	(Portelli et al., 2022)
29	(Porvatov and Semenova, 2022)
30	(Rojas et al., 2022)
31	(Savaliya et al., 2022)
32	(Sinha et al., 2022)
33	(Sultana et al., 2022)
34	(Tamayo et al., 2022)
35	(Tonja et al., 2022)
36	(Trust et al., 2022)
37	(Uludoğan and Yirmibeşoğlu, 2022)
38	(Unnikrishnan et al., 2022)
39	(Verma et al., 2022)
40	(Wei et al., 2022)
41	(Xherija and Choi, 2022)
42	(Yang et al., 2022a)
43	(Yepes and Verspoor, 2022)
44	(Zanwar et al., 2022)
45	(Liu et al., 2022b)
46	(Zhuang and Zhang, 2022)
47	(Zohair et al., 2022)

Author Index

Association for Computational Linguistics
209 N. Eighth Street
Stroudsburg, Pennsylvania 18360

ISBN 978-1-7138-6195-9

CURRAN ASSOCIATES INC.
proceedings
.com